Titu Andreescu • Dorin Andrica

Complex Numbers from A to ... Z

Second Edition

 Birkhäuser

Titu Andreescu
Department of Science and Mathematics
  Education
The University of Texas at Dallas
Richardson, Texas, USA

Dorin Andrica
Department of Mathematics
Babeş-Bolyai University
Cluj-Napoca, Romania

ISBN 978-0-8176-8414-3   ISBN 978-0-8176-8415-0 (eBook)
DOI 10.1007/978-0-8176-8415-0
Springer New York Heidelberg Dordrecht London

Library of Congress Control Number: 2013958316

Mathematics Subject Classification (2010): 00A05, 00A07, 30-99, 30A99, 97U40

Printed on acid-free paper

Springer is part of Springer Science+Business Media (www.birkhauser-science.com)

The shortest path between two truths in the real domain passes through the complex domain.

Jacques Hadamard

Preface

In comparison with the first edition of this work, the main new features of this version of the book are these:

- A number of important new problems have been added.
- Many solutions to the problems presented in the first edition have been revised.
- Alternative solutions to several problems have been provided
- Section 4.6.5, "Blundon's Inequalities," has been added. It contains a brand new geometric proof of the fundamental triangle inequality based on the distance formulas established in the previous subsections.
- A number of typographical errors and LaTeX infelicities have been corrected.
- A few new works have been added to the references section.

We would like to thank all readers who have written to us to express their appreciation of our book, as well as to all who provided pertinent comments and suggestions for future improvement of the text. Special thanks are given to Cătălin Barbu, Dumitru Olteanu, Cosmin Pohoaţă, and Daniel Văcăreţu for their careful proofreading of the material included in this edition.

Happy reading!

Richardson, TX   Titu Andreescu
Cluj-Napoca, Romania   Dorin Andrica

Preface to the First Edition

Solving algebraic equations has been historically one of the favorite topics of mathematicians. While linear equations are always solvable in real numbers, not all quadratic equations have this property. The simplest such equation is $x^2 + 1 = 0$. Until the eighteenth century, mathematicians avoided quadratic equations that were not solvable over $\mathbb{R}$. Leonhard Euler broke the ice by introducing the "number" $\sqrt{-1}$ in his famous book *Elements of Algebra* as "neither nothing, nor greater than nothing, nor less than nothing" and observed the "notwithstanding this, these numbers present themselves to the mind; they exist in our imagination and we still have a sufficient idea of them ... nothing prevents us from making use of these imaginary numbers and employing them in calculation." Euler denoted the number $\sqrt{-1}$ by i and called it the imaginary unit. This became one of the most useful symbols in mathematics. Using this symbol, one defines complex numbers as $z = a + bi$, where a and b are real numbers. The study of complex numbers continues to this day and has been greatly elaborated over the last two and a half centuries; in fact, it is impossible to imagine modern mathematics without complex numbers. All mathematical domains make use of them in some way. This is true of other disciplines as well, including mechanics, theoretical physics, hydrodynamics, and chemistry.

Our main goal is to introduce the reader to this fascinating subject. The book runs smoothly between key concepts and elementary results concerning complex numbers. The reader has the opportunity to learn how complex numbers can be employed in solving algebraic equations and to understand the geometric interpretation of complex numbers and the operations involving them. The theoretical part of the book is augmented by rich exercises and problems of various levels of difficulty. In Chaps. 3 and 4 we cover important applications in Euclidean geometry. Many geometric problems may be solved efficiently and elegantly using complex numbers. The wealth of examples we provide, the presentation of many topics in a personal manner, the presence of numerous original problems, and the attention to detail in the solutions to selected exercises and problems are only some of the key features of this book.

Among the techniques presented, for example, are those for the real and the complex products of complex numbers. In the language of complex numbers, these are the analogues of the scalar and cross products, respectively. Employing these two products turns out to be efficient in solving numerous problems involving complex numbers. After covering this part, the reader will appreciate the use of these techniques.

A special feature of the book is Chap. 5, an outstanding selection of genuine Olympiad and other important mathematical contest problems solved using the methods already presented.

This work does not cover all aspects of complex numbers. It is not a course in complex analysis, but rather a stepping stone toward its study, which is why we have not used the standard notation e^{it} for $z = \cos t + i \sin t$ or the usual power series expansions.

The book reflects the unique experience of the authors. It distills a vast mathematical literature, most of which is unknown to the Western public, capturing the essence of an flourishing problem-solving culture.

Our work is partly based on the Romanian version, *Numere complexe de la A la... Z*, written by D. Andrica and N. Bişboacă and published by Millennium in 2001 (see [17]). We are preserving the title of the Romanian edition and about 35% of the text. Even this 35% has been significantly improved and enhanced with up-to-date material.

The targeted audience includes high school students and their teachers, undergraduates, mathematics contestants such as those training for Olympiads or the W.L. Putnam Mathematical Competition, their coaches, and everyone interested in essential mathematics.

This book might spawn courses such as complex numbers and Euclidean geometry for prospective high school teachers, giving future educators ideas about things they could do with their brighter students or with a math club. This would be quite a welcome development.

Special thanks are offered to Daniel Văcăreţu, Nicolae Bişboacă, Gabriel Dospinescu, and Ioan Şerdean for their careful proofreading of the final version of the manuscript. We would also like to thank the referees, who provided pertinent suggestions that directly contributed to the improvement of the text.

Richardson, TX   Titu Andreescu
Cluj-Napoca, Romania   Dorin Andrica

Contents

About the Authors

Titu Andreescu received his B.A., M.S, and Ph.D. from the West University of Timisoara, Romania. The topic of his doctoral dissertation was "Research on Diophantine Analysis and Applications." Professor Andreescu currently teaches at the University of Texas at Dallas. Titu is past chairman of the USA Mathematical Olympiad, served as director of the MAA American Mathematics Competitions (1998–2003), coach of the USA International Mathematical Olympiad Team (IMO) for 10 years (1993–2002), director of the Mathematical Olympiad Summer Program (1995–2002), and leader of the USA IMO Team (1995–2002). In 2002, Titu was elected member of the IMO Advisory Board, the governing body of the world's most prestigious mathematics competition. Titu received the Edyth May Sliffe Award for Distinguished High School Mathematics Teaching from the MAA in 1994 and a certificate of appreciation from the president of the MAA in 1995 for his outstanding service as coach of the Mathematical Olympiad Summer Program in preparing the US team for its perfect performance in Hong Kong at the 1994 IMO. Titu's contributions to numerous textbooks and problem books are recognized worldwide.

Dorin Andrica received his Ph.D. in 1992 from "Babeş-Bolyai" University in Cluj-Napoca, Romania, with a thesis on critical points and applications to the geometry of differentiable submanifolds. Professor Andrica has been chairman of the Department of Geometry at "Babeş-Bolyai" in the period 1995–2009. Dorin has written and contributed to numerous mathematics textbooks, problem books, articles, and scientific papers at various levels. Dorin is an invited lecturer at university conferences around the world—Austria, Bulgaria, Czech Republic, Egypt, France, Germany, Greece, the Netherlands, Saudi Arabia, Serbia, Turkey, and the USA. He is a member of the Romanian Committee for the Mathematics Olympiad and member of editorial boards of several international journals. Dorin has been a regular faculty member at the Canada–USA Mathcamps in the period 2001–2005 and at AwesomeMath in the period 2006–2011.

Notation

$\mathbb{Z}$	The set of integers
$\mathbb{N}$	The set of positive integers
$\mathbb{Q}$	The set of rational numbers
$\mathbb{R}$	The set of real numbers
$\mathbb{R}^*$	The set of nonzero real numbers
$\mathbb{R}^2$	The set of pairs of real numbers
$\mathbb{C}$	The set of complex numbers
$\mathbb{C}^*$	The set of nonzero complex numbers
$[a,\ b]$	The set of real numbers x such that $a \leq x \leq b$
$(a,\ b)$	The set of real numbers x such that $a < x < b$
$\overline{z}$	The conjugate of the complex number z
$|z|$	The modulus or absolute value of the complex number z
$\overrightarrow{AB}$	The vector AB
(AB)	The open segment determined by A and B
$[AB]$	The closed segment determined by A and B
$(AB$	The open ray with origin A that contains B
area $[F]$	The area of figure F
U_n	The set of nth roots of unity
$\mathcal{C}(P; R)$	The circle centered at point P with radius R

Chapter 1
Complex Numbers in Algebraic Form

1.1 Algebraic Representation of Complex Numbers

1.1.1 Definition of Complex Numbers

In what follows, we assume that the definition and basic properties of the set of real numbers $\mathbb{R}$ are known.

Let us consider the set $\mathbb{R}^2 = \mathbb{R} \times \mathbb{R} = \{(x,\ y)|x,\ y \in \mathbb{R}\}$. Two elements $(x_1,\ y_1)$ and $(x_2,\ y_2)$ of $\mathbb{R}^2$ are equal if and only if $x_1 = x_2$ and $y_1 = y_2$. The operations of addition and multiplication are defined on the set $\mathbb{R}^2$ as follows:

$$z_1 + z_2 = (x_1,\ y_1) + (x_2,\ y_2) = (x_1 + x_2,\ y_1 + y_2) \in \mathbb{R}^2$$

and

$$z_1 \cdot z_2 = (x_1,\ y_1) \cdot (x_2,\ y_2) = (x_1 x_2 - y_1 y_2,\ x_1 y_2 + x_2 y_1) \in \mathbb{R}^2,$$

for all $z_1 = (x_1,\ y_1) \in \mathbb{R}^2$ and $z_2 = (x_2,\ y_2) \in \mathbb{R}^2$.

The element $z_1 + z_2 \in \mathbb{R}^2$ is called the *sum* of z_1 and z_2, and the element $z_1 \cdot z_2 \in \mathbb{R}^2$ is called the *product* of z_1 and z_2 and will often be written simply $z_1 z_2$.

Remarks.

(1) If $z_1 = (x_1, 0) \in \mathbb{R}^2$ and $z_2 = (x_2, 0) \in \mathbb{R}^2$, then $z_1 z_2 = (x_1 x_2, 0)$.
(2) If $z_1 = (0, y_1) \in \mathbb{R}^2$ and $z_2 = (0, y_2) \in \mathbb{R}^2$, then $z_1 z_2 = (-y_1 y_2, 0)$.

Examples.

(1) Let $z_1 = (-5, 6)$ and $z_2 = (1,\ -2)$. Then

$$z_1 + z_2 = (-5, 6) + (1,\ -2) = (-4, 4)$$

T. Andreescu and D. Andrica, *Complex Numbers from A to ... Z*,
DOI 10.1007/978-0-8176-8415-0_1, © Springer Science+Business Media New York 2014

and

$$z_1 z_2 = (-5, 6) \cdot (1, -2) = (-5 + 12, \ 10 + 6) = (7, \ 16).$$

(2) Let $z_1 = \left(-\frac{1}{2}, 1\right)$ and $z_2 = \left(-\frac{1}{3}, \frac{1}{2}\right)$. Then

$$z_1 + z_2 = \left(-\frac{1}{2} - \frac{1}{3}, 1 + \frac{1}{2}\right) = \left(-\frac{5}{6}, \frac{3}{2}\right)$$

and

$$z_1 z_2 = \left(\frac{1}{6} - \frac{1}{2}, \ -\frac{1}{4} - \frac{1}{3}\right) = \left(-\frac{1}{3}, \ -\frac{7}{12}\right).$$

Definition. The set $\mathbb{R}^2$ together with the operations of addition and multiplication is called the *set of complex numbers,* denoted by $\mathbb{C}$. Every element $z = (x, \ y) \in \mathbb{C}$ is called a *complex number.*

The notation $\mathbb{C}^*$ is used to indicate the set $\mathbb{C} \backslash \{(0, 0)\}$.

1.1.2 Properties Concerning Addition

The addition of complex numbers satisfies the following properties:

(a) **Commutative law.**

$$z_1 + z_2 = z_2 + z_1 \text{ for all } z_1, \ z_2 \in \mathbb{C}.$$

(b) **Associative law.**

$$(z_1 + z_2) + z_3 = z_1 + (z_2 + z_3) \text{ for all } z_1, \ z_2, \ z_3 \in \mathbb{C}.$$

Indeed, if $z_1 = (x_1, \ y_1) \in \mathbb{C}$, $z_2 = (x_2, \ y_2) \in \mathbb{C}$, $z_3 = (x_3, \ y_3) \in \mathbb{C}$, then

$$(z_1 + z_2) + z_3 = [(x_1, \ y_1) + (x_2, \ y_2)] + (x_3, \ y_3)$$

$$= (x_1 + x_2, \ y_1 + y_2) + (x_3, \ y_3) = ((x_1 + x_2) + x_3, \ (y_1 + y_2) + y_3),$$

and

$$z_1 + (z_2 + z_3) = (x_1, \ y_1) + [(x_2, \ y_2) + (x_3, \ y_3)]$$

$$= (x_1, \ y_1) + (x_2 + x_3, \ y_2 + y_3) = (x_1 + (x_2 + x_3), \ y_1 + (y_2 + y_3)).$$

The claim holds due to the associativity of the addition of real numbers.

(c) **Additive identity.** There is a unique complex number $0 = (0, \ 0)$ such that

$$z + 0 = 0 + z = z \text{ for all } z = (x, \ y) \in \mathbb{C}.$$

(d) **Additive inverse.** For every complex number $z = (x, y)$, there is a unique $-z = (-x, -y) \in \mathbb{C}$ such that

$$z + (-z) = (-z) + z = 0.$$

The reader can easily prove the claims (a), (c), and (d).

The number $z_1 - z_2 = z_1 + (-z_2)$ is called the *difference* of the numbers z_1 and z_2. The operation that assigns to the numbers z_1 and z_2 the number $z_1 - z_2$ is called *subtraction* and is defined by

$$z_1 - z_2 = (x_1, y_1) - (x_2, y_2) = (x_1 - x_2, y_1 - y_2) \in \mathbb{C}.$$

1.1.3 Properties Concerning Multiplication

The multiplication of complex numbers satisfies the following properties:

(a) **Commutative law.**

$$z_1 \cdot z_2 = z_2 \cdot z_1 \text{ for all } z_1, z_2 \in \mathbb{C}.$$

(b) **Associative law.**

$$(z_1 \cdot z_2) \cdot z_3 = z_1 \cdot (z_2 \cdot z_3) \text{ for all } z_1, z_2, z_3 \in \mathbb{C}.$$

(c) **Multiplicative identity.** There is a unique complex number $1 = (1, 0) \in \mathbb{C}$ such that

$$z \cdot 1 = 1 \cdot z = z \text{ for all } z \in \mathbb{C}.$$

A simple algebraic manipulation is all that is needed to verify these equalities:

$$z \cdot 1 = (x, y) \cdot (1, 0) = (x \cdot 1 - y \cdot 0, \ x \cdot 0 + y \cdot 1) = (x, y) = z$$

and

$$1 \cdot z = (1, 0) \cdot (x, y) = (1 \cdot x - 0 \cdot y, \ 1 \cdot y + 0 \cdot x) = (x, y) = z.$$

(d) **Multiplicative inverse.** For every complex number $z = (x, y) \in \mathbb{C}^*$, there is a unique number $z^{-1} = (x', y') \in \mathbb{C}$ such that

$$z \cdot z^{-1} = z^{-1} \cdot z = 1.$$

To find $z^{-1} = (x', y')$, observe that $(x, y) \neq (0, 0)$ implies $x \neq 0$ or $y \neq 0$, and consequently, $x^2 + y^2 \neq 0$.

The relation $z \cdot z^{-1} = 1$ gives $(x, y) \cdot (x', y') = (1, 0)$, or equivalently,

$$\begin{cases} xx' - yy' = 1, \\ yx' + xy' = 0. \end{cases}$$

Solving this system with respect to x' and y', one obtains

$$x' = \frac{x}{x^2 + y^2} \text{ and } y' = -\frac{y}{x^2 + y^2};$$

hence the multiplicative inverse of the complex number $z = (x, y) \in \mathbb{C}^*$ is

$$z^{-1} = \frac{1}{z} = \left(\frac{x}{x^2 + y^2}, -\frac{y}{x^2 + y^2} \right) \in \mathbb{C}^*.$$

By the commutative law, we also have $z^{-1} \cdot z = 1$.

Two complex numbers $z_1 = (x_1, y_1) \in \mathbb{C}$ and $z = (x, y) \in \mathbb{C}^*$ uniquely determine a third number, called their *quotient*, denoted by $\frac{z_1}{z}$ and defined by

$$\frac{z_1}{z} = z_1 \cdot z^{-1} = (x_1, y_1) \cdot \left(\frac{x}{x^2 + y^2}, -\frac{y}{x^2 + y^2} \right)$$

$$= \left(\frac{x_1 x + y_1 y}{x^2 + y^2}, \frac{-x_1 y + y_1 x}{x^2 + y^2} \right) \in \mathbb{C}.$$

Examples.

(1) If $z = (1, 2)$, then

$$z^{-1} = \left(\frac{1}{1^2 + 2^2}, \frac{-2}{1^2 + 2^2} \right) = \left(\frac{1}{5}, \frac{-2}{5} \right).$$

(2) If $z_1 = (1, 2)$ and $z_2 = (3, 4)$, then

$$\frac{z_1}{z_2} = \left(\frac{3 + 8}{9 + 16}, \frac{-4 + 6}{9 + 16} \right) = \left(\frac{11}{25}, \frac{2}{25} \right).$$

An integer power of a complex number $z \in \mathbb{C}^*$ is defined by

$$z^0 = 1; \quad z^1 = z; \quad z^2 = z \cdot z;$$

$$z^n = \underbrace{z \cdot z \cdots z}_{n \text{ times}} \text{ for all integers } n > 0$$

and $z^n = (z^{-1})^{-n}$ for all integers $n < 0$.

The following properties hold for all complex numbers z, z_1, $z_2 \in \mathbb{C}^*$ and for all integers m, n:

(1) $z^m \cdot z^n = z^{m+n}$;

(2) $\dfrac{z^m}{z^n} = z^{m-n}$;

(3) $(z^m)^n = z^{mn}$;

(4) $(z_1 \cdot z_2)^n = z_1^n \cdot z_2^n$;

(5) $\left(\dfrac{z_1}{z_2}\right)^n = \dfrac{z_1^n}{z_2^n}$.

When $z = 0$, we define $0^n = 0$ for all integers $n > 0$.

(e) **Distributive law.**

$$z_1 \cdot (z_2 + z_3) = z_1 \cdot z_2 + z_1 \cdot z_3 \text{ for all } z_1, \ z_2, \ z_3 \in \mathbb{C}.$$

The above properties of addition and multiplication show that the set $\mathbb{C}$ of all complex numbers, together with these operations, forms a field.

1.1.4 Complex Numbers in Algebraic Form

For algebraic manipulation, it is inconvenient to represent a complex number as an ordered pair. For this reason, another form of writing is preferred.

To introduce this new algebraic representation, consider the set $\mathbb{R} \times \{0\}$, together with the addition and multiplication operations defined on $\mathbb{R}^2$. The function

$$f : \mathbb{R} \to \mathbb{R} \times \{0\}, \ f(x) = (x, \ 0),$$

is bijective, and moreover,

$$(x, \ 0) + (y, \ 0) = (x + y, \ 0) \text{ and } (x, \ 0) \cdot (y, \ 0) = (xy, \ 0).$$

The reader will not fail to notice that the algebraic operations on $\mathbb{R} \times \{0\}$ are similar to the operations on $\mathbb{R}$; therefore, we can identify the ordered pair $(x, \ 0)$ with the number x for all $x \in \mathbb{R}$. Hence we can use, by the above bijection f, the notation $(x, \ 0) = x$.

Setting $i = (0, \ 1)$, we obtain

$$z = (x, \ y) = (x, \ 0) + (0, \ y) = (x, \ 0) + (y, \ 0) \cdot (0, .1)$$
$$= x + yi = (x, \ 0) + (0, \ 1) \cdot (y, \ 0) = x + iy.$$

In this way, we obtain the following result.

Proposition. *Every complex number $z = (x, \ y)$ can be uniquely represented in the form*

$$z = x + yi,$$

where x, y are real numbers. The relation $i^2 = -1$ holds.

The formula $i^2 = -1$ follows directly from the definition of multiplication:

$$i^2 = i \cdot i = (0,1) \cdot (0,1) = (-1,0) = -1.$$

The expression $x + yi$ is called the *algebraic representation* (*form*) of the complex number $z = (x, y)$, so we can write $\mathbb{C} = \{x + yi | x \in \mathbb{R}, \ y \in \mathbb{R}, \ i^2 = -1\}$. From now on, we will denote the complex number $z = (x, y)$ by $x + iy$. The real number $x = \operatorname{Re}(z)$ is called the *real part* of the complex number z, and similarly, $y = \operatorname{Im}(z)$ is called the *imaginary part* of z. Complex numbers of the form iy, $y \in \mathbb{R}^*$, are called *purely imaginary* and the complex number i is called the *imaginary unit*.

The following relations are easy to verify:

(a) $z_1 = z_2$ if and only if $\operatorname{Re}(z_1) = \operatorname{Re}(z_2)$ and $\operatorname{Im}(z_1) = \operatorname{Im}(z_2)$.
(b) $z \in \mathbb{R}$ if and only if $\operatorname{Im}(z) = 0$.
(c) $z \in \mathbb{C} \backslash \mathbb{R}$ if and only if $\operatorname{Im}(z) \neq 0$.

Using the algebraic representation, the usual operations with complex numbers can be performed as follows:

1. Addition

$$z_1 + z_2 = (x_1 + y_1 i) + (x_2 + y_2 i) = (x_1 + x_2) + (y_1 + y_2)i \in \mathbb{C}.$$

It is easy to observe that the sum of two complex numbers is a complex number whose real (imaginary) part is the sum of the real (imaginary) parts of the given numbers:

$$\operatorname{Re}(z_1 + z_2) = \operatorname{Re}(z_1) + \operatorname{Re}(z_2);$$
$$\operatorname{Im}(z_1 + z_2) = \operatorname{Im}(z_1) + \operatorname{Im}(z_2).$$

2. Multiplication

$$z_1 \cdot z_2 = (x_1 + y_1 i)(x_2 + y_2 i) = (x_1 x_2 - y_1 y_2) + (x_1 y_2 + x_2 y_1)i \in \mathbb{C}.$$

In other words,

$$\operatorname{Re}(z_1 z_2) = \operatorname{Re}(z_1) \cdot \operatorname{Re}(z_2) - \operatorname{Im}(z_1) \cdot \operatorname{Im}(z_2)$$

and

$$\operatorname{Im}(z_1 z_2) = \operatorname{Im}(z_1) \cdot \operatorname{Re}(z_2) + \operatorname{Im}(z_2) \cdot \operatorname{Re}(z_1).$$

For a real number λ and a complex number $z = x + yi$,

$$\lambda \cdot z = \lambda(x + yi) = \lambda x + \lambda yi \in \mathbb{C}$$

is the product of a real number and a complex number. The following properties are obvious:

(1) $\lambda(z_1 + z_2) = \lambda z_1 + \lambda z_2$;

(2) $\lambda_1(\lambda_2 z) = (\lambda_1 \lambda_2)z$;

(3) $(\lambda_1 + \lambda_2)z = \lambda_1 z + \lambda_2 z$ for all z, z_1, $z_2 \in \mathbb{C}$ and λ, λ_1, $\lambda_2 \in \mathbb{R}$.

In fact, relations (1) and (3) are special cases of the distributive law, and relation (2) comes from the associative law of multiplication for complex numbers.

3. Subtraction

$$z_1 - z_2 = (x_1 + y_1 i) - (x_2 + y_2 i) = (x_1 - x_2) + (y_1 - y_2)i \in \mathbb{C}.$$

That is,

$$\mathrm{Re}(z_1 - z_2) = \mathrm{Re}(z_1) - \mathrm{Re}(z_2);$$
$$\mathrm{Im}(z_1 - z_2) = = \mathrm{Im}(z_1) - \mathrm{Im}(z_2).$$

1.1.5 Powers of the Number i

The formulas for the powers of a complex number with integer exponents are preserved for the algebraic form $z = x + iy$. Setting $z = i$, we obtain

$$i^0 = 1; \quad i^1 = i; \quad i^2 = -1; \quad i^3 = i^2 \cdot i = -i;$$

$$i^4 = i^3 \cdot i = 1; \quad i^5 = i^4 \cdot i = i; \quad i_6 = i^5 \cdot i = -1; \quad i^7 = i^6 \cdot i = -i.$$

One can prove by induction that for every positive integer n,

$$i^{4n} = 1; \; i^{4n+1} = i; \; i^{4n+2} = -1; \; i^{4n+3} = -i.$$

Hence $i^n \in \{-1, 1, -i, i\}$ for all integers $n \geq 0$. If n is a negative integer, we have

$$i^n = (i^{-1})^{-n} = \left(\frac{1}{i}\right)^{-n} = (-i)^{-n}.$$

Examples.

(1) We have

$$i^{105} + i^{23} + i^{20} - i^{34} = i^{4 \cdot 26 + 1} + i^{4 \cdot 5 + 3} + i^{4 \cdot 5} - i^{4 \cdot 8 + 2} = i - i + 1 + 1 = 2.$$

(2) Let us solve the equation $z^3 = 18 + 26i$, where $z = x + yi$ and x, y are integers. We can write

$$(x + yi)^3 = (x + yi)^2 (x + yi) = (x^2 - y^2 + 2xyi)(x + yi)$$
$$= (x^3 - 3xy^2) + (3x^2 y - y^3)i = 18 + 26i.$$

Using the definition of equality of complex numbers, we obtain

$$\begin{cases} x^3 - 3xy^2 = 18, \\ 3x^2y - y^3 = 26. \end{cases}$$

Setting $y = tx$ in the equality $18(3x^2y - y^3) = 26(x^3 - 3xy^2)$, let us observe that $x \neq 0$ and $y \neq 0$ implies $18(3t - t^3) = 26(1 - 3t^2)$, which is equivalent to $(3t - 1)(3t^2 - 12t - 13) = 0$.

The only rational solution of this equation is $t = \frac{1}{3}$; hence,

$$x = 3, \; y = 1, \text{ and } z = 3 + i.$$

1.1.6 Conjugate of a Complex Number

For a complex number $z = x + yi$ the number $\bar{z} = x - yi$ is called the *complex conjugate* of z or occasionally the *conjugate complex number* of z.

Proposition.

(1) *The relation $z = \bar{z}$ holds if and only if $z \in \mathbb{R}$.*
(2) *For every complex number z, the relation $z = \bar{\bar{z}}$ holds.*
(3) *For every complex number z, the number $z \cdot \bar{z} \in \mathbb{R}$ is a nonnegative real number.*
(4) *$\overline{z_1 + z_2} = \bar{z}_1 + \bar{z}_2$ (the conjugate of a sum is the sum of the conjugates).*
(5) *$\overline{z_1 \cdot z_2} = \bar{z}_1 \cdot \bar{z}_2$ (the conjugate of a product is the product of the conjugates).*
(6) *For every nonzero complex number z, the relation $\overline{z^{-1}} = (\bar{z})^{-1}$ holds.*
(7) *$\overline{\left(\dfrac{z_1}{z_2}\right)} = \dfrac{\bar{z}_1}{\bar{z}_2}$, $z_2 \neq 0$ (the conjugate of a quotient is the quotient of the conjugates).*
(8) *The formulas*

$$\text{Re}(z) = \frac{z + \bar{z}}{2} \text{ and } \text{Im}(z) = \frac{z - \bar{z}}{2i}$$

are valid for all $z \in \mathbb{C}$.

Proof.

(1) If $z = x + yi$, then the relation $z = \bar{z}$ is equivalent to $x + yi = x - yi$. Hence $2yi = 0$, so $y = 0$, and finally, $z = x \in \mathbb{R}$.
(2) We have $\bar{z} = x - yi$ and $\bar{\bar{z}} = x - (-y)i = x + yi = z$.
(3) Observe that $z \cdot \bar{z} = (x + yi)(x - yi) = x^2 + y^2 \geq 0$.
(4) Note that

$$\overline{z_1 + z_2} = \overline{(x_1 + x_2) + (y_1 + y_2)i} = (x_1 + x_2) - (y_1 + y_2)i$$
$$= (x_1 - y_1 i) + (x_2 - y_2 i) = \bar{z}_1 + \bar{z}_2.$$

(5) We can write

$$\overline{z_1 \cdot z_2} = \overline{(x_1 x_2 - y_1 y_2) + (x_1 y_2 + x_2 y_1)i}$$

$$= (x_1 x_2 - y_1 y_2) - (x_1 y_2 + x_2 y_1)i = (x_1 - y_1 i)(x_2 - y_2 i) = \overline{z_1} \cdot \overline{z_2}.$$

(6) Because $z \cdot \dfrac{1}{z} = 1$, we have $\overline{\left(z \cdot \dfrac{1}{z}\right)} = \overline{1}$, and consequently, $\overline{z} \cdot \overline{\left(\dfrac{1}{z}\right)} = 1$,

yielding $\overline{(z^{-1})} = (\overline{z})^{-1}$.

(7) Observe that $\overline{\left(\dfrac{z_1}{z_2}\right)} = \overline{\left(z_1 \cdot \dfrac{1}{z_2}\right)} = \overline{z_1} \cdot \overline{\left(\dfrac{1}{z_2}\right)} = \overline{z_1} \cdot \dfrac{1}{\overline{z_2}} = \dfrac{\overline{z_1}}{\overline{z_2}}.$

(8) From the relations

$$z + \overline{z} = (x + yi) + (x - yi) = 2x,$$
$$z - \overline{z} = (x + yi) - (x - yi) = 2yi,$$

it follows that

$$\mathrm{Re}(z) = \frac{z + \overline{z}}{2} \text{ and } \mathrm{Im}(z) = \frac{z - \overline{z}}{2i},$$

as desired. ☐

Properties (4) and (5) can be easily extended to give

(4') $\overline{\left(\displaystyle\sum_{k=1}^{n} z_k\right)} = \displaystyle\sum_{k=1}^{n} \overline{z_k};$

(5') $\overline{\left(\displaystyle\prod_{k=1}^{n} z_k\right)} = \displaystyle\prod_{k=1}^{n} \overline{z_k}$ for all $z_k \in \mathbb{C}$, $k = 1, 2, \ldots, n$.

As a consequence of (5') and (6), we have

(5'') $\overline{(z^n)} = (\overline{z})^n$ for every integer n and for every $z \in \mathbb{C}$. If $n < 0$, then the formula holds for every $z \neq 0$.

Comments.

(a) To obtain the multiplicative inverse of a complex number $z \in \mathbb{C}^*$, one can use the following approach:

$$\frac{1}{z} = \frac{\overline{z}}{z \cdot \overline{z}} = \frac{x - yi}{x^2 + y^2} = \frac{x}{x^2 + y^2} - \frac{y}{x^2 + y^2}i.$$

(b) The complex conjugate allows us to obtain the quotient of two complex numbers as follows:

$$\frac{z_1}{z_2} = \frac{z_1 \cdot \overline{z_2}}{z_2 \cdot \overline{z_2}} = \frac{(x_1 + y_1 i)(x_2 - y_2 i)}{x_2^2 + y_2^2} = \frac{x_1 x_2 + y_1 y_2}{x_2^2 + y_2^2} + \frac{-x_1 y_2 + x_2 y_1}{x_2^2 + y_2^2}i.$$

Examples.

(1) Compute $z = \frac{5+5i}{3-4i} + \frac{20}{4+3i}$.

Solution. We can write

$$z = \frac{(5+5i)(3+4i)}{9 - 16i^2} + \frac{20(4-3i)}{16 - 9i^2} = \frac{-5+35i}{25} + \frac{80-60i}{25}$$
$$= \frac{75 - 25i}{25} = 3 - i.$$

(2) Let z_1, $z_2 \in \mathbb{C}$. Prove that the number $E = z_1 \cdot \overline{z}_2 + \overline{z}_1 \cdot z_2$ is a real number.

Solution. We have

$$\overline{E} = \overline{z_1 \cdot \overline{z}_2 + \overline{z}_1 \cdot z_2} = \overline{z}_1 \cdot z_2 + z_1 \cdot \overline{z}_2 = E; \text{ so } E \in \mathbb{R}.$$

1.1.7 The Modulus of a Complex Number

The number $|z| = \sqrt{x^2 + y^2}$ is called the *modulus* or the *absolute value* of the complex number $z = x + yi$. For example, the complex numbers

$$z_1 = 4 + 3i, \; z_2 = -3i, \; z_3 = 2$$

have moduli

$$|z_1| = \sqrt{4^2 + 3^2} = 5, \quad |z_2| = \sqrt{0^2 + (-3)^2} = 3, \; \; |z_3| = \sqrt{2^2} = 2.$$

Proposition. *The following properties are satisfied:*

(1) $-|z| \leq \mathrm{Re}(z) \leq |z|$ *and* $-|z| \leq \mathrm{Im}(z) \leq |z|$.
(2) $|z| \geq 0$ *for all* $z \in \mathbb{C}$. *Moreover, we have* $|z| = 0$ *if and only if* $z = 0$.
(3) $|z| = |-z| = |\overline{z}|$.
(4) $z \cdot \overline{z} = |z|^2$.
(5) $|z_1 \cdot z_2| = |z_1| \cdot |z_2|$ *(the modulus of a product is the product of the moduli).*
(6) $|z_1| - |z_2| \leq |z_1 + z_2| \leq |z_1| + |z_2|$.
(7) $|z^{-1}| = |z|^{-1}$, $z \neq 0$.
(8) $\left| \dfrac{z_1}{z_2} \right| = \dfrac{|z_1|}{|z_2|}$, $z_2 \neq 0$ *(the modulus of a quotient is the quotient of the moduli).*
(9) $|z_1| - |z_2| \leq |z_1 - z_2| \leq |z_1| + |z_2|$.

Proof. One can easily check that (1) through (4) hold.

(5) We have $|z_1 \cdot z_2|^2 = (z_1 \cdot z_2)(\overline{z_1 \cdot z_2}) = (z_1 \cdot \overline{z}_1)(z_2 \cdot \overline{z}_2) = |z_1|^2 \cdot |z_2|^2$, and consequently, $|z_1 \cdot z_2| = |z_1| \cdot |z_2|$, since $|z| \geq 0$ for all $z \in \mathbb{C}$.

(6) Observe that

$$|z_1+z_2|^2 = (z_1+z_2)(\overline{z_1+z_2}) = (z_1+z_2)(\overline{z}_1+\overline{z}_2) = |z_1|^2+z_1\overline{z}_2+\overline{z}_1\cdot z_2+|z_2|^2.$$

Because $\overline{z_1\cdot\overline{z}_2} = \overline{z}_1\cdot\overline{\overline{z}}_2 = \overline{z}_1\cdot z_2$, it follows that

$$z_1\overline{z}_2 + \overline{z}_1\cdot z_2 = 2\mathrm{Re}(z_1\,\overline{z}_2) \le 2|z_1\cdot\overline{z}_2| = 2|z_1|\cdot|z_2|,$$

whence

$$|z_1 + z_2|^2 \le (|z_1| + |z_2|)^2,$$

and consequently, $|z_1 + z_2| \le |z_1| + |z_2|$, as desired.
In order to obtain the inequality on the left-hand side, note that

$$|z_1| = |z_1 + z_2 + (-z_2)| \le |z_1 + z_2| + |-z_2| = |z_1 + z_2| + |z_2|,$$

whence

$$|z_1| - |z_2| \le |z_1 + z_2|.$$

(7) Note that the relation $z \cdot \dfrac{1}{z} = 1$ implies $|z| \cdot \left|\dfrac{1}{z}\right| = 1$, or $\left|\dfrac{1}{z}\right| = \dfrac{1}{|z|}$. Hence $|z^{-1}| = |z|^{-1}$.

(8) We have

$$\left|\frac{z_1}{z_2}\right| = \left|z_1 \cdot \frac{1}{z_2}\right| = |z_1 \cdot z_2^{-1}| = |z_1| \cdot |z_2^{-1}| = |z_1| \cdot |z_2|^{-1} = \frac{|z_1|}{|z_2|}.$$

(9) We can write $|z_1| = |z_1 - z_2 + z_2| \le |z_1 - z_2| + |z_2|$, so $|z_1 - z_2| \ge |z_1| - |z_2|$. On the other hand,

$$|z_1 - z_2| = |z_1 + (-z_2)| \le |z_1| + |-z_2| = |z_1| + |z_2|. \qquad \square$$

Remarks.

(1) The inequality $|z_1 + z_2| \le |z_1| + |z_2|$ becomes an equality if and only if $\mathrm{Re}(z_1\overline{z}_2) = |z_1||z_2|$. This is equivalent to $z_1 = tz_2$, where t is a nonnegative real number.

(2) Properties (5) and (6) can be easily extended to give

(5') $\left|\displaystyle\prod_{k=1}^{n} z_k\right| = \displaystyle\prod_{k=1}^{n} |z_k|$;

(6') $\left|\displaystyle\sum_{k=1}^{n} z_k\right| \le \displaystyle\sum_{k=1}^{n} |z_k|$ for all $z_k \in \mathbb{C}$, $k = 1, \ldots, n$.

As a consequence of (5') and (7), we have

(5'') $|z^n| = |z|^n$ for every integer n and complex number z, provided that $z \ne 0$ for $n < 0$.

Problem 1. *Prove the identity*

$$|z_1 + z_2|^2 + |z_1 - z_2|^2 = 2(|z_1|^2 + |z_2|^2)$$

for all complex numbers z_1, z_2.

Solution. Using property (4) in the proposition above, we obtain

$$\begin{aligned}
|z_1 + z_2|^2 + |z_1 - z_2|^2 &= (z_1 + z_2)(\bar{z}_1 + \bar{z}_2) + (z_1 - z_2)(\bar{z}_1 - \bar{z}_2) \\
&= |z_1|^2 + z_1 \cdot \bar{z}_2 + z_2 \cdot \bar{z}_1 + |z_2|^2 + |z_1|^2 - z_1 \cdot \bar{z}_2 \cdot - z_2 \cdot \bar{z}_1 + |z_2|^2 \\
&= 2(|z_1|^2 + |z_2|^2).
\end{aligned}$$

Problem 2. *Prove that if $|z_1| = |z_2| = 1$ and $z_1 z_2 \neq -1$, then $\dfrac{z_1 + z_2}{1 + z_1 z_2}$ is a real number.*

Solution. Using again property (4) in the above proposition, we have

$$z_1 \cdot \bar{z}_1 = |z_1|^2 = 1 \text{ and } \bar{z}_1 = \frac{1}{z_1}.$$

Likewise, $\bar{z}_2 = \frac{1}{z_2}$. Hence denoting by A the number in the problem, we have

$$\bar{A} = \frac{\bar{z}_1 + \bar{z}_2}{1 + \bar{z}_1 \cdot \bar{z}_2} = \frac{\frac{1}{z_1} + \frac{1}{z_2}}{1 + \frac{1}{z_1} \cdot \frac{1}{z_2}} = \frac{z_1 + z_2}{1 + z_1 z_2} = A,$$

so A is a real number.

Problem 3. *Let a be a positive real number and let*

$$M_a = \left\{ z \in \mathbb{C}^* : \left| z + \frac{1}{z} \right| = a \right\}.$$

Find the minimum and maximum values of $|z|$ when $z \in M_a$.

Solution. Squaring both sides of the equality $a = \left| z + \frac{1}{z} \right|$, we get

$$\begin{aligned}
a^2 = \left| z + \frac{1}{z} \right|^2 &= \left(z + \frac{1}{z} \right) \left(\bar{z} + \frac{1}{\bar{z}} \right) = |z|^2 + \frac{z^2 + (\bar{z})^2}{|z|^2} + \frac{1}{|z|^2} \\
&= \frac{|z|^4 + (z + \bar{z})^2 - 2|z|^2 + 1}{|z|^2}.
\end{aligned}$$

Hence

$$|z|^4 - |z|^2 \cdot (a^2 + 2) + 1 = -(z + \bar{z})^2 \leq 0,$$

and consequently,

$$|z|^2 \in \left[\frac{a^2 + 2 - \sqrt{a^4 + 4a^2}}{2}, \frac{a^2 + 2 + \sqrt{a^4 + 4a^2}}{2} \right].$$

It follows that $|z| \in \left[\frac{-a + \sqrt{a^2 + 4}}{2}, \frac{a + \sqrt{a^2 + 4}}{2} \right]$, so

$$\max |z| = \frac{a + \sqrt{a^2 + 4}}{2}, \quad \min |z| = \frac{-a + \sqrt{a^2 + 4}}{2},$$

and the extreme values are obtained for the complex numbers in M satisfying $z = -\bar{z}$.

Problem 4. *Prove that for every complex number z,*

$$|z + 1| \geq \frac{1}{\sqrt{2}} \ or \ |z^2 + 1| \geq 1.$$

Solution. Suppose by way of contradiction that

$$|1 + z| < \frac{1}{\sqrt{2}} \ \text{and} \ |1 + z^2| < 1.$$

Setting $z = a + bi$ with $a, \ b \in \mathbb{R}$ yields $z^2 = a^2 - b^2 + 2abi$. We obtain

$$(1 + a^2 - b^2)^2 + 4a^2b^2 < 1 \ \text{and} \ (1 + a)^2 + b^2 < \frac{1}{2},$$

and consequently,

$$(a^2 + b^2)^2 + 2(a^2 - b^2) < 0 \ \text{and} \ 2(a^2 + b^2) + 4a + 1 < 0.$$

Summing these inequalities implies

$$(a^2 + b^2)^2 + (2a + 1)^2 < 0,$$

which is a contradiction.

Problem 5. *Prove that*

$$\sqrt{3} \leq |1 + z| + |1 - z + z^2| \leq \frac{13}{4}$$

for all complex numbers z such that $|z| = 1$.

Solution. Let $t = |1 + z| \in [0, 2]$. We have

$$t^2 = (1 + z)(1 + \bar{z}) = 2 + 2\mathrm{Re}(z), \quad \text{so} \quad \mathrm{Re}(z) = \frac{t^2 - 2}{2}.$$

Then

$$|1 - z + z^2|^2 = (1 - z + z^2)(1 - \overline{z} + \overline{z}^2) = (1 - z + z^2)\left(1 - \frac{1}{z} + \frac{1}{z^2}\right)$$

$$= \left(\frac{z^2 - z + 1}{z}\right)^2 = \left(z + \frac{1}{z} - 1\right)^2 = (z + \overline{z} - 1)^2 = [2\mathrm{Re}(z) - 1]^2 = (t^2 - 3)^2,$$

and we obtain $|1 - z + z^2| = |t^2 - 3|$. It suffices to find the extreme values of the function (Fig. 1.1).

$$f : [0, 2] \to \mathbb{R}, \quad f(t) = t + |t^2 - 3| = \begin{cases} -t^2 + t + 3 & \text{if } t \in [0, \sqrt{3}], \\ t^2 + t - 3 & \text{if } t \in [\sqrt{3}, 2]. \end{cases}$$

The graph of the function f is shown in Figure 1.1, and we obtain

$$f(\sqrt{3}) = \sqrt{3} \le t + |t^2 - 3| \le f\left(\frac{1}{2}\right) = \frac{13}{4}.$$

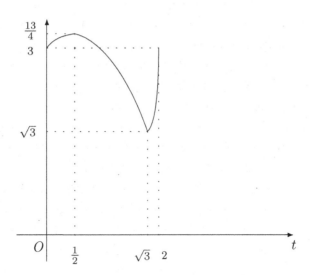

Figure 1.1.

Problem 6. *Consider the set*

$$H = \{z \in \mathbb{C} : z = x - 1 + xi, \quad x \in \mathbb{R}\}.$$

Prove that there is a unique number $z \in H$ such that $|z| \le |w|$ for all $w \in H$.

Solution. Let $\omega = y - 1 + yi$, with $y \in \mathbb{R}$.

It suffices to prove that there is a unique number $x \in \mathbb{R}$ such that

$$(x - 1)^2 + x^2 \le (y - 1)^2 + y^2$$

for all $y \in \mathbb{R}$.

In other words, x is the minimum point of the function

$$f : \mathbb{R} \to \mathbb{R}, \ f(y) = (y - 1)^2 + y^2 = 2y^2 - 2y + 1 = 2\left(y - \frac{1}{2}\right)^2 + \frac{1}{2},$$

whence $x = \frac{1}{2}$ and $z = -\frac{1}{2} + \frac{1}{2}i$.

Problem 7. *Let x, y, z be distinct complex numbers such that*

$$y = tx + (1 - t)z, \quad t \in (0, \ 1).$$

Prove that

$$\frac{|z| - |y|}{|z - y|} \ge \frac{|z| - |x|}{|z - x|} \ge \frac{|y| - |x|}{|y - x|}.$$

Solution. The relation $y = tx + (1 - t)z$ is equivalent to

$$z - y = t(z - x).$$

The inequality

$$\frac{|z| - |y|}{|z - y|} \ge \frac{|z| - |x|}{|z - x|}$$

becomes

$$|z| - |y| \ge t\,(|z| - |x|),$$

and consequently,

$$|y| \le (1 - t)|z| + t|x|.$$

This is the triangle inequality for

$$y = (1 - t)z + tx.$$

The second inequality can be proved similarly, by writing the equality

$$y = tx + (1 - t)z$$

as

$$y - x = (1 - t)(z - x).$$

1.1.8 Solving Quadratic Equations

We are now able to solve the quadratic equation with real coefficients

$$ax^2 + bx + c = 0, \ a \neq 0,$$

in the case that its discriminant $\Delta = b^2 - 4ac$ is negative.

By completing the square, we easily get the equivalent form

$$a \left[\left(x + \frac{b}{2a} \right)^2 + \frac{-\Delta}{4a^2} \right] = 0.$$

Therefore,

$$\left(x + \frac{b}{2a} \right)^2 - i^2 \left(\frac{\sqrt{-\Delta}}{2a} \right)^2 = 0,$$

and so $x_1 = \dfrac{-b + i\sqrt{-\Delta}}{2a}, \ x_2 = \dfrac{-b - i\sqrt{-\Delta}}{2a}.$

Observe that the roots are conjugate complex numbers, and the factorization formula

$$ax^2 + bx + c = a(x - x_1)(x - x_2)$$

holds even in the case $\Delta < 0$.

Let us consider now the general quadratic equation with complex coefficients

$$az^2 + bz + c = 0, \ a \neq 0.$$

Using the same algebraic manipulation as in the case of real coefficients, we get

$$a \left[\left(z + \frac{b}{2a} \right)^2 - \frac{\Delta}{4a^2} \right] = 0.$$

This is equivalent to

$$\left(z + \frac{b}{2a} \right)^2 = \frac{\Delta}{4a^2},$$

or

$$(2az + b)^2 = \Delta,$$

where $\Delta = b^2 - 4ac$ is also called the discriminant of the quadratic equation. If we set $y = 2az + b$, the equation is reduced to

$$y^2 = \Delta = u + vi,$$

where u and v are real numbers.

This equation has the solutions

$$y_{1,2} = \pm \left(\sqrt{\frac{r+u}{2}} + \mathrm{sgn}(v)\sqrt{\frac{r-u}{2}}\, i \right),$$

where $r = |\Delta|$ and $\mathrm{sgn}(v)$ is the sign of the real number v. Indeed, we have $y_{1,2}^2 = \frac{r+u}{2} + 2i\frac{1}{2}\sqrt{r^2 - u^2}\,\mathrm{sgn}(v) = u + i|v|\mathrm{sgn}(v) = u + iv$.
The roots of the initial equation are

$$z_{1,2} = \frac{1}{2a}(-b + y_{1,2}).$$

Observe that the relations

$$z_1 + z_2 = -\frac{b}{a}, \ z_1 z_2 = \frac{c}{a},$$

between roots and coefficients as well as the factorization formula

$$az^2 + bz + c = a(z - z_1)(z - z_2)$$

are also preserved when the coefficients of the equation are elements of the field of complex numbers $\mathbb{C}$.

Problem 1. *Solve, in complex numbers, the quadratic equation*

$$z^2 - 8(1 - i)z + 63 - 16i = 0.$$

Solution. We have

$$\Delta' = (4 - 4i)^2 - (63 - 16i) = -63 - 16i$$

and $r = |\Delta'| = \sqrt{63^2 + 16^2} = 65$, where $\Delta' = \left(\frac{b}{2}\right)^2 - ac$.
The equation

$$y^2 = -63 - 16i$$

has the solution $y_{1,2} = \pm \left(\sqrt{\frac{65-63}{2}} + i\sqrt{\frac{65+63}{2}} \right) = \pm(1 - 8i)$. It follows that $z_{1,2} = 4 - 4i \pm (1 - 8i)$. Hence

$$z_1 = 5 - 12i \text{ and } z_2 = 3 + 4i.$$

Problem 2. *Let p and q be complex numbers with $q \neq 0$. Prove that if the roots of the quadratic equation $x^2 + px + q^2 = 0$ have the same absolute value, then $\frac{p}{q}$ is a real number.*

(1999 Romanian Mathematical Olympiad, Final Round)

Solution. Let x_1 and x_2 be the roots of the equation and let $r = |x_1| = |x_2|$. Then

$$\frac{p^2}{q^2} = \frac{(x_1 + x_2)^2}{x_1 x_2} = \frac{x_1}{x_2} + \frac{x_2}{x_1} + 2 = \frac{x_1 \overline{x_2}}{r^2} + \frac{x_2 \overline{x_1}}{r^2} + 2 = 2 + \frac{2}{r^2} \mathrm{Re}(x_1 \overline{x_2})$$

is a real number. Moreover,

$$\mathrm{Re}(x_1 \overline{x_2}) \geq -|x_1 \overline{x_2}| = -r^2, \text{ so } \frac{p^2}{q^2} \geq 0.$$

Therefore $\frac{p}{q}$ is a real number, as claimed.

Problem 3. *Let* a, b, c *be distinct nonzero complex numbers with* $|a| = |b| = |c|$.

(a) *Prove that if a root of the equation* $az^2 + bz + c = 0$ *has modulus equal to 1, then* $b^2 = ac$.

(b) *If each of the equations*

$$az^2 + bz + c = 0 \text{ and } bz^2 + cz + a = 0$$

has a root of modulus 1, then $|a - b| = |b - c| = |c - a|$.

Solution.

(a) Let z_1, z_2 be the roots of the equation with $|z_1| = 1$. From $z_2 = \frac{c}{a} \cdot \frac{1}{z_1}$, it follows that $|z_2| = |\frac{c}{a}| \cdot \frac{1}{|z_1|} = 1$. Because $z_1 + z_2 = -\frac{b}{a}$ and $|a| = |b|$, we have $|z_1 + z_2|^2 = 1$. This is equivalent to

$$(z_1 + z_2)(\overline{z_1} + \overline{z_2}) = 1, \text{ i.e., } (z_1 + z_2)\left(\frac{1}{z_1} + \frac{1}{z_2}\right) = 1.$$

We find that

$$(z_1 + z_2)^2 = z_1 z_2, \text{ i.e., } \left(\frac{b}{a}\right)^2 = \frac{c}{a},$$

which reduces to $b^2 = ac$, as desired.

(b) As we have already seen, we have $b^2 = ac$ and $c^2 = ab$. Multiplying these relations yields $b^2 c^2 = a^2 bc$, and hence $a^2 = bc$. Therefore,

$$a^2 + b^2 + c^2 = ab + bc + ca. \tag{1}$$

Relation (1) is equivalent to

$$(a - b)^2 + (b - c)^2 + (c - a)^2 = 0,$$

i.e.,

$$(a - b)^2 + (b - c)^2 + 2(a - b)(b - c) + (c - a)^2 = 2(a - b)(b - c).$$

It follows that $(a-c)^2 = (a-b)(b-c)$. Taking absolute values, we obtain $\beta^2 = \gamma\alpha$, where $\alpha = |b-c|$, $\beta = |c-a|$, $\gamma = |a-b|$. In an analogous way, we obtain $\alpha^2 = \beta\gamma$ and $\gamma^2 = \alpha\beta$. Adding these relations yields $\alpha^2 + \beta^2 + \gamma^2 = \alpha\beta + \beta\gamma + \gamma\alpha$, i.e., $(\alpha-\beta)^2 + (\beta-\gamma)^2 + (\gamma-\alpha)^2 = 0$. Hence $\alpha = \beta = \gamma$.

1.1.9 Problems

1. Consider the complex numbers $z_1 = (1, 2)$, $z_2 = (-2, 3)$, and $z_3 = (1, -1)$. Compute the following:

   (a) $z_1 + z_2 + z_3$; (b) $z_1 z_2 + z_2 z_3 + z_3 z_1$; (c) $z_1 z_2 z_3$;

   (d) $z_1^2 + z_2^2 + z_3^2$; (e) $\dfrac{z_1}{z_2} + \dfrac{z_2}{z_3} + \dfrac{z_3}{z_1}$; (f) $\dfrac{z_1^2 + z_2^2}{z_2^2 + z_3^2}$.

2. Solve the following equations:

   (a) $z + (-5, 7) = (2, -1)$; (b) $(2, 3) + z = (-5, -1)$;
   (c) $z \cdot (2, 3) = (4, 5)$; (d) $\dfrac{z}{(-1, 3)} = (3, 2)$.

3. Solve in $\mathbb{C}$ the equations:

   (a) $z^2 + z + 1 = 0$; (b) $z^3 + 1 = 0$.

4. Let $z = (0, 1) \in \mathbb{C}$. Express $\sum\limits_{k=0}^{n} z^k$ in terms of the positive integer n.

5. Solve the following equations:

   (a) $z \cdot (1, 2) = (-1, 3)$; (b) $(1, 1) \cdot z^2 = (-1, 7)$.

6. Let $z = (a, b) \in \mathbb{C}$. Compute z^2, z^3, and z^4.
7. Let $z_0 = (a, b) \in \mathbb{C}$. Find $z \in \mathbb{C}$ such that $z^2 = z_0$.
8. Let $z = (1, -1)$. Compute z^n, where n is a positive integer.
9. Find real numbers x and y in each of the following cases:

   (a) $(1 - 2i)x + (1 + 2i)y = 1 + i$; (b) $\dfrac{x-3}{3+i} + \dfrac{y-3}{3-i} = i$;
   (c) $(4 - 3i)x^2 + (3 + 2i)xy = 4y^2 - \frac{1}{2}x^2 + (3xy - 2y^2)i$.

10. Compute the following:

    (a) $(2 - i)(-3 + 2i)(5 - 4i)$; (b) $(2 - 4i)(5 + 2i) + (3 + 4i)(-6 - i)$;

    (c) $\left(\dfrac{1+i}{1-i}\right)^{16} + \left(\dfrac{1-i}{1+i}\right)^{8}$; (d) $\left(\dfrac{-1+i\sqrt{3}}{2}\right)^{6} + \left(\dfrac{1-i\sqrt{7}}{2}\right)^{6}$;

    (e) $\dfrac{3+7i}{2+3i} + \dfrac{5-8i}{2-3i}$.

11. Compute the following:

(a) $i^{2000} + i^{1999} + i^{201} + i^{82} + i^{47}$;

(b) $E_n = 1 + i + i^2 + i^3 + \cdots + i^n$ for $n \geq 1$;

(c) $i^1 \cdot i^2 \cdot i^3 \ldots i^{2000}$;

(d) $i^{-5} + (-i)^{-7} + (-i)^{13} + i^{-100} + (-i)^{94}$.

12. Solve in $\mathbb{C}$ the following equations:

(a) $z^2 = i$; (b) $z^2 = -i$; (c) $z^2 = \frac{1}{2} - i\frac{\sqrt{2}}{2}$.

13. Find all complex numbers $z \neq 0$ such that $z + \dfrac{1}{z} \in \mathbb{R}$.

14. Prove the following:

(a) $E_1 = (2 + i\sqrt{5})^7 + (2 - i\sqrt{5})^7 \in \mathbb{R}$;

(b) $E_2 = \left(\dfrac{19 + 7i}{9 - i}\right)^n + \left(\dfrac{20 + 5i}{7 + 6i}\right)^n \in \mathbb{R}$.

15. Prove the following identities:

(a) $|z_1 + z_2|^2 + |z_2 + z_3|^2 + |z_3 + z_1|^2 = |z_1|^2 + |z_2|^2 + |z_3|^2 + |z_1 + z_2 + z_3|^2$;

(b) $|1 + z_1\bar{z}_2|^2 + |z_1 - z_2|^2 = (1 + |z_1|^2)(1 + |z_2|^2)$;

(c) $|1 - z_1\bar{z}_2|^2 - |z_1 - z_2|^2 = (1 - |z_1|^2)(1 - |z_2|^2)$;

(d) $|z_1 + z_2 + z_3|^2 + |-z_1 + z_2 + z_3|^2 + |z_1 - z_2 + z_3|^2 + |z_1 + z_2 - z_3|^2$

$$= 4(|z_1|^2 + |z_2|^2 + |z_3|^2).$$

16. Let $z \in \mathbb{C}^*$ be such that $\left|z^3 + \dfrac{1}{z^3}\right| \leq 2$. Prove that $\left|z + \dfrac{1}{z}\right| \leq 2$.

17. Find all complex numbers z such that

$$|z| = 1 \text{ and } |z^2 + \bar{z}^2| = 1.$$

18. Find all complex numbers z such that

$$4z^2 + 8|z|^2 = 8.$$

19. Find all complex numbers z such that $z^3 = \bar{z}$.

20. Consider $z \in \mathbb{C}$ with $\operatorname{Re}(z) > 1$. Prove that

$$\left|\frac{1}{z} - \frac{1}{2}\right| < \frac{1}{2}.$$

21. Let a, b, c be real numbers and $\omega = -\frac{1}{2} + i\frac{\sqrt{3}}{2}$. Compute

$$(a + b\omega + c\omega^2)(a + b\omega^2 + c\omega).$$

22. Solve the following equations:

(a) $|z| - 2z = 3 - 4i$;

  (b) $|z| + z = 3 + 4i$;
  (c) $z^3 = 2 + 11i$, where $z = x + yi$ and $x, y \in \mathbb{Z}$;
  (d) $iz^2 + (1 + 2i)z + 1 = 0$;
  (e) $z^4 + 6(1 + i)z^2 + 5 + 6i = 0$;
  (f) $(1 + i)z^2 + 2 + 11i = 0$.

23. Find all real numbers m for which the equation

$$z^3 + (3 + i)z^2 - 3z - (m + i) = 0$$

has at least one real root.

24. Find all complex numbers z such that

$$z' = (z - 2)(\overline{z} + i)$$

is a real number.

25. Find all complex numbers z such that $|z| = |\frac{1}{z}|$.

26. Let z_1, $z_2 \in \mathbb{C}$ be complex numbers such that $|z_1 + z_2| = \sqrt{3}$ and $|z_1| = |z_2| = 1$. Compute $|z_1 - z_2|$.

27. Find all positive integers n such that

$$\left(\frac{-1 + i\sqrt{3}}{2}\right)^n + \left(\frac{-1 - i\sqrt{3}}{2}\right)^n = 2.$$

28. Let $n > 2$ be an integer. Find the number of solutions to the equation

$$z^{n-1} = i\overline{z}.$$

29. Let z_1, z_2, z_3 be complex numbers with

$$|z_1| = |z_2| = |z_3| = R > 0.$$

Prove that

$$|z_1 - z_2| \cdot |z_2 - z_3| + |z_3 - z_1| \cdot |z_1 - z_2| + |z_2 - z_3| \cdot |z_3 - z_1| \leq 9R^2.$$

30. Let u, v, w, z be complex numbers such that $|u| < 1$, $|v| = 1$, and $w = \dfrac{v(u - z)}{\overline{u} \cdot z - 1}$. Prove that $|w| \leq 1$ if and only if $|z| \leq 1$.

31. Let z_1, z_2, z_3 be complex numbers such that

$$z_1 + z_2 + z_3 = 0 \text{ and } |z_1| = |z_2| = |z_3| = 1.$$

Prove that

$$z_1^2 + z_2^2 + z_3^2 = 0.$$

32. Consider the complex numbers z_1, z_2, $\ldots$, z_n with

$$|z_1| = |z_2| = \cdots = |z_n| = r > 0.$$

Prove that the number

$$E = \frac{(z_1 + z_2)(z_2 + z_3) \cdots (z_{n-1} + z_n)(z_n + z_1)}{z_1 \cdot z_2 \ldots z_n}$$

is real.

33. Let z_1, z_2, z_3 be distinct complex numbers such that

$$|z_1| = |z_2| = |z_3| > 0.$$

If $z_1 + z_2 z_3$, $z_2 + z_1 z_3$, and $z_3 + z_1 z_2$ are real numbers, prove that $z_1 z_2 z_3 = 1$.

34. Let x_1 and x_2 be the roots of the equation $x^2 - x + 1 = 0$. Compute the following:

(a) $x_1^{2000} + x_2^{2000}$; (b) $x_1^{1999} + x_2^{1999}$; (c) $x_1^n + x_2^n$, for $n \in \mathbb{N}$.

35. Factorize (in linear polynomials) the following polynomials:

(a) $x^4 + 16$; (b) $x^3 - 27$; (c) $x^3 + 8$; (d) $x^4 + x^2 + 1$.

36. Find all quadratic equations with real coefficients that have one of the following roots:

(a) $(2 + i)(3 - i)$; (b) $\dfrac{5 + i}{2 - i}$; (c) $i^{51} + 2i^{80} + 3i^{45} + 4i^{38}$.

37. (Hlawka's inequality) Prove that the inequality

$$|z_1 + z_2| + |z_2 + z_3| + |z_3 + z_1| \leq |z_1| + |z_2| + |z_3| + |z_1 + z_2 + z_3|$$

holds for all complex numbers z_1, z_2, z_3.

38. Suppose that complex numbers x_i, y_i, $i = 1, 2, \ldots, n$, satisfy $|x_i| = |y_i| = 1$. Let

$$x = \frac{1}{n} \sum_{i=1}^{n} x_i, \; y = \frac{1}{n} \sum_{i=1}^{n} y_i, \text{ and } z_i = xy_i + yx_i - x_i y_i.$$

Prove that $\displaystyle\sum_{i=1}^{n} |z_i| \leq n$.

1.2 Geometric Interpretation of the Algebraic Operations

1.2.1 Geometric Interpretation of a Complex Number

We have defined a complex number $z = (x, y) = x + yi$ to be an ordered pair of real numbers $(x, y) \in \mathbb{R} \times \mathbb{R}$, so it is natural to let a complex number $z = x + yi$ correspond to a point $M(x, y)$ in the plane $\mathbb{R} \times \mathbb{R}$.

For a formal introduction, let us consider $\mathcal{P}$ to be the set of points of a given plane Π equipped with a coordinate system xOy. Consider the bijective function $\varphi : \mathbb{C} \to \mathcal{P}$, $\varphi(z) = M(x, y)$.

Definition. The point $M(x, y)$ is called the *geometric image* of the complex number $z = x + yi$.

The complex number $z = x + yi$ is called the *complex coordinate* of the point $M(x, y)$. We will use the notation $M(z)$ to indicate that the complex coordinate of M is the complex number z.

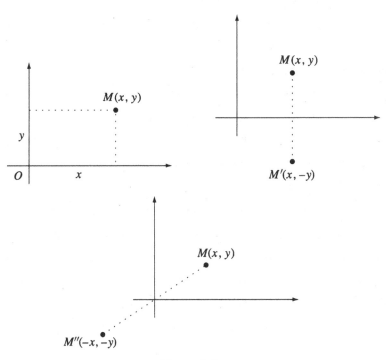

Figure 1.2.

The geometric image of the complex conjugate $\bar{z}$ of a complex number $z = x + yi$ is the reflection point $M'(x, -y)$ across the x-axis of the point $M(x, y)$ (see Fig. 1.2).

The geometric image of the additive inverse $-z$ of a complex number $z = x + yi$ is the reflection $M''(-x, -y)$ across the origin of the point $M(x, y)$ (see Fig. 1.2).

The bijective function φ maps the set $\mathbb{R}$ onto the x-axis, which is called the *real axis*. On the other hand, the imaginary complex numbers correspond to the y-axis, which is called the *imaginary axis*. The plane Π, whose points are identified with complex numbers, is called the *complex plane*.

On the other hand, we can also identify a complex number $z = x + yi$ with the vector $\vec{v} = \overrightarrow{OM}$, where $M(x, y)$ is the geometric image of the complex number z (Fig. 1.3).

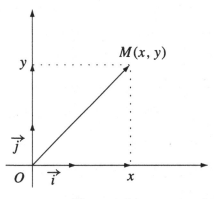

$M(x, y)$

Figure 1.3.

Let V_0 be the set of vectors whose initial points are the origin O. Then we can define the bijective function

$$\varphi' : \mathbb{C} \to V_0, \ \varphi'(z) = \overrightarrow{OM} = \vec{v} = x\,\vec{i} + y\,\vec{j},$$

where $\vec{i}, \vec{j}$ are the unit vectors of the x-axis and y-axis, respectively.

1.2.2 Geometric Interpretation of the Modulus

Let us consider a complex number $z = x + yi$ and the geometric image $M(x, y)$ in the complex plane. The Euclidean distance OM is given by the formula

$$OM = \sqrt{(x_M - x_O)^2 + (y_M - y_O)^2};$$

hence $OM = \sqrt{x^2 + y^2} = |z| = |\overrightarrow{v}|$. In other words, the absolute value $|z|$ of a complex number $z = x + yi$ is the length of the segment OM or the magnitude of the vector $\overrightarrow{v} = x\,\overrightarrow{i} + y\,\overrightarrow{j}$.

Remarks.

(a) For a positive real number r, the set of complex numbers with modulus r corresponds in the complex plane to $\mathcal{C}(O; r)$, our notation for the circle $\mathcal{C}$ with center O and radius r.

(b) The complex numbers z with $|z| < r$ correspond to the interior points of circle $\mathcal{C}$; on the other hand, the complex numbers z with $|z| > r$ correspond to the points in the exterior of circle $\mathcal{C}$.

Example. The numbers $z_k = \pm\dfrac{1}{2} \pm \dfrac{\sqrt{3}}{2}i$, $k = 1, 2, 3, 4$, are represented in the complex plane by four points on the unit circle centered at the origin, since

$$|z_1| = |z_2| = |z_3| = |z_4| = 1.$$

1.2.3 Geometric Interpretation of the Algebraic Operations

(a) **Addition and subtraction.** Consider the complex numbers $z_1 = x_1 + y_1 i$ and $z_2 = x_2 + y_2 i$ and the corresponding vectors $\overrightarrow{v}_1 = x_1\,\overrightarrow{i} + y_1\,\overrightarrow{j}$ and $\overrightarrow{v}_2 = x_2\,\overrightarrow{i} + y_2\,\overrightarrow{j}$. Observe that the sum of the complex numbers is

$$z_1 + z_2 = (x_1 + x_2) + (y_1 + y_2)i,$$

and the sum of the vectors is

$$\overrightarrow{v}_1 + \overrightarrow{v}_2 = (x_1 + x_2)\,\overrightarrow{i} + (y_1 + y_2)\,\overrightarrow{j}.$$

Therefore, the sum $z_1 + z_2$ corresponds to the sum $\overrightarrow{v}_1 + \overrightarrow{v}_2$ (Fig. 1.4).

Examples.

(1) We have $(3 + 5i) + (6 + i) = 9 + 6i$; hence the geometric image of the sum is given in Fig. 1.5.

(2) Observe that $(6 - 2i) + (-2 + 5i) = 4 + 3i$. Therefore, the geometric image of the sum of these two complex numbers is the point $M(4, 3)$ (see Fig. 1.6).

On the other hand, the difference of the complex numbers z_1 and z_2 is

$$z_1 - z_2 = (x_1 - x_2) + (y_1 - y_2)i,$$

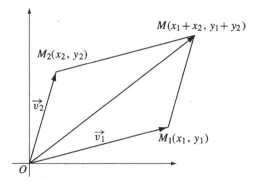

Figure 1.4.

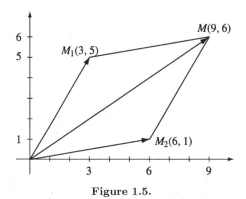

Figure 1.5.

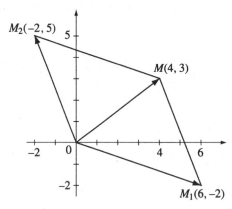

Figure 1.6.

and the difference of the vectors v_1 and v_2 is

$$\vec{v}_1 - \vec{v}_2 = (x_1 - x_2)\,\vec{i} + (y_1 - y_2)\,\vec{j}\,.$$

Hence, the difference $z_1 - z_2$ corresponds to the difference $\vec{v}_1 - \vec{v}_2$.

(3) We have $(-3+i) - (2+3i) = (-3+i) + (-2-3i) = -5 - 2i$; hence the geometric image of the difference of these two complex numbers is the point $M(-5,\ -2)$ given in Fig. 1.7.

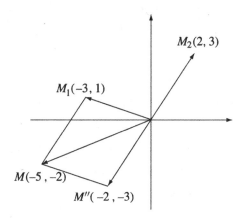

Figure 1.7.

(4) Note that $(3\ - 2i) - (-2 - 4i) = (3 - 2i) + (2 + 4i) = 5 + 2i$; we obtain the point $M_2(-2,\ -4)$ as the geometric image of the difference of these two complex numbers (see Fig. 1.8).

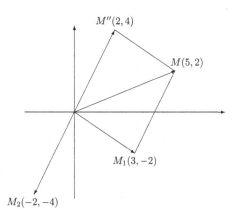

Figure 1.8.

Remark. The distance between $M_1(x_1, \ y_1)$ and $M_2(x_2, \ y_2)$ is equal to the modulus of the complex number $z_1 - z_2$, or to the length of the vector $\overrightarrow{v}_1 - \overrightarrow{v}_2$. Indeed,

$$|M_1 M_2| = |z_1 - z_2| = |\overrightarrow{v}_1 - \overrightarrow{v}_2| = \sqrt{(x_2 - x_1)^2 + (y_2 - y_1)^2}.$$

(b) **Real multiples of a complex number.** Consider a complex number $z = x + iy$ and the corresponding vector $\overrightarrow{v} = x \overrightarrow{i} + y \overrightarrow{j}$. If λ is a real number, then the real multiple $\lambda z = \lambda x + i \lambda y$ corresponds to the vector

$$\lambda \overrightarrow{v} = \lambda x \overrightarrow{i} + \lambda y \overrightarrow{j}.$$

Note that if $\lambda > 0$, then the vectors $\lambda \overrightarrow{v}$ and $\overrightarrow{v}$ have the same orientation and

$$|\lambda \overrightarrow{v}| = \lambda |\overrightarrow{v}|.$$

When $\lambda < 0$, the vector $\lambda \overrightarrow{v}$ changes to the opposite orientation, and $|\lambda \overrightarrow{v}| = -\lambda |\overrightarrow{v}|$. Of course, if $\lambda = 0$, then $\lambda \overrightarrow{v} = \overrightarrow{0}$ (Fig. 1.9).

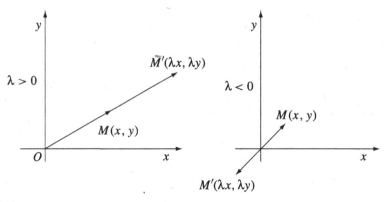

Figure 1.9.

Examples.

(1) We have $3(1 + 2i) = 3 + 6i$; therefore, $M'(3, \ 6)$ is the geometric image of the product of 3 and $z = 1 + 2i$.

(2) Observe that $-2(-3 + 2i) = 6 - 4i$; we obtain the point $M'(6, \ -4)$ as the geometric image of the product of -2 and $z = -3 + 2i$ (Fig. 1.10).

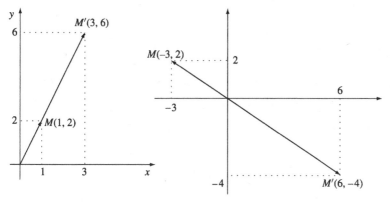

Figure 1.10.

1.2.4 Problems

1. Represent the geometric images of the following complex numbers:

$$z_1 = 3 + i; \ z_2 = -4 + 2i; \ z_3 = -5 - 4i; \ z_4 = 5 - i;$$
$$z_5 = 1; \ z_6 = -3i; \ z_7 = 2i; \ z_8 = -4.$$

2. Find the geometric interpretation for the following equalities:

(a) $(-5 + 4i) + (2 - 3i) = -3 + i$;
(b) $(4 - i) + (-6 + 4i) = -2 + 3i$;
(c) $(-3 - 2i) - (-5 + i) = 2 - 3i$;
(d) $(8 - i) - (5 + 3i) = 3 - 4i$;
(e) $2(-4 + 2i) = -8 + 4i$;
(f) $-3(-1 + 2i) = 3 - 6i$.

3. Find the geometric image of the complex number z in each of the following cases:

(a) $|z - 2| = 3$; (b) $|z + i| < 1$; (c) $|z - 1 + 2i| > 3$;
(d) $|z - 2| - |z + 2| < 2$; (e) $0 < \mathrm{Re}(iz) < 1$; (f) $-1 < \mathrm{Im}(z) < 1$;
(g) $\mathrm{Re}\left(\dfrac{z-2}{z-1}\right) = 0$; (h) $\dfrac{1+\bar{z}}{z} \in \mathbb{R}$.

4. Find the set of points $P(x, \ y)$ in the complex plane such that

$$\left|\sqrt{x^2 + 4} + i\sqrt{y - 4}\right| = \sqrt{10}.$$

5. Let $z_1 = 1 + i$ and $z_2 = -1 - i$. Find $z_3 \in \mathbb{C}$ such that the triangle z_1, z_2, z_3 is equilateral.

6. Find the geometric images of the complex numbers z such that the triangle with vertices at z, z^2, and z^3 is a right triangle.

7. Find the geometric images of the complex numbers z such that

$$\left| z + \frac{1}{z} \right| = 2.$$

Chapter 2
Complex Numbers in Trigonometric Form

2.1 Polar Representation of Complex Numbers

2.1.1 Polar Coordinates in the Plane

Let us consider a coordinate plane and a point $M(x, y)$ that is not the origin.

The real number $r = \sqrt{x^2 + y^2}$ is called the *polar radius* of the point M. The direct angle $t^* \in [0, 2\pi)$ between the vector $\overrightarrow{OM}$ and the positive x-axis is called the *polar argument* of the point M. The pair (r, t^*) are called the *polar coordinates* of the point M. We will write $M(r, t^*)$. Note that the function $h : \mathbb{R} \times \mathbb{R}\backslash\{(0,0)\} \to (0, \infty) \times [0, 2\pi)$, $h((x, y)) = (r, t^*)$ is bijective.

The origin O is the unique point such that $r = 0$; the argument t^* of the origin is not defined.

For each point M in the plane there is a unique intersection point P of the ray $(OM$ with the unit circle centered at the origin. The point P has the same polar argument t^*. Using the definition of the sine and cosine functions, we find that

$$x = r \cos t^* \text{ and } y = r \sin t^*.$$

Therefore, it is easy to obtain the Cartesian coordinates of a point from its polar coordinates (Fig. 2.1).

Conversely, let us consider a point $M(x, y)$. The polar radius is

$$r = \sqrt{x^2 + y^2}.$$

To determine the polar argument, we study the following cases:

(a) If $x \neq 0$, from $\tan t^* = \dfrac{y}{x}$ we deduce that

$$t^* = \arctan \frac{y}{x} + k\pi,$$

T. Andreescu and D. Andrica, *Complex Numbers from A to ... Z,*
DOI 10.1007/978-0-8176-8415-0_2, © Springer Science+Business Media New York 2014

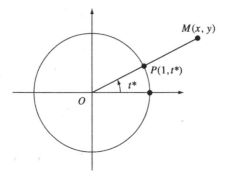

Figure 2.1.

where

$$k = \begin{cases} 0, & \text{for} \quad x > 0 \quad \text{and } y \ge 0, \\ 1, & \text{for} \quad x < 0 \quad \text{and any } y, \\ 2, & \text{for} \quad x > 0 \quad \text{and } y < 0. \end{cases}$$

(b) If $x = 0$ and $y \neq 0$, then

$$t^* = \begin{cases} \pi/2, & \text{for} \quad y > 0, \\ 3\pi/2, & \text{for} \quad y < 0. \end{cases}$$

Examples.

1. Let us find the polar coordinates of the points $M_1(2, -2)$, $M_2(-1, 0)$, $M_3(-2\sqrt{3}, -2)$, $M_4(\sqrt{3}, 1)$, $M_5(3, 0)$, $M_6(-2, 2)$, $M_7(0, 1)$ and $M_8(0, -4)$. In this case, we have $r_1 = \sqrt{2^2 + (-2)^2} = 2\sqrt{2}$; $t_1^* = \arctan(-1) + 2\pi = -\frac{\pi}{4} + 2\pi = \frac{7\pi}{4}$, so $M_1\left(2\sqrt{2}, \frac{7\pi}{4}\right)$.

Observe that $r_2 = 1$, $t_2^* = \arctan 0 + \pi = \pi$, so $M_2(1, \pi)$.

We have $r_3 = 4$, $t_3^* = \arctan\frac{\sqrt{3}}{3} + \pi = \frac{\pi}{6} + \pi = \frac{7\pi}{6}$, so $M_3\left(4, \frac{7\pi}{6}\right)$.

Note that $r_4 = 2$, $t_4^* = \arctan\frac{\sqrt{3}}{3} = \frac{\pi}{6}$, so $M_4\left(2, \frac{\pi}{6}\right)$.

We have $r_5 = 3$, $t_5^* = \arctan 0 + 0 = 0$, so $M_5(3, 0)$.

We have $r_6 = 2\sqrt{2}$, $t_6^* = \arctan(-1) + \pi = -\frac{\pi}{4} + \pi = \frac{3\pi}{4}$, so $M_6\left(2\sqrt{2}, \frac{3\pi}{4}\right)$.

Note that $r_7 = 1$, $t_7^* = \frac{\pi}{2}$, so $M_7\left(1, \frac{\pi}{2}\right)$.

Observe that $r_8 = 4$, $t_8^* = \frac{3\pi}{2}$, so $M_8\left(1, \frac{3\pi}{2}\right)$.

2. Let us find the Cartesian coordinates of the following points given in polar coordinates: $M_1\left(2, \frac{2\pi}{3}\right)$, $M_2\left(3, \frac{7\pi}{4}\right)$, and $M_3(1, 1)$.

We have $x_1 = 2\cos\dfrac{2\pi}{3} = 2\left(-\dfrac{1}{2}\right) = -1$, $y_1 = 2\sin\dfrac{2\pi}{3} = 2\dfrac{\sqrt{3}}{2} = \sqrt{3}$, so
$M_1(-1,\ \sqrt{3})$.
Note that

$$x_2 = 3\cos\frac{7\pi}{4} = \frac{3\sqrt{2}}{2},\quad y_2 = 3\sin\frac{7\pi}{4} = -\frac{3\sqrt{2}}{2},$$

so

$$M_2\left(\frac{3\sqrt{2}}{2}, -\frac{3\sqrt{2}}{2}\right).$$

Observe that $x_3 = \cos 1$, $y_3 = \sin 1$, so $M_3(\cos 1,\ \sin 1)$.

2.1.2 Polar Representation of a Complex Number

For a complex number $z = x + yi$, we can write the polar representation

$$z = r(\cos t^* + i\sin t^*),$$

where $r \in [0,\ \infty)$ and $t^* \in [0,\ 2\pi)$ are the polar coordinates of the geometric image of z.

The polar argument t^* of the geometric image of z is called the *principal* (or *reduced*) *argument* of z, denoted by $\arg z$. The polar radius r of the geometric image of z is equal to the modulus of z. For $z \neq 0$, the modulus and argument of z are uniquely determined.

Consider $z = r(\cos t^* + i\sin t^*)$ and let $t = t^* + 2k\pi$ for an integer k. Then

$$z = r[\cos(t - 2k\pi) + i\sin(t - 2k\pi)] = r(\cos t + i\sin t),$$

i.e., every complex number z can be represented as $z = r(\cos t + i\sin t)$, where $r \geq 0$ and $t \in \mathbb{R}$. The set $\operatorname{Arg} z = \{t : t^* + 2k\pi,\ k \in \mathbb{Z}\}$ is called the *extended argument* of the complex number z.

Therefore, two complex numbers z_1, $z_2 \neq 0$ represented as

$$z_1 = r_1(\cos t_1 + i\sin t_1)\ \text{and}\ z_2 = r_2(\cos t_2 + i\sin t_2)$$

are equal if and only if $r_1 = r_2$ and $t_1 - t_2 = 2k\pi$, for some integer k.

Example 1. Let us find the polar representation of the following numbers:

(a) $z_1 = -1 - i$,
(b) $z_2 = 2 + 2i$,
(c) $z_3 = -1 + i\sqrt{3}$,
(d) $z_4 = 1 - i\sqrt{3}$,

and determine their extended arguments (Fig. 2.2).

(a) As in Figure 2.2, the geometric image $P_1(-1, -1)$ lies in the third quadrant. Then $r_1 = \sqrt{(-1)^2 + (-1)^2} = \sqrt{2}$ and

$$t_1^* = \arctan \frac{y}{x} + \pi = \arctan 1 + \pi = \frac{\pi}{4} + \pi = \frac{5\pi}{4}.$$

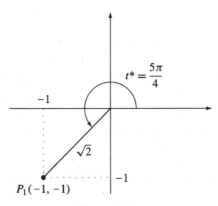

Figure 2.2.

Hence

$$z_1 = \sqrt{2}\left(\cos\frac{5\pi}{4} + i\sin\frac{5\pi}{4}\right)$$

and

$$\operatorname{Arg} z_1 = \left\{\frac{5\pi}{4} + 2k\pi \mid k \in \mathbb{Z}\right\}.$$

(b) The point $P_2(2, 2)$ lies in the first quadrant, so we can write

$$r_2 = \sqrt{2^2 + 2^2} = 2\sqrt{2} \text{ and } t_2^* = \arctan 1 = \frac{\pi}{4}.$$

Hence

$$z_2 = 2\sqrt{2}\left(\cos\frac{\pi}{4} + i\sin\frac{\pi}{4}\right)$$

and

$$\operatorname{Arg} z = \left\{\frac{\pi}{4} + 2k\pi \mid k \in \mathbb{Z}\right\}.$$

(c) The point $P_3(-1, \sqrt{3})$ lies in the second quadrant, so (Fig. 2.3)

$$r_3 = 2 \text{ and } t_3^* = \arctan(-\sqrt{3}) + \pi = -\frac{\pi}{3} + \pi = \frac{2\pi}{3}.$$

Therefore,

$$z_3 = 2\left(\cos\frac{2\pi}{3} + i\sin\frac{2\pi}{3}\right)$$

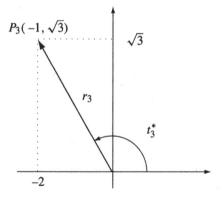

Figure 2.3.

and

$$\operatorname{Arg} z_3 = \left\{ \frac{2\pi}{3} + 2k\pi \,\middle|\, k \in \mathbb{Z} \right\}.$$

(d) The point $P_4(1, -\sqrt{3})$ lies in the fourth quadrant (Fig. 2.4), so

$$r_4 = 2 \text{ and } t_4^* = \arctan(-\sqrt{3}) + 2\pi = -\frac{\pi}{3} + 2\pi = \frac{5\pi}{3}.$$

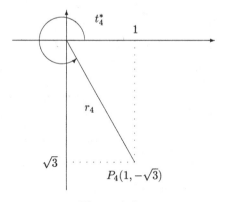

Figure 2.4.

Hence

$$z_4 = 2 \left(\cos \frac{5\pi}{3} + i \sin \frac{5\pi}{3} \right)$$

and

$$\operatorname{Arg} z_4 = \left\{ \frac{5\pi}{3} + 2k\pi \,\middle|\, k \in \mathbb{Z} \right\}.$$

Example 2. Let us find the polar representation of the following numbers:

(a) $z_1 = 2i$,
(b) $z_2 = -1$,
(c) $z_3 = 2$,
(d) $z_4 = -3i$,

and determine their extended arguments.

(a) The point $P_1(0,\ 2)$ lies on the positive y-axis, so

$$r_1 = 2, \quad t_1^* = \frac{\pi}{2}, \quad z_1 = 2\left(\cos\frac{\pi}{2} + i\sin\frac{\pi}{2}\right)$$

and

$$\operatorname{Arg} z_1 = \left\{\frac{\pi}{2} + 2k\pi \,\middle|\, k \in \mathbb{Z}\right\}.$$

(b) The point $P_2(-1,0)$ lies on the negative x-axis, so

$$r_2 = 1, \quad t_2^* = \pi, \quad z_2 = \cos\pi + i\sin\pi$$

and

$$\operatorname{Arg} z_2 = \{\pi + 2k\pi \,|\, k \in \mathbb{Z}\}.$$

(c) The point $P_3(2,\ 0)$ lies on the positive x-axis, so

$$r_3 = 2, \quad t_3^* = 0, \quad z_3 = 2(\cos 0 + i\ \sin\ 0)$$

and

$$\operatorname{Arg} z_3 = \{2k\pi \,|\, k \in \mathbb{Z}\}.$$

(d) The point $P_4(0,\ -3)$ lies on the negative y-axis, so

$$r_4 = 3, \quad t_4^* = \frac{3\pi}{2}, \quad z_4 = 3\left(\cos\frac{3\pi}{2} + i\sin\frac{3\pi}{2}\right)$$

and

$$\operatorname{Arg} z_4 = \left\{\frac{3\pi}{2} + 2k\pi \,\middle|\, k \in \mathbb{Z}\right\}.$$

Remark. The following formulas should be memorized:

$$1 = \cos 0 + i\sin 0; \quad i = \cos\frac{\pi}{2} + i\sin\frac{\pi}{2};$$

$$-1 = \cos\pi + i\sin\pi; \quad -i = \cos\frac{3\pi}{2} + i\sin\frac{3\pi}{2}.$$

Problem 1. *Find the polar representation of the complex number*

$$z = 1 + \cos a + i\sin a, \quad a \in (0,\ 2\pi).$$

Solution. The modulus is

$$|z| = \sqrt{(1 + \cos a)^2 + \sin^2 a} = \sqrt{2(1 + \cos a)} = \sqrt{4 \cos^2 \frac{a}{2}} = 2 \left| \cos \frac{a}{2} \right|.$$

The argument of z is determined as follows:

(a) If $a \in (0, \pi)$, then $\frac{a}{2} \in \left(0, \frac{\pi}{2}\right)$, and the point $P(1 + \cos a, \sin a)$ lies in the first quadrant. Hence

$$t^* = \arctan \frac{\sin a}{1 + \cos a} = \arctan \left(\tan \frac{a}{2}\right) = \frac{a}{2},$$

and in this case,

$$z = 2 \cos \frac{a}{2} \left(\cos \frac{a}{2} + i \sin \frac{a}{2}\right).$$

(b) If $a \in (\pi, 2\pi)$, then $\frac{a}{2} \in \left(\frac{\pi}{2}, \pi\right)$, and the point $P(1 + \cos a, \sin a)$ lies in the fourth quadrant. Hence

$$t^* = \arctan \left(\tan \frac{a}{2}\right) + 2\pi = \frac{a}{2} - \pi + 2\pi = \frac{a}{2} + \pi$$

and

$$z = -2 \cos \frac{a}{2} \left(\cos \left(\frac{a}{2} + \pi\right) + i \sin \left(\frac{a}{2} + \pi\right)\right).$$

(c) If $a = \pi$, then $z = 0$.

Problem 2. *Find all complex numbers z such that $|z| = 1$ and*

$$\left| \frac{z}{\overline{z}} + \frac{\overline{z}}{z} \right| = 1.$$

Solution. Let $z = \cos x + i \sin x$, $x \in [0, 2\pi)$. Then

$$1 = \left| \frac{z}{\overline{z}} + \frac{\overline{z}}{z} \right| = \frac{|z^2 + \overline{z}^2|}{|z|^2} = |\cos 2x + i \sin 2x + \cos 2x - i \sin 2x| = 2|\cos 2x|,$$

whence

$$\cos 2x = \frac{1}{2} \text{ or } \cos 2x = -\frac{1}{2}.$$

If $\cos 2x = \frac{1}{2}$, then

$$x_1 = \frac{\pi}{6}, \quad x_2 = \frac{5\pi}{6}, \quad x_3 = \frac{7\pi}{6}, \quad x_4 = \frac{11\pi}{6}.$$

If $\cos 2x = -\frac{1}{2}$, then

$$x_5 = \frac{\pi}{3}, \quad x_6 = \frac{2\pi}{3}, \quad x_7 = \frac{4\pi}{3}, \quad x_8 = \frac{5\pi}{3}.$$

Hence there are eight solutions:

$$z_k = \cos x_k + i \sin x_k, \ k = 1, 2, \ldots, 8.$$

2.1.3 Operations with Complex Numbers in Polar Representation

1. Multiplication

Proposition. *Suppose that*

$$z_1 = r_1(\cos t_1 + i \sin t_1) \, and \, z_2 = r_2(\cos t_2 + i \sin t_2).$$

Then

$$z_1 z_2 = r_1 r_2(\cos(t_1 + t_2) + i \sin(t_1 + t_2)). \tag{1}$$

Proof. Indeed,

$$z_1 z_2 = r_1 r_2(\cos t_1 + i \sin t_1)(\cos t_2 + i \sin t_2)$$

$$= r_1 r_2((\cos t_1 \cos t_2 - \sin t_1 \sin t_2) + i(\sin t_1 \cos t_2 + \sin t_2 \cos t_1))$$

$$= r_1 r_2(\cos(t_1 + t_2) + i \sin(t_1 + t_2)). \qquad \square$$

Remarks.

(a) We find again that $|z_1 z_2| = |z_1| \cdot |z_2|$.
(b) We have $\arg(z_1 z_2) = \arg z_1 + \arg z_2 - 2k\pi$, where

$$k = \begin{cases} 0, & \text{for} \quad \arg z_1 + \arg z_2 < 2\pi, \\ 1, & \text{for} \quad \arg z_1 + \arg z_2 \geq 2\pi. \end{cases}$$

(c) Also we can write $\text{Arg}(z_1 z_2) = \{\arg z_1 + \arg z_2 + 2k\pi : k \in \mathbb{Z}\}$.
(d) Formula (1) can be extended to $n \geq 2$ complex numbers. If $z_k = r_k(\cos t_k + i \sin t_k), \ k = 1, \ldots, n$, then

$$z_1 z_2 \ldots z_n = r_1 r_2 \cdots r_n(\cos(t_1 + t_2 + \cdots + t_n) + i \sin(t_1 + t_2 + \cdots + t_n)).$$

The proof by induction is immediate. This formula can be written as

$$\prod_{k=1}^{n} z_k = \prod_{k=1}^{n} r_k \left(\cos \sum_{k=1}^{n} t_k + i \sin \sum_{k=1}^{n} t_k \right). \tag{2}$$

Example. Let $z_1 = 1 - i$ and $z_2 = \sqrt{3} + i$. Then

$$z_1 = \sqrt{2} \left(\cos \frac{7\pi}{4} + i \sin \frac{7\pi}{4} \right), \quad z_2 = 2 \left(\cos \frac{\pi}{6} + i \sin \frac{\pi}{6} \right),$$

and

$$z_1 z_2 = 2\sqrt{2} \left[\cos\left(\frac{7\pi}{4} + \frac{\pi}{6} \right) + i \sin\left(\frac{7\pi}{4} + \frac{\pi}{6} \right) \right]$$
$$= 2\sqrt{2} \left(\cos \frac{23\pi}{12} + i \sin \frac{23\pi}{12} \right).$$

2. The power of a complex number

Proposition. (De Moivre[1]) *For $z = r(\cos t + i \sin t)$ and $n \in \mathbb{N}$, we have*

$$z^n = r^n (\cos nt + i \sin nt). \tag{3}$$

Proof. Apply formula (2) with $z = z_1 = z_2 = \cdots = z_n$ to obtain

$$z^n = \underbrace{r \cdot r \cdots r}_{n \text{ times}} (\cos(\underbrace{t + t + \cdots + t}_{n \text{ times}}) + i \ \sin(\underbrace{t + t + \cdots + t}_{n \text{ times}}))$$

$$= r^n \ (\cos nt + i \sin nt). \qquad \square$$

Remarks.

(a) We find again that $|z^n| = |z|^n$.
(b) If $r = 1$, then $(\cos t + i \sin t)^n = \cos nt + i \sin nt$.
(c) We can write $\operatorname{Arg} z^n = \{ n \arg z + 2k\pi : k \in \mathbb{Z} \}$.

Example. Let us compute $(1 + i)^{1000}$.
The polar representation of $1 + i$ is $\sqrt{2} \left(\cos \frac{\pi}{4} + i \sin \frac{\pi}{4} \right)$. Applying de Moivre's formula, we obtain

$$(1 + i)^{1000} = (\sqrt{2})^{1000} \left(\cos 1000 \frac{\pi}{4} + i \sin 1000 \frac{\pi}{4} \right)$$
$$= 2^{500} (\cos 250\pi + i \sin 250\pi) = 2^{500}.$$

Problem. *Prove that*

$$\sin 5t = 16 \sin^5 t - 20 \sin^3 t + 5 \sin t;$$
$$\cos 5t = 16 \cos^5 t - 20 \cos^3 t + 5 \cos t.$$

Solution. Using de Moivre's theorem to expand $(\cos t + i \sin t)^5$, then using the binomial theorem, we have

$$\cos 5t + i \sin 5t = \cos^5 t + 5i \cos^4 t \ \sin t + 10 i^2 \cos^3 t \ \sin^2 t$$
$$+ 10 i^3 \cos^2 t \ \sin^3 t + 5 i^4 \cos t \ \sin^4 t + i^5 \sin^5 t.$$

---

[1] Abraham de Moivre (1667–1754), French mathematician, a pioneer in probability theory and trigonometry.

Hence

$$\cos 5t + i \sin 5t = \cos^5 t - 10 \cos^3 t \, (1 - \cos^2 t) + 5 \cos t \, (1 - \cos^2 t)^2$$
$$+ i(5(1 - \sin^2 t)^2 \sin t - 10(1 - \sin^2 t) \sin^3 t + \sin^5 t).$$

Simple algebraic manipulation leads to the desired result.

3. Division

Proposition. *Suppose that*

$$z_1 = r_1(\cos t_1 + i \sin t_1), \ z_2 = r_2(\cos t_2 + i \sin t_2) \neq 0.$$

Then

$$\frac{z_1}{z_2} = \frac{r_1}{r_2}[\cos(t_1 - t_2) + i \sin(t_1 - t_2)].$$

Proof. We have

$$\frac{z_1}{z_2} = \frac{r_1(\cos t_1 + i \sin t_1)}{r_2(\cos t_2 + i \sin t_2)}$$

$$= \frac{r_1(\cos t_1 + i \sin t_1)(\cos t_2 - i \sin t_2)}{r_2(\cos^2 t_2 + \sin^2 t_2)}$$

$$= \frac{r_1}{r_2}[(\cos t_1 \cos t_2 + \sin t_1 \sin t_2) + i(\sin t_1 \cos t_2 - \sin t_2 \cos t_1)]$$

$$= \frac{r_1}{r_2}(\cos(t_1 - t_2) + i \sin(t_1 - t_2)). \qquad \square$$

Remarks.

(a) We have again $\left|\dfrac{z_1}{z_2}\right| = \dfrac{r_1}{r_2} = \dfrac{|z_1|}{|z_2|}$.

(b) We can write $\text{Arg}\left(\dfrac{z_1}{z_2}\right) = \{\arg z_1 - \arg z_2 + 2k\pi : k \in \mathbb{Z}\}$.

(c) For $z_1 = 1$ and $z_2 = z$,

$$\frac{1}{z} = z^{-1} = \frac{1}{r}(\cos(-t) + i \sin(-t)).$$

(d) De Moivre's formula also holds for negative integer exponents n, i.e., we have

$$z^n = r^n \, (\cos nt + i \sin nt).$$

Problem. *Compute*

$$z = \frac{(1 - i)^{10}(\sqrt{3} + i)^5}{(-1 - i\sqrt{3})^{10}}.$$

Solution. We can write

$$z = \frac{(\sqrt{2})^{10}\left(\cos\dfrac{7\pi}{4} + i\sin\dfrac{7\pi}{4}\right)^{10} \cdot 2^5\left(\cos\dfrac{\pi}{6} + i\sin\dfrac{\pi}{6}\right)^5}{2^{10}\left(\cos\dfrac{4\pi}{3} + i\sin\dfrac{4\pi}{3}\right)^{10}}$$

$$= \frac{2^{10}\left(\cos\dfrac{35\pi}{2} + i\sin\dfrac{35\pi}{2}\right)\left(\cos\dfrac{5\pi}{6} + i\sin\dfrac{5\pi}{6}\right)}{2^{10}\left(\cos\dfrac{40\pi}{3} + i\sin\dfrac{40\pi}{3}\right)}$$

$$= \frac{\cos\dfrac{55\pi}{3} + i\sin\dfrac{55\pi}{3}}{\cos\dfrac{40\pi}{3} + i\sin\dfrac{40\pi}{3}} = \cos 5\pi + i\sin 5\pi = -1.$$

2.1.4 Geometric Interpretation of Multiplication

Consider the complex numbers

$$z_1 = r_1(\cos t_1^* + i\sin t_1^*), \quad z_2 = r_2(\cos t_2^* + i\sin t_2^*),$$

and their geometric images $M_1(r_1, t_1^*)$, $M_2(r_2, t_2^*)$. Let P_1, P_2 be the intersection points of the circle $\mathcal{C}(O; 1)$ with the rays $(OM_1$ and $(OM_2$. Construct the point $P_3 \in \mathcal{C}(O; 1)$ with the polar argument $t_1^* + t_2^*$ and choose the point $M_3 \in (OP_3$ such that $OM_3 = OM_1 \cdot OM_2$. Let z_3 be the complex coordinate of M_3. The point $M_3(r_1r_2, t_1^* + t_2^*)$ is the geometric image of the product $z_1 \cdot z_2$.

Let A be the geometric image of the complex number 1. Because

$$\frac{OM_3}{OM_1} = \frac{OM_2}{1}, \quad \text{i.e.,} \quad \frac{OM_3}{OM_1} = \frac{OM_2}{OA}$$

and $\widehat{M_2OM_3} = \widehat{AOM_1}$, it follows that triangles $O\,AM_1$ and $O\,M_2M_3$ are similar (see Fig. 2.5).

In order to construct the geometric image of the quotient, note that the image of $\dfrac{z_3}{z_2}$ is M_1.

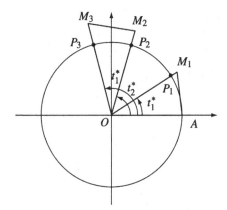

Figure 2.5.

2.1.5 Problems

1. Find the polar coordinates for the following points, given their Cartesian coordinates:

   (a) $M_1(-3, 3)$;   (b) $M_2(-4\sqrt{3}, -4)$;   (c) $M_3(0, -5)$;
   (d) $M_4(-2, -1)$;   (e) $M_5(4, -2)$.

2. Find the Cartesian coordinates for the following points, given their polar coordinates:

   (a) $P_1\left(2, \dfrac{\pi}{3}\right)$;   (b) $P_2\left(4, \ 2\pi - \arcsin\dfrac{3}{5}\right)$;   (c) $P_3(2, \ \pi)$;

   (d) $P_4(3, -\pi)$;   (e) $P_5(1, \dfrac{\pi}{2})$;   (f) $P_6(4, \dfrac{3\pi}{2})$.

3. Express $\arg(\bar{z})$ and $\arg(-z)$ in terms of $\arg(z)$.

4. Find the geometric images for the complex numbers z in each of the following cases:

   (a) $|z| = 2$;   (b) $|z + i| \geq 2$;   (c) $|z - i| \leq 3$;

   (d) $\pi < \arg z < \dfrac{5\pi}{4}$;   (e) $\arg z \geq \dfrac{3\pi}{2}$;   (f) $\arg z < \dfrac{\pi}{2}$;

   (g) $\arg(-z) \in \left(\dfrac{\pi}{6}, \dfrac{\pi}{3}\right)$;   (h) $|z + 1 + i| < 3$ and $0 < \arg z < \dfrac{\pi}{6}$.

5. Find polar representations for the following complex numbers:

   (a) $z_1 = 6 + 6i\sqrt{3}$;   (b) $z_2 = -\dfrac{1}{4} + i\dfrac{\sqrt{3}}{4}$;   (c) $z_3 = -\dfrac{1}{2} - i\dfrac{\sqrt{3}}{2}$;
   (d) $z_4 = 9 - 9i\sqrt{3}$;   (e) $z_5 = 3 - 2i$;   (f) $z_6 = -4i$.

6. Find polar representations for the following complex numbers:

   (a) $z_1 = \cos a - i \sin a$, $\ \cdot \ a \in [0, \ 2\pi)$;
   (b) $z_2 = \sin a + i(1 + \cos a)$, $a \in [0, \ 2\pi)$;

(c) $z_3 = \cos a + \sin a + i(\sin a - \cos a)$, $\quad a \in [0, 2\pi)$;

(d) $z_4 = 1 - \cos a + i \sin a$, $\quad a \in [0, 2\pi)$.

7. Compute the following products using the polar representation of a complex number:

(a) $\left(\dfrac{1}{2} - i \dfrac{\sqrt{3}}{2} \right)(-3 + 3i)(2\sqrt{3} + 2i)$; (b) $(1+i)(-2-2i) \cdot i$;

(c) $-2i \cdot (-4 + 4\sqrt{3}i) \cdot (3 + 3i)$; (d) $3 \cdot (1-i)(-5+5i)$.

Verify your results using the algebraic form.

8. Find $|z|$, $\arg z$, $\mathrm{Arg} z$, $\arg \overline{z}$, $\arg(-z)$ for

(a) $z = (1-i)(6+6i)$; (b) $z = (7 - 7\sqrt{3}i)(-1 - i)$.

9. Find $|z|$ and $\arg z$ for

(a) $z = \dfrac{(2\sqrt{3} + 2i)^8}{(1-i)^6} + \dfrac{(1+i)^6}{(2\sqrt{3} - 2i)^8}$;

(b) $z = \dfrac{(-1+i)^4}{(\sqrt{3} - i)^{10}} + \dfrac{1}{(2\sqrt{3} + 2i)^4}$;

(c) $z = (1 + i\sqrt{3})^n + (1 - i\sqrt{3})^n$.

10. Prove that de Moivre's formula holds for negative integer exponents.

11. Compute the following:

(a) $(1 - \cos a + i \sin a)^n$ for $a \in [0, 2\pi)$ and $n \in \mathbb{N}$;

(b) $z^n + \dfrac{1}{z^n}$, if $z + \dfrac{1}{z} = \sqrt{3}$.

12. Given that z is a complex number such that $z + \dfrac{1}{z} = 2 \cos 3°$, find the least integer that is greater than $z^{2000} + \dfrac{1}{z^{2000}}$.

(2000 AIME II, Problem 9)

13. For how many positive integers n less than or equal to 1000 is

$$(\sin t + i \cos t)^n = \sin nt + i \cos nt$$

true for all real t?

(2005 AIME II, Problem 9)

14. Let $(1 - \sqrt{3}i)^n = x_n + iy_n$, where x_n, y_n are real for $n = 1, 2, 3, \ldots$.

(a) Show that $x_n y_{n-1} - x_{n-1} y_n = 4^{n-1}\sqrt{3}$.

(b) Compute $x_n x_{n-1} + y_n y_{n-1}$.

2.2 The nth Roots of Unity

2.2.1 Defining the nth Roots of a Complex Number

Consider a positive integer $n \geq 2$ and a complex number $z_0 \neq 0$. As in the field of real numbers, the equation

$$Z^n - z_0 = 0 \qquad (1)$$

is used for defining the nth roots of the number z_0. Hence we call any solution Z of equation (1) an *nth root of the complex number z_0.*

Theorem. *Let $z_0 = r(\cos t^* + i \sin t^*)$ be a complex number with $r > 0$ and $t^* \in [0, 2\pi)$.*
The number z_0 has n distinct nth roots, given by the formulas

$$Z_k = \sqrt[n]{r}\left(\cos\frac{t^* + 2k\pi}{n} + i\sin\frac{t^* + 2k\pi}{n}\right),$$

$k = 0, 1, \ldots, n - 1$.

Proof. We use the polar representation of the complex number Z with the extended argument:

$$Z = \rho(\cos\varphi + i\sin\varphi).$$

By definition, we have $Z^n = z_0$, or equivalently,

$$\rho^n(\cos n\varphi + i\sin n\varphi) = r(\cos t^* + i\sin t^*).$$

We obtain $\rho^n = r$ and $n\varphi = t^* + 2k\pi$ for $k \in \mathbb{Z}$; hence $\rho = \sqrt[n]{r}$ and $\varphi_k = \frac{t^*}{n} + k \cdot \frac{2\pi}{n}$ for $k \in \mathbb{Z}$.
   So far, the roots of equation (1) are

$$Z_k = \sqrt[n]{r}(\cos\varphi_k + i\sin\varphi_k) \text{ for } k \in \mathbb{Z}.$$

Now observe that $0 \leq \varphi_0 < \varphi_1 < \cdots < \varphi_{n-1} < 2\pi$, so the numbers φ_k, $k \in \{0, 1, \ldots, n-1\}$, are reduced arguments, i.e., $\varphi_k^* = \varphi_k$. Until now, we had n distinct roots of z_0:

$$Z_0, Z_1, \ldots, Z_{n-1}.$$

Consider some integer k and let $r \in \{0, 1, \ldots, n-1\}$ be the residue of k modulo n. Then $k = nq + r$ for $q \in \mathbb{Z}$, and

$$\varphi_k = \frac{t^*}{n} + (nq + r)\frac{2\pi}{n} = \frac{t^*}{n} + r\frac{2\pi}{n} + 2q\pi = \varphi_r + 2q\pi.$$

It is clear that $Z_k = Z_r$. Hence

$$\{Z_k : k \in \mathbb{Z}\} = \{Z_0, Z_1, \ldots, Z_{n-1}\}.$$

In other words, there are exactly n distinct nth roots of z_0, as claimed. □

The geometric images of the nth roots of a complex number $z_0 \neq 0$ are the vertices of a regular n-gon inscribed in a circle with center at the origin and radius $\sqrt[n]{r}$.

To prove this, denote by M_0, M_1, ..., M_{n-1} the points with complex coordinates Z_0, Z_1, ..., Z_{n-1}. Because $OM_k = |Z_k| = \sqrt[n]{r}$ for $k \in \{0, 1, \ldots, n-1\}$, it follows that the points M_k lie on the circle $\mathcal{C}(O; \sqrt[n]{r})$. On the other hand, the measure of the arc $\overparen{M_k M_{k+1}}$ is equal to

$$\arg Z_{k+1} - \arg Z_k = \frac{t^* + 2(k+1)\pi - (t^* + 2k\pi)}{n} = \frac{2\pi}{n},$$

for all $k \in \{0, 1, \ldots, n-2\}$, and the remaining arc $\overparen{M_{n-1} M_0}$ is

$$\frac{2\pi}{n} = 2\pi - (n-1)\frac{2\pi}{n}.$$

Because all of the arcs $\overparen{M_0 M_1}$, $\overparen{M_1 M_2}$, ..., $\overparen{M_{n-1} M_0}$ are equal, the polygon $M_0 M_1 \cdots M_{n-1}$ is regular.

Example. Let us find the cube roots of the number $z = 1 + i$ and represent them in the complex plane.

The polar representation of $z = 1 + i$ is

$$z = \sqrt{2}\left(\cos\frac{\pi}{4} + i\sin\frac{\pi}{4}\right).$$

The cube roots of the number z are

$$Z_k = \sqrt[6]{2}\left(\cos\left(\frac{\pi}{12} + k\frac{2\pi}{3}\right) + i\sin\left(\frac{\pi}{12} + k\frac{2\pi}{3}\right)\right), \quad k = 0, 1, 2,$$

or in explicit form,

$$Z_0 = \sqrt[6]{2}\left(\cos\frac{\pi}{12} + i\sin\frac{\pi}{12}\right),$$

$$Z_1 = \sqrt[6]{2}\left(\cos\frac{3\pi}{4} + i\sin\frac{3\pi}{4}\right),$$

and

$$Z_2 = \sqrt[6]{2}\left(\cos\frac{17\pi}{12} + i\sin\frac{17\pi}{12}\right).$$

In polar coordinates, the geometric images of the numbers Z_0, Z_1, Z_2 are

$$M_0\left(\sqrt[6]{2},\ \frac{\pi}{12}\right),\quad M_1\left(\sqrt[6]{2},\ \frac{3\pi}{4}\right),\quad M_2\left(\sqrt[6]{2},\ \frac{17\pi}{12}\right).$$

The resulting equilateral triangle $M_0M_1M_2$ is shown in Fig. 2.6.

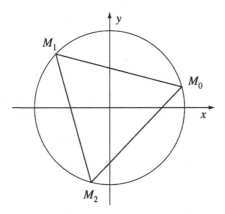

Figure 2.6.

2.2.2 The nth Roots of Unity

The roots of the equation $Z^n - 1 = 0$ are called the nth *roots of unity*. Since $1 = \cos 0 + i\sin 0$, from the formulas for the nth roots of a complex number, we derive that the nth roots of unity are

$$\varepsilon_k = \cos\frac{2k\pi}{n} + i\sin\frac{2k\pi}{n},\quad k \in \{0, 1, 2, \ldots, n-1\}.$$

Explicitly, we have

$$\varepsilon_0 = \cos 0 + i\sin 0 = 1;$$

$$\varepsilon_1 = \cos\frac{2\pi}{n} + i\sin\frac{2\pi}{n} = \varepsilon;$$

$$\varepsilon_2 = \cos\frac{4\pi}{n} + i\sin\frac{4\pi}{n} = \varepsilon^2;$$

$$\cdots$$

$$\varepsilon_{n-1} = \cos\frac{2(n-1)\pi}{n} + i\sin\frac{2(n-1)\pi}{n} = \varepsilon^{n-1}.$$

The set $\{1,\ \varepsilon,\ \varepsilon^2,\ \ldots,\ \varepsilon^{n-1}\}$ is denoted by U_n. Observe that the set U_n is generated by the element ε, i.e., the elements of U_n are the powers of ε.

As stated before, the geometric images of the nth roots of unity are the vertices of a regular polygon with n sides inscribed in the unit circle with one of the vertices at 1.

We take a brief look at some particular values of n:

1. For $n = 2$, the equation $Z^2 - 1 = 0$ has the roots -1 and 1, which are the square roots of unity.
2. For $n = 3$, the cube roots of unity, i.e., the roots of equation $Z^3 - 1 = 0$, are given by

$$\varepsilon_k = \cos \frac{2k\pi}{3} + i \sin \frac{2k\pi}{3} \text{ for } k \in \{0, 1, 2\}.$$

Hence

$$\varepsilon_0 = 1, \quad \varepsilon_1 = \cos \frac{2\pi}{3} + i \sin \frac{2\pi}{3} = -\frac{1}{2} + i \frac{\sqrt{3}}{2} = \varepsilon$$

and

$$\varepsilon_2 = \cos \frac{4\pi}{3} + i \sin \frac{4\pi}{3} = -\frac{1}{2} - i \frac{\sqrt{3}}{2} = \varepsilon^2.$$

They form an equilateral triangle inscribed in the circle $\mathcal{C}(O; 1)$ as in Fig. 2.7.

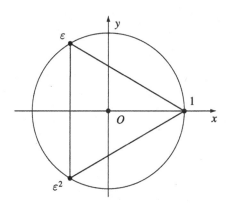

Figure 2.7.

3. For $n = 4$, the fourth roots of unity are

$$\varepsilon_k = \cos \frac{2k\pi}{4} + i \sin \frac{2k\pi}{4} \text{ for } k = 0, 1, 2, 3.$$

In explicit form, we have

$$\varepsilon_0 = \cos 0 + i \sin 0 = 1; \quad \varepsilon_1 = \cos \frac{\pi}{2} + i \sin \frac{\pi}{2} = i;$$

$$\varepsilon_2 = \cos \pi + i \sin \pi = -1 \text{ and } \varepsilon_3 = \cos \frac{3\pi}{2} + i \sin \frac{3\pi}{2} = -i.$$

Observe that $U_4 = \{1,\ i,\ i^2,\ i^3\} = \{1,\ i,\ -1,\ -i\}$. The geometric images of the fourth roots of unity are the vertices of a square inscribed in the circle $\mathcal{C}(O; 1)$ (Fig. 2.8).

The root $\varepsilon_k \in U_n$ is called *primitive* if for all positive integer, $m < n$ we have $\varepsilon_k^m \neq 1$.

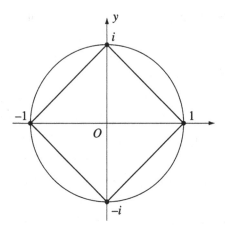

Figure 2.8.

Proposition 1.

(a) If $n | q$, then every root of $Z^n - 1 = 0$ is a root of $Z^q - 1 = 0$.

(b) The common roots of $Z^m - 1 = 0$ and $Z^n - 1 = 0$ are the roots of $Z^d - 1 = 0$, where $d = \gcd(m,\ n)$, i.e., $U_m \cap U_n = U_d$.

(c) The primitive roots of $Z^m - 1 = 0$ are $\varepsilon_k = \cos\frac{2k\pi}{m} + i\sin\frac{2k\pi}{m}$, where $0 \leq k \leq m$ and $\gcd(k,\ m) = 1$.

Proof.

(a) If $q = pn$, then $Z^q - 1 = (Z^n)^p - 1 = (Z^n - 1)(Z^{(p-1)n} + \cdots + Z^n + 1)$, and the conclusion follows.

(b) Consider $\varepsilon_p = \cos\frac{2p\pi}{m} + i\sin\frac{2p\pi}{m}$ a root of $Z^m - 1 = 0$ and $\varepsilon_q' = \cos\frac{2q\pi}{n} + i\sin\frac{2q\pi}{n}$ a root of $Z^n - 1 = 0$. Since $|\varepsilon_p| = |\varepsilon_q'| = 1$, we have $\varepsilon_p = \varepsilon_q'$ if and only if $\arg\varepsilon_p = \arg\varepsilon_q'$, i.e., $\frac{2p\pi}{m} = \frac{2q\pi}{n} + 2r\pi$ for some integer r. The last relation is equivalent to $\frac{p}{m} - \frac{q}{n} = r$, that is, $pn - qm = rmn$.

On the other hand we have $m = m'd$ and $n = n'd$, where $\gcd(m',\ n') = 1$. From the relation $pn - qm = rmn$, we obtain $n'p - m'q = rm'n'd$. Hence $m' | n'p$, so $m' | p$. That is, $p = p'm'$ for some positive integer p' and

$$\arg\varepsilon_p = \frac{2p\pi}{m} = \frac{2p'm'\pi}{m'd} = \frac{2p'\pi}{d} \text{ and } \varepsilon_p^d = 1.$$

Conversely, since $d | m$ and $d | n$ (from property a), every root of $Z^d - 1 = 0$ is a root of $Z^m - 1 = 0$ and $Z^n - 1 = 0$.

(c) First we will find the smallest positive integer p such that $\varepsilon_k^p = 1$. From the relation $\varepsilon_k^p = 1$, it follows that $\frac{2kp\pi}{m} = 2q\pi$ for some positive integer q. That is, $\frac{kp}{m} = q \in \mathbb{Z}$. Consider $d = \gcd(k, m)$ and $k = k'd$, $m = m'd$, where $\gcd(k', m') = 1$. We obtain $\frac{k'pd}{m'd} = \frac{k'p}{m'} \in \mathbb{Z}$. Since k' and m' are relatively primes, we get $m'|p$. Therefore, the smallest positive integer p with $\varepsilon_k^p = 1$ is $p = m'$. Substituting in the relation $m = m'd$ yields that $p = \frac{m}{d}$, where $d = \gcd(k, m)$.

If ε_k is a primitive root of unity, then from relation $\varepsilon_k^p = 1$, $p = \frac{m}{\gcd(k,m)}$, it follows that $p = m$, i.e., $\gcd(k, m) = 1$.   □

Remark. From Proposition 1(b) in Sect. 2.2.2, one obtains that the equations $Z^m - 1 = 0$ and $Z^n - 1 = 0$ have the unique common root 1 if and only if $\gcd(m, n) = 1$.

Proposition 2. *If $\varepsilon \in U_n$ is a primitive root of unity, then the roots of the equation $z^n - 1 = 0$ are ε^r, ε^{r+1}, ..., ε^{r+n-1}, where r is an arbitrary positive integer.*

Proof. Let r be a positive integer and consider $h \in \{0, 1, \ldots, n-1\}$. Then $(\varepsilon^{r+h})^n = (\varepsilon^n)^{r+h} = 1$, i.e., ε^{r+h} is a root of $Z^n - 1 = 0$.

We need only prove that ε^r, ε^{r+1}, ..., ε^{r+n-1} are distinct. Assume by way of contradiction that for $r + h_1 \neq r + h_2$ and $h_1 > h_2$, we have $\varepsilon^{r+h_1} = \varepsilon^{r+h_2}$. Then $\varepsilon^{r+h_2}(\varepsilon^{h_1-h_2} - 1) = 0$. But $\varepsilon^{r+h_2} \neq 0$ implies $\varepsilon^{h_1-h_2} = 1$. Taking into account that $h_1 - h_2 < n$ and ε is a primitive root of $Z^n - 1 = 0$, we obtain a contradiction.   □

Proposition 3. *Let $\varepsilon_0, \varepsilon_1, \ldots, \varepsilon_{n-1}$ be the nth roots of unity. For every positive integer k, the following relation holds:*

$$\sum_{j=0}^{n-1} \varepsilon_j^k = \begin{cases} n, & \text{if } n|k, \\ 0, & \text{otherwise.} \end{cases}$$

Proof. Consider $\varepsilon = \cos\frac{2\pi}{n} + i\sin\frac{2\pi}{n}$. Then $\varepsilon \in U_n$ is a primitive root of unity, whence $\varepsilon^m = 1$ if and only if $n|m$. Assume that n does not divide k. We have

$$\sum_{j=0}^{n-1} \varepsilon_j^k = \sum_{j=0}^{n-1} (\varepsilon^j)^k = \sum_{j=0}^{n-1} (\varepsilon^k)^j = \frac{1 - (\varepsilon^k)^n}{1 - \varepsilon^k} = \frac{1 - (\varepsilon^n)^k}{1 - \varepsilon^k} = 0.$$

If $n|k$, then $k = qn$ for some positive integer q, and we obtain

$$\sum_{j=0}^{n-1} \varepsilon_j^k = \sum_{j=0}^{n-1} \varepsilon_j^{qn} = \sum_{j=0}^{n-1} (\varepsilon_j^n)^q = \sum_{j=0}^{n-1} 1 = n.$$ □

Proposition 4. *Let p be a prime number and let $\varepsilon = \cos\frac{2\pi}{p} + i\sin\frac{2\pi}{p}$. If $a_0, a_1, \ldots, a_{p-1}$ are nonzero integers, the relation*

$$a_0 + a_1\varepsilon + \cdots + a_{p-1}\varepsilon^{p-1} = 0$$

holds if and only if $a_0 = a_1 = \cdots = a_{p-1}$.

Proof. If $a_0 = a_1 = \cdots = a_{p-1}$, then the above relation is clearly true.

Conversely, define the polynomials f, $g \in \mathbb{Z}[X]$ by $f = a_0 + a_1 X + \cdots + a_{p-1}X^{p-1}$ and $g = 1 + X + \cdots + X^{p-1}$. If the polynomials f, g have common zeros, then $\gcd(f, g)$ divides g. But it is well known (for example by Eisenstein's irreducibility criterion) that g is irreducible over $\mathbb{Z}$. Hence $\gcd(f, g) = g$, so $g|f$, and we obtain $g = kf$ for some nonzero integer k, i.e., $a_0 = a_1 = \cdots = a_{n-1}$. $\qquad\square$

Problem 1. *Find the number of ordered pairs (a, b) of real numbers such that $(a + bi)^{2002} = a - bi$.*

(American Mathematics Contest 12A, 2002, Problem 24)

Solution. Let $z = a + bi$, $\bar{z} = a - bi$, and $|z| = \sqrt{a^2 + b^2}$. The given relation becomes $z^{2002} = \bar{z}$. Note that

$$|z|^{2002} = |z^{2002}| = |\bar{z}| = |z|,$$

from which it follows that

$$|z|(|z|^{2001} - 1) = 0.$$

Hence $|z| = 0$ and $(a, b) = (0, 0)$, or $|z| = 1$. In the case $|z| = 1$, we have $z^{2002} = \bar{z}$, which is equivalent to $z^{2003} = \bar{z} \cdot z = |z|^2 = 1$. Since the equation $z^{2003} = 1$ has 2003 distinct solutions, there are altogether $1 + 2003 = 2004$ ordered pairs that meet the required conditions.

Problem 2. *Two regular polygons are inscribed in the same circle. The first polygon has 1982 sides and the second has 2973 sides. If the polygons have any common vertices, how many such vertices will there be?*

Solution. The number of common vertices is given by the number of common roots of $z^{1982} - 1 = 0$ and $z^{2973} - 1 = 0$. Applying Proposition 1(b) in Sect. 2.2.2 yields that the desired number is $d = \gcd(1982, 2973) = 991$.

Problem 3. *Let $\varepsilon \in U_n$ be a primitive root of unity and let z be a complex number such that $|z - \varepsilon^k| \leq 1$ for all $k = 0, 1, \ldots, n - 1$. Prove that $z = 0$.*

Solution. From the given condition, it follows that $(z - \varepsilon^k)\overline{(z - \varepsilon^k)} \leq 1$, yielding $|z|^2 \leq z\overline{(\varepsilon^k)} + \bar{z} \cdot \varepsilon^k$, $k = 0, 1, \ldots, n-1$. By summing these relations, we obtain

$$n|z|^2 \le z \left(\overline{\sum_{k=0}^{n-1} \varepsilon^k} \right) + \overline{z} \cdot \sum_{k=0}^{n-1} \varepsilon^k = 0.$$

Thus $z = 0$.

Problem 4. Let $P_0 P_1 \cdots P_{n-1}$ be a regular polygon inscribed in a circle of radius 1. Prove the following:

(a) $P_0 P_1 \cdot P_0 P_2 \cdots P_0 P_{n-1} = n$;

(b) $\sin \dfrac{\pi}{n} \sin \dfrac{2\pi}{n} \cdots \sin \dfrac{(n-1)\pi}{n} = \dfrac{n}{2^{n-1}}$;

(c) $\sin \dfrac{\pi}{2n} \sin \dfrac{3\pi}{2n} \cdots \sin \dfrac{(2n-1)\pi}{2n} = \dfrac{1}{2^{n-1}}$.

Solution.

(a) Without loss of generality, we may assume that the vertices of the polygon are the geometric images of the nth roots of unity, and $P_0 = 1$. Consider the polynomial $f = z^n - 1 = (z - 1)(z - \varepsilon) \cdots (z - \varepsilon^{n-1})$, where $\varepsilon = \cos \frac{2\pi}{n} + i \sin \frac{2\pi}{n}$. Then it is clear that

$$n = f'(1) = (1 - \varepsilon)(1 - \varepsilon^2) \cdots (1 - \varepsilon^{n-1}).$$

Taking the modulus of each side yields the desired result.

(b) We have

$$1 - \varepsilon^k = 1 - \cos \frac{2k\pi}{n} - i \sin \frac{2k\pi}{n} = 2 \sin^2 \frac{k\pi}{n} - 2i \sin \frac{k\pi}{n} \cos \frac{k\pi}{n}$$

$$= 2 \sin \frac{k\pi}{n} \left(\sin \frac{k\pi}{n} - i \cos \frac{k\pi}{n} \right),$$

whence $|1 - \varepsilon^k| = 2 \sin \dfrac{k\pi}{n}$, $k = 1, 2, \ldots, n-1$, and the desired trigonometric identity follows from part (a).

(c) Consider the regular polygon $Q_0 Q_1 \cdots Q_{2n-1}$ inscribed in the same circle whose vertices are the geometric images of the $2n$th roots of unity. According to (a),

$$Q_0 Q_1 \cdot Q_0 Q_2 \cdots Q_0 Q_{2n-1} = 2n.$$

Now taking into account that $Q_0 Q_2 \cdots Q_{2n-2}$ is also a regular polygon, we deduce from (a) that

$$Q_0 Q_2 \cdot Q_0 Q_4 \cdots Q_0 Q_{2n-2} = n.$$

Combining the last two relations yields

$$Q_0 Q_1 \cdot Q_0 Q_3 \cdots Q_0 Q_{2n-1} = 2.$$

A similar computation to that in (b) leads to

$$Q_0 Q_{2k-1} = 2\sin\frac{(2k-1)\pi}{2n}, \quad k = 1, 2, \ldots, n,$$

and the desired result follows.

Let n be a positive integer and let $\varepsilon_n = \cos\dfrac{2\pi}{n} + i\sin\dfrac{2\pi}{n}$. The nth *cyclotomic polynomial* is defined by

$$\phi_n(x) = \prod_{\substack{1 \le k \le n-1 \\ \gcd(k,n)=1}} (x - \varepsilon_n^k).$$

Clearly, the degree of ϕ_n is $\varphi(n)$, where φ is the Euler totient function. The polynomial ϕ_n is monic with integer coefficients and is irreducible over $\mathbb{Q}$. The first sixteen cyclotomic polynomials are given below:

$$\phi(x) = x - 1,$$
$$\phi_2(x) = x + 1,$$
$$\phi_3(x) = x^2 + x + 1,$$
$$\phi_4(x) = x^2 + 1,$$
$$\phi_5(x) = x^4 + x^3 + x^2 + x + 1,$$
$$\phi_6(x) = x^2 - x + 1,$$
$$\phi_7(x) = x^6 + x^5 + x^4 + x^3 + x^2 + x + 1,$$
$$\phi_8(x) = x^4 + 1,$$
$$\phi_9(x) = x^6 + x^3 + 1,$$
$$\phi_{10}(x) = x^4 - x^3 + x^2 - x + 1,$$
$$\phi_{11}(x) = x^{10} + x^9 + x^8 + \cdots + x + 1,$$
$$\phi_{12}(x) = x^4 - x^2 + 1,$$
$$\phi_{13}(x) = x^{12} + x^{11} + x^{10} + \cdots + x + 1,$$
$$\phi_{14}(x) = x^6 - x^5 + x^4 - x^3 + x^2 - x + 1,$$
$$\phi_{15}(x) = x^8 - x^7 + x^5 - x^4 + x^3 - x + 1,$$
$$\phi_{16}(x) = x^8 + 1.$$

The following properties of cyclotomic polynomials are well known:

(1) If $q > 1$ is an odd integer, then $\phi_{2q}(x) = \phi_q(-x)$.
(2) If $n > 1$, then

$$\phi_n(1) = \begin{cases} p, & \text{when } n \text{ is a power of a prime } p, \\ 1, & \text{otherwise.} \end{cases}$$

The next problem extends the trigonometric identity in Problem 4(b) in Sect. 2.2.2.

Problem 5. *The following identities hold:*

(a) $\displaystyle\prod_{\substack{1\le k\le n-1 \\ \gcd(k,\,n)=1}} \sin\frac{k\pi}{n} = \frac{1}{2^{\varphi(n)}},$ *whenever n is not a power of a prime;*

(b) $\displaystyle\prod_{\substack{1\le k\le n-1 \\ \gcd(k,\,n)=1}} \cos\frac{k\pi}{n} = \frac{(-1)^{\frac{\varphi(n)}{2}}}{2^{\varphi(n)}},$ *for all odd positive integers n.*

Solution.

(a) As we have seen in Problem 4(b) in Sect. 2.2.2,

$$1 - \varepsilon_n^k = 2\sin\frac{k\pi}{n}\left(\sin\frac{k\pi}{n} - i\cos\frac{k\pi}{n}\right) = \frac{2}{i}\sin\frac{k\pi}{n}\left(\cos\frac{k\pi}{n} + i\sin\frac{k\pi}{n}\right).$$

We have

$$1 = \phi_n(1) = \prod_{\substack{1\le k\le n-1 \\ \gcd(k,\,n)=1}} (1 - \varepsilon_n^k) = \prod_{\substack{1\le k\le n-1 \\ \gcd(k,\,n)=1}} \frac{2}{i}\sin\frac{k\pi}{n}\left(\cos\frac{k\pi}{n} + i\sin\frac{k\pi}{n}\right)$$

$$= \frac{2^{\varphi(n)}}{i^{\varphi(n)}}\left(\prod_{\substack{1\le k\le n-1 \\ \gcd(k,\,n)=1}} \sin\frac{k\pi}{n}\right)\left(\cos\frac{\varphi(n)}{2}\pi + i\sin\frac{\varphi(n)}{2}\pi\right)$$

$$= \frac{2^{\varphi(n)}}{(-1)^{\frac{\varphi(n)}{2}}}\left(\prod_{\substack{1\le k\le n-1 \\ \gcd(k,\,n)=1}} \sin\frac{k\pi}{n}\right)(-1)^{\frac{\varphi(n)}{2}},$$

where we have used the fact that $\varphi(n)$ is even, and also the well-known relation

$$\sum_{\substack{1\le k\le n-1 \\ \gcd(k,\,n)=1}} k = \frac{1}{2}n\varphi(n).$$

The conclusion follows.

(b) We have

$$1 + \varepsilon_n^k = 1 + \cos\frac{2k\pi}{n} + i\sin\frac{2k\pi}{n} = 2\cos^2\frac{k\pi}{n} + 2i\sin\frac{k\pi}{n}\cos\frac{k\pi}{n}$$

$$= 2\cos\frac{k\pi}{n}\left(\cos\frac{k\pi}{n} + i\sin\frac{k\pi}{n}\right), \quad k = 0, 1, \ldots, n-1.$$

Because n is odd, it follows from the relation $\phi_{2n}(x) = \phi_n(-x)$ that $\phi_n(-1) = \phi_{2n}(1) = 1$. Then

$$1 = \phi_n(-1) = \prod_{\substack{1 \le k \le n-1 \\ \gcd(k,\, n)=1}} (1 - \varepsilon_n^k) = (-1)^{\varphi(n)} \prod_{\substack{1 \le k \le n-1 \\ \gcd(k,\, n)=1}} (1 + \varepsilon_n^k)$$

$$= (-1)^{\varphi(n)} \prod_{\substack{1 \le k \le n-1 \\ \gcd(k,\, n)=1}} 2\cos\frac{k\pi}{n}\left(\cos\frac{k\pi}{n} + i\sin\frac{k\pi}{n}\right)$$

$$= (-1)^{\varphi(n)} 2^{\varphi(n)} \left(\prod_{\substack{1 \le k \le n-1 \\ \gcd(k,\, n)=1}} \cos\frac{k\pi}{n} \right) \left(\cos\frac{\varphi(n)}{2}\pi + i\sin\frac{\varphi(n)}{2}\pi \right)$$

$$= (-1)^{\frac{\varphi(n)}{2}} 2^{\varphi(n)} \prod_{\substack{1 \le k \le n-1 \\ \gcd(k,\, n)=1}} \cos\frac{k\pi}{n},$$

yielding the desired identity.

2.2.3 Binomial Equations

A binomial equation is an equation of the form $Z^n + a = 0$, where $a \in \mathbb{C}^*$ and $n \ge 2$ is an integer.

Solving for Z means finding the nth roots of the complex number $-a$. This is, in fact, a simple polynomial equation of degree n with complex coefficients. From the well-known fundamental theorem of algebra, it follows that it has exactly n complex roots, and it is obvious that the roots are distinct.

Example.

(1) Let us find the roots of $Z^3 + 8 = 0$.
   We have $-8 = 8(\cos\pi + i\sin\pi)$, so the roots are

$$Z_k = 2\left(\cos\frac{\pi + 2k\pi}{3} + i\sin\frac{\pi + 2k\pi}{3}\right), \quad k \in \{0, 1, 2\}.$$

(2) Let us solve the equation $Z^6 - Z^3(1+i) + i = 0$.
   Observe that the equation is equivalent to

$$(Z^3 - 1)(Z^3 - i) = 0.$$

Solving the binomial equations $Z^3 - 1 = 0$ and $Z^3 - i = 0$ for Z, we obtain the solutions

$$\varepsilon_k = \cos \frac{2k\pi}{3} + i \sin \frac{2k\pi}{3} \quad \text{for } k \in \{0, 1, \ 2\}$$

and

$$Z_k = \cos \frac{\frac{\pi}{2} + 2k\pi}{3} + i \sin \frac{\frac{\pi}{2} + 2k\pi}{3} \quad \text{for } k \in \{0, 1, \ 2\}.$$

2.2.4 Problems

1. Find the square roots of the following complex numbers:

    (a) $z = 1 + i$; (b) $z = i$; (c) $z = \dfrac{1}{\sqrt{2}} + \dfrac{i}{\sqrt{2}}$;

    (d) $z = -2(1 + i\sqrt{3})$; (e) $z = 7 - 24i$.

2. Find the cube roots of the following complex numbers:

    (a) $z = -i$; (b) $z = -27$; (c) $z = 2 + 2i$;

    (d) $z = \dfrac{1}{2} - i\dfrac{\sqrt{3}}{2}$; (e) $z = 18 + 26i$.

3. Find the fourth roots of the following complex numbers:

    (a) $z = 2 - i\sqrt{12}$; (b) $z = \sqrt{3} + i$; (c) $z = i$;
    (d) $z = -2i$; (e) $z = -7 + 24i$.

4. Find the 5th, 6th, 7th, 8th, and 12th roots of the complex numbers given above.

5. Let $U_n = \{\varepsilon_0, \varepsilon_1, \varepsilon_2, \ldots, \varepsilon_{n-1}\}$. Prove the following:

    (a) $\varepsilon_j \cdot \varepsilon_k \in U_n$, for all $j, \ k \in \{0, 1, \ldots, n - 1\}$;
    (b) $\varepsilon_j^{-1} \in U_n$, for all $j \in \{0, 1, \ldots, n - 1\}$.

6. Solve the following equations:

    (a) $z^3 - 125 = 0$; (b) $z^4 + 16 = 0$;
    (c) $z^3 + 64i = 0$; (d) $z^3 - 27i = 0$.

7. Solve the following equations:

    (a) $z^7 - 2iz^4 - iz^3 - 2 = 0$; (b) $z^6 + iz^3 + i - 1 = 0$;
    (c) $(2 - 3i)z^6 + 1 + 5i = 0$; (d) $z^{10} + (-2 + i)z^5 - 2i = 0$.

8. Solve the equation

$$z^4 = 5(z - 1)(z^2 - z + 1).$$

9. Let z be a complex number such that $z^n + z^{n-1} + \ldots + 1 = 0$. Prove that

$$nz^{n-1} + \ldots + 2z + 1 = \frac{n+1}{z^2 - z}.$$

10. Let z be a complex number such that

$$\left(z + \frac{1}{z}\right)\left(z + \frac{1}{z} + 1\right) = 1.$$

For an arbitrary integer n, evaluate

$$\left(z^n + \frac{1}{z^n}\right)\left(z^n + \frac{1}{z^n} + 1\right).$$

11. Let v and w be distinct randomly chosen roots of the equation

$$z^{1997} - 1 = 0.$$

Let $\dfrac{m}{n}$ be the probability that $\sqrt{2 + \sqrt{3}} \le |v + w|$, where m and n are relatively prime positive integers. Find $m + n$.

(1997 AIME, Problem 14)

12. Let z_1, z_2, z_3, z_4 be the roots of $\left(\dfrac{z - i}{2z - i}\right)^4 = 1$. Determine the value of

$$(z_1^2 + 1)(z_2^2 + 1)(z_3^2 + 1)(z_4^2 + 1).$$

13. The equation $x^{10} + (13x - 10^{10}) = 0$ has 10 complex roots $r_1, \overline{r_1}, r_2, \overline{r_2}, r_3, \overline{r_3}, r_4, \overline{r_4}, r_5, \overline{r_5}$, where the bar denotes the complex conjugate. Find the value of

$$\frac{1}{r_1\overline{r_1}} + \frac{1}{r_2\overline{r_2}} + \frac{1}{r_3\overline{r_3}} + \frac{1}{r_4\overline{r_4}} + \frac{1}{r_5\overline{r_5}}.$$

(1994 AIME, Problem 13)

14. For certain real values of a, b, c, and d, the equation

$$x^4 + ax^3 + bx^2 + cx + d = 0$$

has four nonreal roots. The product of two of these roots is $13 + i$, and the sum of the other two roots is $3 + 4i$, where $i = \sqrt{-1}$. Find b.

(1995 AIME, Problem 5)

Chapter 3
Complex Numbers and Geometry

3.1 Some Simple Geometric Notions and Properties

3.1.1 The Distance Between Two Points

Suppose that the complex numbers z_1 and z_2 have the geometric images M_1 and M_2. Then the distance between the points M_1 and M_2 is given by

$$M_1 M_2 = |z_1 - z_2|.$$

The distance function $d : \mathbb{C} \times \mathbb{C} \to [0, \infty)$ is defined by

$$d(z_1, z_2) = |z_1 - z_2|,$$

and it satisfies the following properties:

(a) (positivity and nondegeneracy):

$$d(z_1, \ z_2) \geq 0 \text{ for all } z_1, \ z_2 \in \mathbb{C};$$

$$d(z_1, \ z_2) = 0 \text{ if and only if } z_1 = z_2.$$

(b) (symmetry):

$$d(z_1, \ z_2) = d(z_2, \ z_1) \text{ for all } z_1, \ z_2 \in \mathbb{C}.$$

(c) (triangle inequality):

$$d(z_1, \ z_2) \leq d(z_1, \ z_3) + d(z_3, \ z_2) \text{ for all } z_1, \ z_2, \ z_3 \in \mathbb{C}.$$

To justify (c), let us observe that

$$|z_1 - z_2| = |(z_1 - z_3) + (z_3 - z_2)| \leq |z_1 - z_3| + |z_3 - z_2|,$$

T. Andreescu and D. Andrica, *Complex Numbers from A to ... Z*,
DOI 10.1007/978-0-8176-8415-0_3, © Springer Science+Business Media New York 2014

from the modulus property. Equality holds if and only if there is a positive real number k such that

$$z_3 - z_1 = k(z_2 - z_3).$$

3.1.2 Segments, Rays, and Lines

Let A and B be two distinct points with complex coordinates a and b. We say that the point M with complex coordinate z is between the points A and B if $z \neq a$, $z \neq b$, and the following relation holds:

$$|a - z| + |z - b| = |a - b|.$$

We use the notation $A - M - B$ when the point M is between A and B.

The set $(AB) = \{M : A - M - B\}$ is called the *open segment* determined by the points A and B. The set $[AB] = (AB) \cup \{A, B\}$ represents the *closed segment* defined by the points A and B.

Theorem 1. *Suppose $A(a)$ and $B(b)$ are two distinct points. The following statements are equivalent:*

(1) $M \in (AB)$.
(2) There is a positive real number k such that $z - a = k(b - z)$.
(3) There is a real number $t \in (0, 1)$ such that $z = (1 - t)a + tb$, where z is the complex coordinate of M.

Proof. We first prove that (1) and (2) are equivalent. Indeed, we have $M \in (AB)$ if and only if $|a - z| + |z - b| = |a - b|$. That is, $d(a, z) + d(z, b) = d(a, b)$, or equivalently, there is a real number $k > 0$ such that $z - a = k(b - z)$.

To prove (2) $\Leftrightarrow$ (3), set $t = \frac{k}{k+1} \in (0, 1)$, or $k = \frac{t}{1-t} > 0$. Then we have $z - a = k(b - z)$ if and only if $z = \frac{1}{k+1}a + \frac{k}{k+1}b$. That is, $z = (1 - t)a + tb$, and we are done. $\square$

The set $(AB = \{M | A - M - B \text{ or } A - B - M\}$ is called the *open ray* with endpoint A that contains B.

Theorem 2. *Suppose $A(a)$ and $B(b)$ are two distinct points. The following statements are equivalent:*

(1) $M \in (AB$.
(2) There is a positive real number t such that $z = (1 - t)a + tb$, where z is the complex coordinate of M.
(3) $\arg(z - a) = \arg(b - a)$.
(4) $\dfrac{z - a}{b - a} \in \mathbb{R}^+$.

Proof. It suffices to prove that (1) $\Rightarrow$ (2) $\Rightarrow$ (3) $\Rightarrow$ (4) $\Rightarrow$ (1).

(1) $\Rightarrow$ (2). Since $M \in (AB$, we have $A - M - B$ or $A - B - M$. There are numbers t, $l \in (0, 1)$ such that

$$z = (1 - t)a + tb \text{ or } b = (1 - l)a + lz.$$

In the first case, we are done; for the second case, set $t = \dfrac{1}{l}$, and therefore,

$$z = tb - (t - 1)a = (1 - t)a + tb,$$

as claimed.

(2) $\Rightarrow$ (3). From $z = (1 - t)a + tb$, $t > 0$, we obtain

$$z - a = t (b - a), \ t > 0.$$

Hence

$$\arg(z - a) = \arg(b - a).$$

(3) $\Rightarrow$ (4). The relation

$$\arg \frac{z - a}{b - a} = \arg(z - a) - \arg(b - a) + 2k\pi \text{ for some } k \in \mathbb{Z}$$

implies $\arg \dfrac{z - a}{b - a} = 2k\pi$, $k \in \mathbb{Z}$. Since $\arg \dfrac{z - a}{b - a} \in [0, \ 2\pi)$, it follows that $k = 0$ and $\arg \dfrac{z - a}{b - a} = 0$. Thus $\dfrac{z - a}{b - a} \in \mathbb{R}^+$, as desired.

(4) $\Rightarrow$ (1). Let $t = \dfrac{z - a}{b - a} \in \mathbb{R}^+$. Hence

$$z = a + t (b - a) = (1 - t)a + tb, \ t > 0.$$

If $t \in (0, 1)$, then $M \in (AB) \subset (AB$.

If $t = 1$, then $z = b$ and $M = B \in (AB$. Finally, if $t > 1$, then setting $l = \dfrac{1}{t} \in (0, 1)$, we have

$$b = lz + (1 - l)a.$$

It follows that $A - B - M$ and $M \in (AB$.

The proof is now complete. $\square$

Theorem 3. *Suppose $A(a)$ and $B(b)$ are two distinct points. The following statements are equivalent:*

(1) $M(z)$ lies on the line AB.
(2) $\frac{z-a}{b-a} \in \mathbb{R}$.
(3) There is a real number t such that $z = (1 - t)a + tb$.

(4)
$$\begin{vmatrix} z-a & \bar{z}-\bar{a} \\ b-a & \bar{b}-\bar{a} \end{vmatrix} = 0.$$

(5)
$$\begin{vmatrix} z & \bar{z} & 1 \\ a & \bar{a} & 1 \\ b & \bar{a} & 1 \end{vmatrix} = 0.$$

Proof. To obtain the equivalences (1) $\Leftrightarrow$ (2) $\Leftrightarrow$ (3), observe that for a point C such that $C - A - B$, the line AB is the union $(AB \cup \{A\} \cup (AC$. Then apply Theorem 2.

Next we prove the equivalences (2) $\Leftrightarrow$ (4) $\Leftrightarrow$ (5).

Indeed, we have $\dfrac{z-a}{b-a} \in \mathbb{R}$ if and only if $\dfrac{z-a}{b-a} = \overline{\left(\dfrac{z-a}{b-a}\right)}$.

That is, $\dfrac{z-a}{b-a} = \dfrac{\bar{z}-\bar{a}}{\bar{b}-\bar{a}}$, or equivalently, $\begin{vmatrix} z-a & \bar{z}-\bar{a} \\ b-a & \bar{b}-\bar{a} \end{vmatrix} = 0$, so we obtain

that (2) is equivalent to (4).

Moreover, we have

$$\begin{vmatrix} z & \bar{z} & 1 \\ a & \bar{a} & 1 \\ b & \bar{b} & 1 \end{vmatrix} = 0 \text{ if and only if } \begin{vmatrix} z-a & \bar{z}-\bar{a} & 0 \\ a & \bar{a} & 1 \\ b-a & \bar{b}-\bar{a} & 0 \end{vmatrix} = 0.$$

The last relation is equivalent to

$$\begin{vmatrix} z-a & \bar{z}-\bar{a} \\ b-a & \bar{b}-\bar{a} \end{vmatrix} = 0,$$

so we obtain that (4) is equivalent to (5), and we are done. $\qquad\square$

Problem 1. *Let z_1, z_2, z_3 be complex numbers such that $|z_1| = |z_2| = |z_3| = R$ and $z_2 \neq z_3$. Prove that*

$$\min_{a \in \mathbb{R}} |az_2 + (1-a)z_3 - z_1| = \frac{1}{2R}|z_1 - z_2| \cdot |z_1 - z_3|.$$

(Romanian Mathematical Olympiad—Final Round, 1984)

Solution. Let $z = az_2 + (1-a)z_3$, $a \in \mathbb{R}$, and consider the points A_1, A_2, A_3, A with complex coordinates z_1, z_2, z_3, z, respectively. From the hypothesis, it follows that the circumcenter of triangle $A_1 A_2 A_3$ is the origin of the complex plane. Notice that point A lies on the line $A_2 A_3$, so $A_1 A = |z - z_1|$ is greater than or equal to the altitude $A_1 B$ of the triangle $A_1 A_2 A_3$ (Fig. 3.1).

It suffices to prove that

$$A_1 B = \frac{1}{2R}|z_1 - z_2||z_1 - z_3| = \frac{1}{2R}A_1 A_2 \cdot A_1 A_3.$$

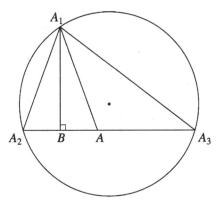

Figure 3.1.

Indeed, since R is the circumradius of the triangle $A_1 A_2 A_3$, we have

$$A_1 B = \frac{2\text{area}[A_1 A_2 A_3]}{A_2 A_3} = \frac{2\dfrac{A_1 A_2 \cdot A_2 A_3 \cdot A_3 A_1}{4R}}{A_2 A_3} = \frac{A_1 A_2 \cdot A_3 A_1}{2R},$$

as claimed.

3.1.3 Dividing a Segment into a Given Ratio

Consider two distinct points $A(a)$ and $B(b)$. A point $M(z)$ on the line AB divides the segment AB into the ratio $k \in \mathbb{R}\backslash\{1\}$ if the following vectorial relation holds:

$$\overrightarrow{MA} = k \cdot \overrightarrow{MB}.$$

In terms of complex numbers, this relation can be written as

$$a - z = k(b - z) \text{ or } (1 - k)z = a - kb.$$

Hence, we obtain

$$z = \frac{a - kb}{1 - k}.$$

Observe that for $k < 0$, the point M lies on the line segment joining the points A and B. If $k \in (0, 1)$, then $M \in (AB\backslash[AB]$. Finally, if $k > 1$, then $M \in (BA\backslash[AB]$.

As a consequence, note that for $k = -1$, we obtain that the coordinate of the midpoint of segment AB is given by $z_M = \dfrac{a + b}{2}$.

Example. Let $A(a)$, $B(b)$, $C(c)$ be noncollinear points in the complex plane. Then the midpoint M of segment AB has the complex coordinate $z_M = \dfrac{a+b}{2}$. The centroid G of triangle ABC divides the median CM in the proportion $2:1$ internally; hence its complex coordinate is given by $k = -2$, i.e.,

$$z_G = \frac{c + 2z_M}{1 + 2} = \frac{a+b+c}{3}.$$

3.1.4 Measure of an Angle

Recall that a triangle is oriented if an ordering of its vertices is specified. It is positively, or directly, oriented if the vertices are oriented counterclockwise. Otherwise, we say that the triangle is negatively oriented. Consider two distinct points $M_1(z_1)$ and $M_2(z_2)$ other than the origin in the complex plane. The angle $\widehat{M_1 O M_2}$ is positively, or directly, oriented if the points M_1 and M_2 are ordered counterclockwise (Fig. 3.2).

Proposition. *The measure of the directly oriented angle $\widehat{M_1 O M_2}$ is equal to* $\arg \dfrac{z_2}{z_1}$.

Proof. In order to simplify the presentation, we will use the same notation for the measure of an angle as for the angle. We consider the following two cases.

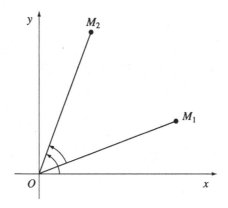

Figure 3.2.

(a) If the triangle $M_1 O M_2$ is negatively oriented (Fig. 3.2), then

$$\widehat{M_1 O M_2} = \widehat{x O M_2} - \widehat{x O M_1} = \arg(z_2) - \arg(z_1) = \arg \frac{z_2}{z_1}.$$

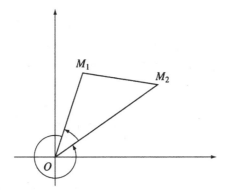

Figure 3.3.

(b) If the triangle M_1OM_2 is positively oriented (Fig. 3.3), then

$$\widehat{M_1OM_2} = 2\pi - \widehat{M_2OM_1} = 2\pi - \arg\frac{z_1}{z_2},$$

since the triangle M_2OM_1 is negatively oriented. Thus

$$\widehat{M_1OM_2} = 2\pi - \arg\frac{z_1}{z_2} = 2\pi - \left(2\pi - \arg\frac{z_2}{z_1}\right) = \arg\frac{z_2}{z_1},$$

as claimed. $\qquad\square$

Remark. The result also holds if the points O, M_1, M_2 are collinear.

Example.

(a) Suppose that $z_1 = 1 + i$ and $z_2 = -1 + i$. Then (see Fig. 3.4)

$$\frac{z_2}{z_1} = \frac{-1+i}{1+i} = \frac{(-1+i)(1-i)}{2} = i,$$

so

$$\widehat{M_1OM_2} = \arg i = \frac{\pi}{2} \quad \text{and} \quad \widehat{M_2OM_1} = \arg(-i) = \frac{3\pi}{2}.$$

(b) Suppose that $z_1 = i$ and $z_2 = 1$. Then $\dfrac{z_2}{z_1} = \dfrac{1}{i} = -i$, so (see Fig. 3.5)

$$\widehat{M_1OM_2} = \arg(-i) = \frac{3\pi}{2} \quad \text{and} \quad \widehat{M_2OM_1} = \arg(i) = \frac{\pi}{2}.$$

Theorem. *Consider three distinct points $M_1(z_1)$, $M_2(z_2)$, and $M_3(z_3)$. The measure of the oriented angle $\widehat{M_2M_1M_3}$ is* $\arg\dfrac{z_3 - z_1}{z_2 - z_1}$.

Proof. Translation by the vector $-z_1$ maps the points M_1, M_2, M_3 into the points O, M_2', M_3', with complex coordinates O, $z_2 - z_1$, $z_3 - z_1$. Moreover, we have $\widehat{M_2 M_1 M_3} = \widehat{M_2' O M_3'}$. By the previous result, we obtain

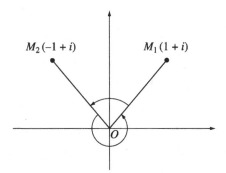

Figure 3.4.

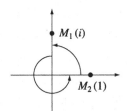

Figure 3.5.

$$\widehat{M_2' O M_3'} = \arg \frac{z_3 - z_1}{z_2 - z_1},$$

as claimed.   □

Example. Suppose that $z_1 = 4 + 3i$, $z_2 = 4 + 7i$, $z_3 = 8 + 7i$. Then

$$\frac{z_2 - z_1}{z_3 - z_1} = \frac{4i}{4 + 4i} = \frac{i(1 - i)}{2} = \frac{1 + i}{2},$$

so

$$\widehat{M_3 M_1 M_2} = \arg \frac{1 + i}{2} = \frac{\pi}{4}$$

and

$$\widehat{M_2 M_1 M_3} = \arg \frac{2}{1 + i} = \arg(1 - i) = \frac{7\pi}{4}.$$

Remark. Using polar representation, from the above result we have

$$\frac{z_3 - z_1}{z_2 - z_1} = \left| \frac{z_3 - z_1}{z_2 - z_1} \right| \left(\cos \left(\arg \frac{z_3 - z_1}{z_2 - z_1} \right) + i \sin \left(\arg \frac{z_3 - z_1}{z_2 - z_1} \right) \right)$$

$$= \left| \frac{z_3 - z_1}{z_2 - z_1} \right| \left(\cos \widehat{M_2 M_1 M_3} + i \sin \widehat{M_2 M_1 M_3} \right).$$

3.1.5 Angle Between Two Lines

Consider four distinct points $M_i(z_i)$, $i \in \{1, 2, 3, 4\}$. The measure of the angle determined by the lines $M_1 M_3$ and $M_2 M_4$ equals $\arg \dfrac{z_3 - z_1}{z_4 - z_2}$ or $\arg \dfrac{z_4 - z_2}{z_3 - z_1}$. The proof is obtained following the same ideas as in the previous subsection.

3.1.6 Rotation of a Point

Consider an angle α and the complex number given by

$$\varepsilon = \cos \alpha + i \sin \alpha.$$

Let $z = r(\cos t + i \sin t)$ be a complex number and M its geometric image.

Form the product $z\varepsilon = r(\cos(t + \alpha) + i \sin(t + \alpha))$ and let us observe that $|z\varepsilon| = r$ and

$$\arg(z\varepsilon) = \arg z + \alpha.$$

It follows that the geometric image M' of $z\varepsilon$ is the rotation of M with respect to the origin through the angle α (Fig. 3.6).

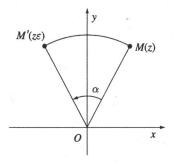

Figure 3.6.

Now we have all the ingredients to establish the following result:

Proposition. *Suppose that the point C is the rotation of B with respect to A through the angle α.*

If a, b, c are the coordinates of the points A, B, C, respectively, then

$$c = a + (b - a)\varepsilon, \ where \ \varepsilon = \cos\alpha + i\sin\alpha.$$

Proof. Translation by the vector $-a$ maps the points A, B, C into the points O, B', C', with complex coordinates O, $b-a$, $c-a$, respectively (see Fig. 3.7). The point C' is the image of B' under rotation about the origin through the angle α, so $c - a = (b - a)\varepsilon$, or $c = a + (b - a)\varepsilon$, as desired.   □

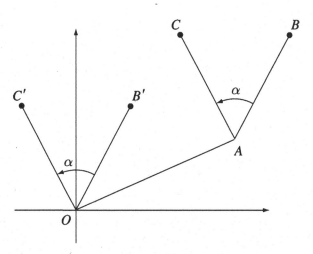

Figure 3.7.

We will call the formula in the above proposition the *rotation formula*.

Problem 1. *Let $ABCD$ and $BNMK$ be two nonoverlapping squares and let E be the midpoint of AN. If point F is the foot of the perpendicular from B to the line CK, prove that points E, F, B are collinear.*

Solution. Consider the complex plane with origin at F and the axis CK and FB, where FB is the imaginary axis (Fig. 3.8).

Let c, k, bi be the complex coordinates of points C, K, B with c, k, $b \in \mathbb{R}$. Rotation with center B through the angle $\theta = \dfrac{\pi}{2}$ maps point C to A, so A has the complex coordinate $a = b(1 - i) + ci$. Similarly, point N is obtained by rotating point K around B through the angle $\theta = -\dfrac{\pi}{2}$, and its complex coordinate is

$$n = b(1 + i) - ki.$$

The midpoint E of segment AN has complex coordinate

$$e = \frac{a + n}{2} = b + \frac{c - k}{2}i,$$

so E lies on the line FB, as desired.

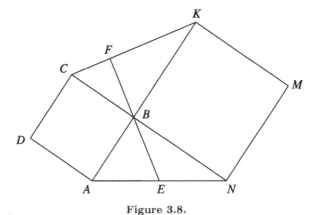

Figure 3.8.

Problem 2. *On the sides AB, BC, CD, DA of quadrilateral ABCD, and exterior to the quadrilateral, we construct squares of centers O_1, O_2, O_3, O_4, respectively. Prove that*

$$O_1O_3 \perp O_2O_4 \quad and \quad O_1O_3 = O_2O_4.$$

(Van Aubel)

Solution. Let $ABMM'$, $BCNN'$, $CDPP'$, and $DAQQ'$ be the constructed squares with centers O_1, O_2, O_3, O_4, respectively (Fig. 3.9).

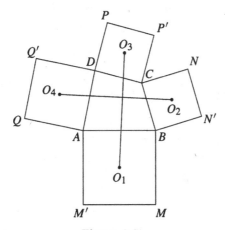

Figure 3.9.

Denote by the corresponding lowercase letter the coordinate of each of the points denoted by an uppercase letter, i.e., o_1 is the coordinate of O_1, etc.

Point M is obtained from point A by a rotation about B through the angle $\theta = \dfrac{\pi}{2}$; hence $m = b + (a - b)i$. Likewise,

$$n = c + (b - c)i, \; p = d + (c - d)i \text{ and } q = a + (d - a)i.$$

It follows that

$$o_1 = \frac{a + m}{2} = \frac{a + b + (a - b)i}{2}, \quad o_2 = \frac{b + c + (b - c)i}{2},$$

$$o_3 = \frac{c + d + (c - d)i}{2}, \quad o_4 = \frac{d + a + (d - a)i}{2}.$$

Then

$$\frac{o_3 - o_1}{o_4 - o_2} = \frac{c + d - a - b + i(c - d - a + b)}{a + d - b - c + i(d - a - b + c)} = -i \in i\mathbb{R}^*,$$

so $O_1 O_3 \perp O_2 O_4$. Moreover,

$$\left| \frac{o_3 - o_1}{o_4 - o_2} \right| = |-i| = 1;$$

hence $O_1 O_3 = O_2 O_4$, as desired.

Problem 3. *In the exterior of the triangle ABC we construct triangles ABR, BCP, and CAQ such that*

$$m(\widehat{PBC}) = m(\widehat{CAQ}) = 45°,$$
$$m(\widehat{BCP}) = m(\widehat{QCA}) = 30°,$$

and

$$m(\widehat{ABR}) = m(\widehat{RAB}) = 15°.$$

Prove that

$$m(\widehat{QRP}) = 90° \text{ and } RQ = RP.$$

Solution. Consider the complex plane with origin at point R and let M be the foot of the perpendicular from P to the line BC (Fig. 3.10).

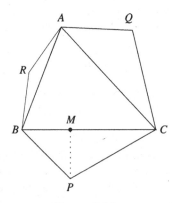

Figure 3.10.

Denote by the corresponding lowercase letter the coordinate of a point denoted by an uppercase letter. From $MP = MB$ and $\dfrac{MC}{MP} = \sqrt{3}$, it follows that

$$\frac{p-m}{b-m} = i \text{ and } \frac{c-m}{p-m} = i\sqrt{3},$$

whence

$$p = \frac{c+\sqrt{3}b}{1+\sqrt{3}} + \frac{b-c}{1+\sqrt{3}}i.$$

Likewise,

$$q = \frac{c+\sqrt{3}a}{1+\sqrt{3}} + \frac{a-c}{1+\sqrt{3}}i.$$

Point B is obtained from point A by a rotation about R through the angle $\theta = 150°$, so

$$b = a\left(-\frac{\sqrt{3}}{2} + \frac{1}{2}i\right).$$

Simple algebraic manipulations show that $\dfrac{p}{q} = i \in \mathbb{R}^*$, whence $QR \perp PR$. Moreover, $|p| = |iq| = |q|$, $RP = RQ$, and we are done.

Problem 4. *The points $(0,0)$, $(a, 11)$, and $(b, 37)$ are the vertices of an equilateral triangle. Find the value of ab.*

(1994 AIME, Problem 8)

Solution. Consider the points as lying in the complex plane. The point $b + 37i$ is then a rotation of $60°$ of $a + 11i$ about the origin, so

$$(a + 11i)(\cos 60° + i\sin 60°) = (a + 11i)\left(\frac{1}{2} + \frac{\sqrt{3}i}{2}\right) = b + 37i.$$

Equating the real and imaginary parts, we have

$$b = \frac{a}{2} - \frac{11\sqrt{3}}{2},$$

$$37 = \frac{11}{2} + \frac{a\sqrt{3}}{2}.$$

By solving this system, we find that $a = 21\sqrt{3}$, $b = 5\sqrt{3}$. Thus, the answer is 315.

Remark. There is another solution in which the point $b + 37i$ is a rotation of $a + 11i$ through $-60°$. However, this triangle is just a reflection of the first triangle in the y-axis, and the signs of a and b are reversed. However, the product ab is unchanged.

Problem 5. *Let ABCD be a convex quadrilateral. Let P be the point outside ABCD such that angle APB is a right angle and P is equidistant from A and B. Let points Q, R, and S be given by the same conditions with respect to the other three edges of ABCD. Let J, K, L, and M be the midpoints of PQ, QR, RS, and SP, respectively. Prove that JKLM is a square.*

(American Mathematical Monthly)

Solution. By Van Aubel's theorem, the lines PR and QS are perpendicular, and the segments PR and QS are equal. Let O be the intersection point of the lines PR and QS. Without loss of generality, assume that $|PR| = |QS| = 1$. Consider now the Cartesian coordinate system centered at O with axes PR and QS. In this case,

$$Q = (u, 0), \ S = (u - 1, 0), \ R = (0, v), \ P = (0, v - 1),$$

for some positive real numbers u, v less than 1. Hence

$$J = \left(\frac{u}{2}, \frac{v-1}{2} \right), \ K = \left(\frac{u}{2}, \frac{v}{2} \right), \ L = \left(\frac{u-1}{2}, \frac{v}{2} \right), \ M = \left(\frac{u-1}{2}, \frac{v-1}{2} \right),$$

from which we can deduce, by making use of the distance formula combined with the Pythagorean theorem, that the quadrilateral $JKLM$ is a square.

3.2 Conditions for Collinearity, Orthogonality, and Concyclicity

In this section, we consider four distinct points $M_i(z_i)$, $i \in \{1, 2, 3, \ 4\}$.

Proposition 1. *The points M_1, M_2, M_3 are collinear if and only if*

$$\frac{z_3 - z_1}{z_2 - z_1} \in \mathbb{R}^*.$$

Proof. The collinearity of the points M_1, M_2, M_3 is equivalent to $\widehat{M_2 M_1 M_3} \in \{0, \ \pi\}$. It follows that

$$\arg \frac{z_3 - z_1}{z_2 - z_1} \in \{0, \ \pi\},$$

or equivalently,

$$\frac{z_3 - z_1}{z_2 - z_1} \in \mathbb{R}^*,$$

as claimed.   □

Proposition 2. *The lines $M_1 M_2$ and $M_3 M_4$ are orthogonal if and only if*

$$\frac{z_1 - z_2}{z_3 - z_4} \in i\mathbb{R}^*.$$

Proof. We have $M_1M_2 \perp M_3M_4$ if and only if $(\widehat{M_1M_2, M_3M_4}) \in \{\frac{\pi}{2}, \frac{3\pi}{2}\}$. This is equivalent to $\arg \frac{z_1-z_2}{z_3-z_4} \in \{\frac{\pi}{2}, \frac{3\pi}{2}\}$. We obtain $\frac{z_1-z_2}{z_3-z_4} \in i\mathbb{R}^*$.   □

Remark. Suppose that $M_2 = M_4$. Then $M_1M_2 \perp M_3M_2$ if and only if $\frac{z_1-z_2}{z_3-z_2} \in i\mathbb{R}^*$.

Example.

(1) Consider the points $M_1(2 - i)$, $M_2(-1 + 2i)$, $M_3(-2 - i)$, $M_4(1 + 2i)$. Simple algebraic manipulation shows that

$$\frac{z_1 - z_2}{z_3 - z_4} = i; \text{ hence } M_1M_2 \perp M_3M_4.$$

(2) Consider the points $M_1(2 - i)$, $M_2(-1 + 2i)$, $M_3(1 + 2i)$, $M_4(-2 - i)$. Then we have $\frac{z_1-z_2}{z_3-z_4} = -i$, and hence $M_1M_2 \perp M_3M_4$.

Problem 1. *Let z_1, z_2, z_3 be the coordinates of vertices A, B, C of a triangle. If $w_1 = z_1 - z_2$ and $w_2 = z_3 - z_1$, prove that $\hat{A} = 90°$ if and only if $\mathrm{Re}(w_1 \cdot \overline{w}_2) = 0$.*

Solution. We have $\hat{A} = 90°$ if and only if $\frac{z_2-z_1}{z_3-z_1} \in i\mathbb{R}$, which is equivalent to $\frac{w_1}{-w_2} \in i\mathbb{R}$, i.e., $\mathrm{Re}\left(\frac{w_1}{-w_2}\right) = 0$. The last relation is equivalent to $\mathrm{Re}\left(\frac{w_1 \cdot \overline{w}_2}{-|w_2|^2}\right) = 0$, i.e., $\mathrm{Re}(w_1 \cdot \overline{w}_2) = 0$, as desired.

Proposition 3. *The distinct points $M_1(z_1)$, $M_2(z_2)$, $M_3(z_3)$, $M_4(z_4)$ are concyclic or collinear if and only if*

$$k = \frac{z_3 - z_2}{z_1 - z_2} : \frac{z_3 - z_4}{z_1 - z_4} \in \mathbb{R}^*.$$

Proof. Assume that the points are collinear. We can arrange four points on a circle in $(4 - 1)! = 3! = 6$ different ways. Consider the case in which M_1, M_2, M_3, M_4 are given in this order. Then M_1, M_2, M_3, M_4 are concyclic if and only if

$$\widehat{M_1M_2M_3} + \widehat{M_3M_4M_1} \in \{3\pi, \pi\}.$$

That is,

$$\arg \frac{z_3 - z_2}{z_1 - z_2} + \arg \frac{z_1 - z_4}{z_3 - z_4} \in \{3\pi, \pi\}.$$

Because

$$\arg \frac{1}{z} = \begin{cases} 2\pi - \arg z & \text{if } z \in \mathbb{C}^* \setminus \mathbb{R}_+, \\ 0 & \text{if } z \in \mathbb{R}_+^*, \end{cases}$$

we obtain

$$\arg \frac{z_3 - z_2}{z_1 - z_2} - \arg \frac{z_3 - z_4}{z_1 - z_4} \in \{-\pi, \pi\},$$

i.e., $k < 0$.

For all other arrangements of the four points, the proof is similar. Note that $k > 0$ in three cases and $k < 0$ in the other three.   □

The number k is called the *cross ratio* of the four points $M_1(z_1)$, $M_2(z_2)$, $M_3(z_3)$, and $M_4(z_4)$.

Remarks.

(1) The points M_1, M_2, M_3, M_4 are collinear if and only if

$$\frac{z_3 - z_2}{z_1 - z_2} \in \mathbb{R}^* \text{ and } \frac{z_3 - z_4}{z_1 - z_4} \in \mathbb{R}^*.$$

(2) The points M_1, M_2, M_3, M_4 are concyclic if and only if

$$k = \frac{z_3 - z_2}{z_1 - z_2} : \frac{z_3 - z_4}{z_1 - z_4} \in \mathbb{R}^*, \text{ but } \frac{z_3 - z_2}{z_1 - z_2} \notin \mathbb{R} \text{ and } \frac{z_3 - z_4}{z_1 - z_4} \notin \mathbb{R}.$$

Example.

(1) The geometric images of the complex numbers $1, i, -1, -i$ are concyclic. Indeed, we have the cross ratio $k = \frac{-1-i}{1-i} : \frac{-1+i}{1+i} = -1 \in \mathbb{R}^*$, and clearly $\frac{-1-i}{1-i} \notin \mathbb{R}$ and $\frac{-1+i}{1+i} \notin \mathbb{R}$.

(2) The points $M_1(2 - i)$, $M_2(3 - 2i)$, $M_3(-1 + 2i)$, and $M_4(-2 + 3i)$ are collinear. Indeed, $k = \frac{-4+4i}{-1+i} : \frac{1-i}{4-4i} = 1 \in \mathbb{R}^*$ and $\frac{-4+4i}{-1+i} = 4 \in \mathbb{R}^*$.

Problem 2. *Find all complex numbers z such that the points with complex coordinates z, z^2, z^3, z^4, in this order, are the vertices of a cyclic quadrilateral.*

Solution. If the points of complex coordinates z, z^2, z^3, z^4, in this order, are the vertices of a cyclic quadrilateral, then

$$\frac{z^3 - z^2}{z - z^2} : \frac{z^3 - z^4}{z - z^4} \in \mathbb{R}^*.$$

It follows that

$$-\frac{1 + z + z^2}{z} \in \mathbb{R}^*, \text{ i.e., } -1 - \left(z + \frac{1}{z}\right) \in \mathbb{R}^*.$$

We obtain $z + \frac{1}{z} \in \mathbb{R}$, i.e., $z + \frac{1}{z} = \bar{z} + \frac{1}{\bar{z}}$. Hence $(z - \bar{z})(|z|^2 - 1) = 0$, whence $z \in \mathbb{R}$ or $|z| = 1$.

If $z \in \mathbb{R}$, then the points with complex coordinates z, z^2, z^3, z^4 are collinear; hence it is left to consider the case $|z| = 1$.

Let $t = \arg z \in [0, 2\pi)$. We prove that the points with complex coordinates z, z^2, z^3, z^4 lie in this order on the unit circle if and only if $t \in \left(0, \frac{2\pi}{3}\right) \cup \left(\frac{4\pi}{3}, 2\pi\right)$. Indeed,

(a) If $t \in \left(0, \frac{\pi}{2}\right)$, then $0 < t < 2t < 3t < 4t < 2\pi$ or

$$0 < \arg z < \arg z^2 < \arg z^3 < \arg z^4 < 2\pi.$$

(b) If $t \in \left[\frac{\pi}{2}, \frac{2\pi}{3}\right)$, then $0 \le 4t - 2\pi < t < 2t < 3t < 2\pi$ or

$$0 \le \arg z^4 < \arg z < \arg z^2 < \arg z^3 < 2\pi.$$

(c) If $t \in \left[\frac{2\pi}{3}, \pi\right)$, then $0 \le 3t - 2\pi < t \le 4t - 2\pi < 2t < 2\pi$ or

$$0 \le \arg z^3 < \arg z \le \arg z^4 < \arg z^2.$$

In the same manner, we can analyze the case $t \in [\pi, 2\pi)$.

To conclude, the complex numbers satisfying the desired property are

$$z = \cos t + i \sin t, \quad \text{with} \quad t \in \left(0, \frac{2\pi}{3}\right) \cup \left(\frac{4\pi}{3}, \pi\right).$$

3.3 Similar Triangles

Consider six points $A_1(a_1)$, $A_2(a_2)$, $A_3(a_3)$, $B_1(b_1)$, $B_2(b_2)$, $B_3(b_3)$ in the complex plane. We say that the triangles $A_1A_2A_3$ and $B_1B_2B_3$ are similar if the angle at A_k is equal to the angle at B_k, $k \in \{1, 2, 3\}$.

Proposition 1. *The triangles $A_1A_2A_3$ and $B_1B_2B_3$ are similar, with the same orientation, if and only if*

$$\frac{a_2 - a_1}{a_3 - a_1} = \frac{b_2 - b_1}{b_3 - b_1}. \tag{1}$$

Proof. We have $\triangle A_1A_2A_3 \sim \triangle B_1B_2B_3$ if and only if $\frac{A_1A_2}{A_1A_3} = \frac{B_1B_2}{B_1B_3}$ and $\widehat{A_3A_1A_2} \equiv \widehat{B_3B_1B_2}$. This is equivalent to $\frac{|a_2-a_1|}{|a_3-a_1|} = \frac{|b_2-b_1|}{|b_3-b_1|}$ and $\arg \frac{a_2-a_1}{a_3-a_1} = \arg \frac{b_2-b_1}{b_3-b_1}$. We obtain

$$\frac{a_2 - a_1}{a_3 - a_1} = \frac{b_2 - b_1}{b_3 - b_1}.$$

$\square$

Remarks.

(1) The condition (1) is equivalent to

$$\begin{vmatrix} 1 & 1 & 1 \\ a_1 & a_2 & a_3 \\ b_1 & b_2 & b_3 \end{vmatrix} = 0.$$

(2) The triangles $A_1(0)$, $A_2(1)$, $A_3(2i)$ and $B_1(0)$, $B_2(-i)$, $B_3(-2)$ are similar, but oppositely oriented. In this case, the condition (1) is not satisfied. Indeed,

$$\frac{a_2 - a_1}{a_3 - a_1} = \frac{1 - 0}{2i - 0} = \frac{1}{2i} \neq \frac{b_2 - b_1}{b_3 - b_1} = \frac{-i - 0}{-2 - 0} = \frac{i}{2}.$$

Proposition 2. *The triangles $A_1A_2A_3$ and $B_1B_2B_3$ are similar, having opposite orientations, if and only if*

$$\frac{a_2 - a_1}{a_3 - a_1} = \frac{\bar{b}_2 - \bar{b}_1}{\bar{b}_3 - \bar{b}_1}.$$

Proof. Reflection across the x-axis maps the points B_1, B_2, B_3 into the points $M_1(\bar{b}_1)$, $M_2(\bar{b}_2)$, $M_3(\bar{b}_3)$. The triangles $B_1B_2B_3$ and $M_1M_2M_3$ are similar and have opposite orientations; hence triangles $A_1A_2A_3$ and $M_1M_2M_3$ are similar with the same orientation. The conclusion follows from the previous proposition.   □

Problem 1. *On sides AB, BC, CA of a triangle ABC we draw similar triangles ADB, BEC, CFA, having the same orientation. Prove that triangles ABC and DEF have the same centroid.*

Solution. Denote by the corresponding lowercase letter the coordinate of a point denoted by an uppercase letter.

Triangles ADB, BEC, CFA are similar with the same orientation, whence

$$\frac{d - a}{b - a} = \frac{e - b}{c - b} = \frac{f - c}{a - c} = z,$$

and consequently,

$$d = a + (b - a)z, \quad e = b + (c - b)z, \quad f = c + (a - c)z.$$

Then

$$\frac{d + e + f}{3} = \frac{a + b + c}{3},$$

so triangles ABC and DEF have the same centroid.

Problem 2. *Let M, N, P be the midpoints of sides AB, BC, CA of triangle ABC. On the perpendicular bisectors of segments $[AB]$, $[BC]$, $[CA]$, points C', A', B' are chosen inside the triangle such that*

$$\frac{MC'}{AB} = \frac{NA'}{BC} = \frac{PB'}{CA}.$$

Prove that ABC and $A'B'C'$ have the same centroid.

Solution. Note that from

$$\frac{MC'}{AB} = \frac{NA'}{BC} = \frac{PB'}{CA},$$

it follows that $\tan(\widehat{C'AB}) = \tan(\widehat{A'BC}) = \tan(\widehat{B'CA})$. Hence triangles $AC'B$, $BA'C$, $CB'A$ are similar, and so from Problem 2, it follows that ABC and $A'B'C'$ have the same centroid.

Problem 3. *Let ABO be an equilateral triangle with center S and let $A'B'O$ be another equilateral triangle with the same orientation and $S \neq A'$, $S \neq B'$. Consider M and N the midpoints of the segments $A'B$ and AB'.*
  Prove that triangles $SB'M$ and $SA'N$ are similar.

(30th IMO-Shortlist)

Solution. Let R be the circumradius of the triangle ABO and let

$$\varepsilon = \cos \frac{2\pi}{3} + i \sin \frac{2\pi}{3}.$$

Consider the complex plane with origin at point S such that point O lies on the positive real axis. Then the coordinates of points O, A, B are R, $R\varepsilon$, $R\varepsilon^2$, respectively (Fig. 3.11).

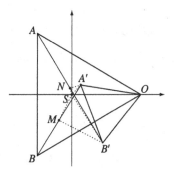

Figure 3.11.

Let $R + z$ be the coordinate of point B', so $R - z\varepsilon$ is the coordinate of point A'. It follows that the midpoints M, N have the coordinates

$$z_M = \frac{z_B + z_{A'}}{2} = \frac{R\varepsilon^2 + R - z\varepsilon}{2} = \frac{R(\varepsilon^2 + 1) - z\varepsilon}{2}$$
$$= \frac{-R\varepsilon - z\varepsilon}{2} = \frac{-\varepsilon(R + z)}{2}$$

and

$$z_N = \frac{z_A + z_{B'}}{2} = \frac{R\varepsilon + R + z}{2} = \frac{R(\varepsilon + 1) + z}{2} = \frac{-R\varepsilon^2 + z}{2}$$

$$= \frac{z - \dfrac{R}{\varepsilon}}{2} = \frac{R - z\varepsilon}{-2\varepsilon}.$$

Now we have

$$\frac{z_{B'} - z_S}{z_M - z_S} = \frac{\overline{z_{A'}} - \overline{z_S}}{\overline{z_N} - \overline{z_S}}$$

if and only if

$$\frac{R + z}{\dfrac{-\varepsilon(R + z)}{2}} = \frac{\overline{R - z\varepsilon}}{\dfrac{\overline{R - z\varepsilon}}{-2\overline{\varepsilon}}}.$$

The last relation is equivalent to $\varepsilon \cdot \overline{\varepsilon} = 1$, i.e., $|\varepsilon|^2 = 1$. Hence the triangles $SB'M$ and $SA'N$ are similar, with opposite orientations.

3.4 Equilateral Triangles

Proposition 1. *Suppose z_1, z_2, z_3 are the coordinates of the vertices of the triangle $A_1 A_2 A_3$. The following statements are equivalent:*

(a) $A_1 A_2 A_3$ is an equilateral triangle.
(b) $|z_1 - z_2| = |z_2 - z_3| = |z_3 - z_1|$.
(c) $z_1^2 + z_2^2 + z_3^2 = z_1 z_2 + z_2 z_3 + z_3 z_1$.
(d) $\dfrac{z_2 - z_1}{z_3 - z_1} = \dfrac{z_3 - z_2}{z_1 - z_2}$.
(e) $\dfrac{1}{z - z_1} + \dfrac{1}{z - z_2} + \dfrac{1}{z - z_3} = 0$, where $z = \dfrac{z_1 + z_2 + z_3}{3}$.
(f) $(z_1 + \varepsilon z_2 + \varepsilon^2 z_3)(z_1 + \varepsilon^2 z_2 + \varepsilon z_3) = 0$, where $\varepsilon = \cos\dfrac{2\pi}{3} + i\sin\dfrac{2\pi}{3}$.
(g)

$$\begin{vmatrix} 1 & 1 & 1 \\ z_1 & z_2 & z_3 \\ z_2 & z_3 & z_1 \end{vmatrix} = 0.$$

Proof. The triangle $A_1 A_2 A_3$ is equilateral if and only if $A_1 A_2 A_3$ is similar to $A_2 A_3 A_1$ with the same orientation, or

$$\begin{vmatrix} 1 & 1 & 1 \\ z_1 & z_2 & z_3 \\ z_2 & z_3 & z_1 \end{vmatrix} = 0;$$

thus (a) $\Leftrightarrow$ (g).

Computing the determinant, we obtain

$$0 = \begin{vmatrix} 1 & 1 & 1 \\ z_1 & z_2 & z_3 \\ z_2 & z_3 & z_1 \end{vmatrix}.$$

$$= z_1 z_2 + z_2 z_3 + z_3 z_1 - (z_1^2 + z_2^2 + z_3^2)$$
$$= -(z_1 + \varepsilon z_2 + \varepsilon^2 z_3)(z_1 + \varepsilon^2 z_2 + \varepsilon z_3);$$

hence (g) $\Leftrightarrow$ (c) $\Leftrightarrow$ (f).

Simple algebraic manipulation shows that (d) $\Leftrightarrow$ (c). Since (a) $\Leftrightarrow$ (b) is obvious, we leave for the reader to prove that (a) $\Leftrightarrow$ (e). $\square$

The next results bring some refinements to this issue.

Proposition 2. *Let z_1, z_2, z_3 be the coordinates of the vertices A_1, A_2, A_3 of a positively oriented triangle. The following statements are equivalent.*

(a) $A_1 A_2 A_3$ is an equilateral triangle.

(b) $z_3 - z_1 = \varepsilon(z_2 - z_1)$, where $\varepsilon = \cos \dfrac{\pi}{3} + i \sin \dfrac{\pi}{3}$.

(c) $z_2 - z_1 = \varepsilon(z_3 - z_1)$, where $\varepsilon = \cos \dfrac{5\pi}{3} + i \sin \dfrac{5\pi}{3}$.

(d) $z_1 + \varepsilon z_2 + \varepsilon^2 z_3 = 0$, where $\varepsilon = \cos \dfrac{2\pi}{3} + i \sin \dfrac{2\pi}{3}$.

Proof. $A_1 A_2 A_3$ is equilateral and positively oriented if and only if A_3 is obtained from A_2 by rotation about A_1 through the angle $\frac{\pi}{3}$. That is,

$$z_3 = z_1 + \left(\cos \frac{\pi}{3} + i \sin \frac{\pi}{3} \right)(z_2 - z_1);$$

hence (a) $\Leftrightarrow$ (b) (Fig. 3.12).

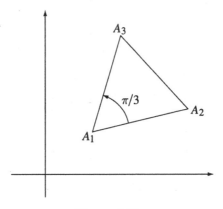

Figure 3.12.

The rotation about A_1 through the angle $\dfrac{5\pi}{3}$ maps A_3 into A_2. Similar considerations show that (a) $\Leftrightarrow$ (c).

To prove that (b) $\Leftrightarrow$ (d), observe that (b) is equivalent to (b')

$$z_3 = z_1 + \left(\frac{1}{2} + i\frac{\sqrt{3}}{2}\right)(z_2 - z_1) = \left(\frac{1}{2} - i\frac{\sqrt{3}}{2}\right)z_1 + \left(\frac{1}{2} + i\frac{\sqrt{3}}{2}\right)z_2.$$

Hence

$$z_1 + \varepsilon z_2 + \varepsilon^2 z_3 = z_1 + \left(-\frac{1}{2} + i\frac{\sqrt{3}}{2}\right)z_2 + \left(-\frac{1}{2} - i\frac{\sqrt{3}}{2}\right)z_3$$

$$= z_1 + \left(-\frac{1}{2} + i\frac{\sqrt{3}}{2}\right)z_2$$

$$+ \left(-\frac{1}{2} - i\frac{\sqrt{3}}{2}\right)\left[\left(\frac{1}{2} - i\frac{\sqrt{3}}{2}\right)z_1 + \left(\frac{1}{2} + i\frac{\sqrt{3}}{2}\right)z_2\right]$$

$$= z_1 + \left(-\frac{1}{2} + i\frac{\sqrt{3}}{2}\right)z_2 - z_1 + \left(\frac{1}{2} - i\frac{\sqrt{3}}{2}\right)z_2 = 0,$$

so (b) $\Leftrightarrow$ (d). $\qquad\qquad\square$

Proposition 3. *Let* z_1, z_2, z_3 *be the coordinates of the vertices* A_1, A_2, A_3 *of a negatively oriented triangle.*

The following statements are equivalent:

(a) $A_1A_2A_3$ *is an equilateral triangle.*

(b) $z_3 - z_1 = \varepsilon(z_2 - z_1)$, *where* $\varepsilon = \cos\dfrac{5\pi}{3} + i\sin\dfrac{5\pi}{3}$.

(c) $z_2 - z_1 = \varepsilon(z_3 - z_1)$, *where* $\varepsilon = \cos\dfrac{\pi}{3} + i\sin\dfrac{\pi}{3}$.

(d) $z_1 + \varepsilon^2 z_2 + \varepsilon z_3 = 0$, *where* $\varepsilon = \cos\dfrac{2\pi}{3} + i\sin\dfrac{2\pi}{3}$.

Proof. Equilateral triangle $A_1A_2A_3$ is negatively oriented if and only if $A_1A_3A_2$ is a positively oriented equilateral triangle. The rest follows from the previous proposition. $\qquad\square$

Proposition 4. *Let* z_1, z_2, z_3 *be the coordinates of the vertices of equilateral triangle* $A_1A_2A_3$. *Consider the following statements:*

(1) $A_1A_2A_3$ *is an equilateral triangle.*

(2) $z_1 \cdot \bar{z}_2 = z_2 \cdot \bar{z}_3 = z_3 \cdot \bar{z}_1$.

(3) $z_1^2 = z_2 \cdot z_3$ *and* $z_2^2 = z_1 \cdot z_3$.

Then (2) $\Rightarrow$ *(1), (3)* $\Rightarrow$ *(1), and (2)* $\Leftrightarrow$ *(3).*

Proof. $(2) \Rightarrow (1)$. Taking the moduli of the terms in the given relation, we obtain

$$|z_1| \cdot |\bar{z}_2| = |z_2| : |\bar{z}_3| = |z_3| \cdot |\bar{z}_1|,$$

or equivalently,

$$|z_1| \cdot |z_2| = |z_2| \cdot |z_3| = |z_3| \cdot |z_1|.$$

This implies

$$r = |z_1| = |z_2| = |z_3|$$

and

$$\bar{z}_1 = \frac{r^2}{z_1}, \ \bar{z}_2 = \frac{r^2}{z_2}, \ \bar{z}_3 = \frac{r^2}{z_3}.$$

Returning to the given relation, we have

$$\frac{z_1}{z_2} = \frac{z_2}{z_3} = \frac{z_3}{z_1},$$

or

$$z_1^2 = z_2 z_3, \quad z_2^2 = z_3 z_1, \quad z_3^2 = z_1 z_2.$$

Summing up these relations yields

$$z_1^2 + z_2^2 + z_3^2 = z_1 z_2 + z_2 z_3 + z_3 z_1,$$

so triangle $A_1 A_2 A_3$ is equilateral.

Observe that we have also proved that $(2) \Rightarrow (3)$ and that the arguments are reversible; hence $(2) \Leftrightarrow (3)$. As a consequence, $(3) \Rightarrow (1)$, and we are done. $\qquad \square$

Problem 1. *Let z_1, z_2, z_3 be nonzero complex coordinates of the vertices of the triangle $A_1 A_2 A_3$. If $z_1^2 = z_2 z_3$ and $z_2^2 = z_1 z_3$, show that triangle $A_1 A_2 A_3$ is equilateral.*

Solution. Multiplying the relations $z_1^2 = z_2 z_3$ and $z_2^2 = z_1 z_3$ yields $z_1^2 z_2^2 = z_1 z_2 z_3^2$, and consequently $z_1 z_2 = z_3^2$. Thus

$$z_1^2 + z_2^2 + z_3^2 = z_1 z_2 + z_2 z_3 + z_3 z_1,$$

so triangle $A_1 A_2 A_3$ is equilateral, by Proposition 1 in this section.

Problem 2. *Let z_1, z_2, z_3 be the coordinates of the vertices of triangle $A_1 A_2 A_3$. If $|z_1| = |z_2| = |z_3|$ and $z_1 + z_2 + z_3 = 0$, prove that triangle $A_1 A_2 A_3$ is equilateral.*

Solution 1. The following identity holds for all complex numbers z_1 and z_2 (see Problem 1 in Sect. 1.1.7):

$$|z_1 - z_2|^2 + |z_1 + z_2|^2 = 2(|z_1|^2 + |z_2|^2). \tag{1}$$

From $z_1 + z_2 + z_3 = 0$, it follows that $z_1 + z_2 = -z_3$, so $|z_1 + z_2| = |z_3|$. Using the relations $|z_1| = |z_2| = |z_3|$ and (1), we get $|z_1 - z_2|^2 = 3|z_1|^2$. Analogously, we obtain the relations $|z_2 - z_3|^2 = 3|z_1|^2$ and $|z_3 - z_1|^2 = 3|z_1|^2$. Therefore, $|z_1 - z_2| = |z_2 - z_3| = |z_3 - z_1|$, i.e., triangle $A_1 A_2 A_3$ is equilateral.

Solution 2. If we pass to conjugates, then we obtain $\dfrac{1}{z_1} + \dfrac{1}{z_2} + \dfrac{1}{z_3} = 0$.

Combining this with the hypothesis yields $z_1^2 + z_2^2 + z_3^2 = z_1 z_2 + z_2 z_3 + z_3 z_1 = 0$, from which the desired conclusion follows by Proposition 1.

Solution 3. Taking into account the hypotheses $|z_1| = |z_2| = |z_3|$, it follows that we can consider the complex plane with its origin at the circumcenter of triangle $A_1 A_2 A_3$. Then the coordinate of the orthocenter H is $z_H = z_1 + z_2 + z_3 = 0 = z_0$. Hence $H = O$, and triangle $A_1 A_2 A_3$ is equilateral.

Problem 3. *In the exterior of triangle ABC, three positively oriented equilateral triangles $AC'B$, $BA'C$, and $CB'A$ are constructed. Prove that the centroids of these triangles are the vertices of an equilateral triangle.*

(Napoleon's problem)

Solution. Let a, b, c be the coordinates of vertices A, B, C, respectively (Fig. 3.13).

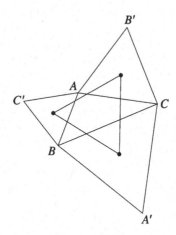

Figure 3.13.

Using Proposition 2, we have

$$a + c'\varepsilon + b\varepsilon^2 = 0, \quad b + a'\varepsilon + c\varepsilon^2 = 0, \quad c + b'\varepsilon + a\varepsilon^2 = 0, \qquad (1)$$

where a', b', c' are the coordinates of points A', B', C'.

The centroids of triangles $A'BC$, $AB'C$, ABC' have the coordinates

$$a'' = \frac{1}{3}(a' + b + c), \quad b'' = \frac{1}{3}(a + b' + c), \quad c'' = \frac{1}{3}(a + b + c'),$$

respectively. We have to check that $c'' + a''\varepsilon + b''\varepsilon^2 = 0$. Indeed,

$$3(c'' + a''\varepsilon + b''\varepsilon^2) = (a + b + c') + (a' + b + c)\varepsilon + (a + b' + c)\varepsilon^2$$
$$= (b + a'\varepsilon + c\varepsilon^2) + (c + b'\varepsilon + a\varepsilon^2)\varepsilon + (a + c'\varepsilon + b\varepsilon^2)\varepsilon^2 = 0.$$

Problem 4. *On the sides of the triangle ABC, we draw three regular n-gons, external to the triangle. Find all values of n for which the centers of the n-gons are the vertices of an equilateral triangle.*

(Balkan Mathematical Olympiad 1990—Shortlist)

Solution. Let A_0, B_0, C_0 be the centers of the regular n-gons constructed externally on the sides BC, CA, AB, respectively (Fig. 3.14).

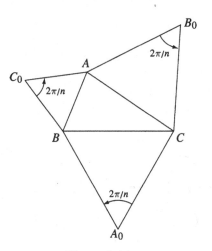

Figure 3.14.

The angles $\widehat{AC_0B}$, $\widehat{BA_0C}$, $\widehat{AB_0C}$ have measure $\dfrac{2\pi}{n}$. Let

$$\varepsilon = \cos\frac{2\pi}{n} + i\sin\frac{2\pi}{n}$$

and denote by a, b, c, a_0, b_0, c_0 the coordinates of the points A, B, C, A_0, B_0, C_0, respectively.

Using the rotation formula, we obtain

$$a = c_0 + (b - c_0)\varepsilon,$$
$$b = a_0 + (c - a_0)\varepsilon,$$
$$c = b_0 + (a - b_0)\varepsilon.$$

Thus

$$a_0 = \frac{b - c\varepsilon}{1 - \varepsilon}, \quad b_0 = \frac{c - a\varepsilon}{1 - \varepsilon}, \quad c_0 = \frac{a - b\varepsilon}{1 - \varepsilon}.$$

Triangle $A_0\, B_0\, C_0$ is equilateral if and only if

$$a_0^2 + b_0^2 + c_0^2 = a_0 b_0 + b_0 c_0 + c_0 a_0.$$

Substituting the above values of $a_0,\ b_0,\ c_0$, we obtain

$$(b - c\varepsilon)^2 + (c - a\varepsilon)^2 + (a - b\varepsilon)^2$$

$$= (b - c\varepsilon)(c - a\varepsilon) + (c - a\varepsilon)(a - b\varepsilon) + (a - b\varepsilon)(c - a\varepsilon).$$

This is equivalent to

$$(1 + \varepsilon + \varepsilon^2)[(a - b)^2 + (b - c)^2 + (c - a)^2] = 0.$$

It follows that $1 + \varepsilon + \varepsilon^2 = 0$, i.e., $\dfrac{2\pi}{n} = \dfrac{2\pi}{3}$, and we get $n = 3$. Therefore, $n = 3$ is the only value with the desired property.

3.5 Some Analytic Geometry in the Complex Plane

3.5.1 Equation of a Line

Proposition 1. *The equation of a line in the complex plane is*

$$\overline{\alpha} \cdot \overline{z} + \alpha z + \beta = 0,$$

where $\alpha \in \mathbb{C}^*$, $\beta \in \mathbb{R}$ and $z = x + iy \in \mathbb{C}$.

Proof. The equation of a line in the Cartesian plane is

$$Ax + By + C = 0,$$

where $A,\ B,\ C \in \mathbb{R}$ and $A^2 + B^2 \neq 0$. If we set $z = x + iy$, then $x = \dfrac{z + \overline{z}}{2}$ and $y = \dfrac{z - \overline{z}}{2i}$. Thus

$$A\frac{z+\overline{z}}{2} - Bi\frac{z-\overline{z}}{2} + C = 0,$$

or equivalently,

$$\overline{z}\left(\frac{A+Bi}{2}\right) + z\frac{A-Bi}{2} + C = 0.$$

Let $\alpha = \dfrac{A-Bi}{2} \in \mathbb{C}^*$ and $\beta = C \in \mathbb{R}$. Then $\alpha \neq 0$, because $|\alpha|^2 = \dfrac{A^2+B^2}{4} \neq 0$, and

$$\overline{\alpha}\cdot\overline{z} + \alpha z + \beta = 0,$$

as claimed. $\qquad\qquad\qquad\qquad\qquad\qquad\qquad\qquad\qquad\qquad\qquad\square$

If $\alpha = \overline{\alpha}$, then $B = 0$, and we have a vertical line. If $\alpha \neq \overline{\alpha}$, then we define the *angular coefficient* of the line as

$$m = -\frac{A}{B} = \frac{\alpha+\overline{\alpha}}{\dfrac{\alpha-\overline{\alpha}}{i}} = \frac{\alpha+\overline{\alpha}}{\alpha-\overline{\alpha}}i.$$

Proposition 2. *Consider the lines d_1 and d_2 with equations*

$$\overline{\alpha}_1\cdot\overline{z} + \alpha_1\cdot z + \beta_1 = 0$$

and

$$\overline{\alpha}_2\cdot\overline{z} + \alpha_2\cdot z + \beta_2 = 0,$$

respectively.
Then the lines d_1 and d_2 are:

(1) parallel if and only if $\dfrac{\overline{\alpha}_1}{\alpha_1} = \dfrac{\overline{\alpha}_2}{\alpha_2}$;

(2) perpendicular if and only if $\dfrac{\overline{\alpha}_1}{\alpha_1} + \dfrac{\overline{\alpha}_2}{\alpha_2} = 0$;

(3) concurrent if and only if $\dfrac{\overline{\alpha}_1}{\alpha_1} \neq \dfrac{\overline{\alpha}_2}{\alpha_2}$.

Proof.

(1) We have $d_1\|d_2$ if and only if $m_1 = m_2$. Therefore, $\dfrac{\alpha_1+\overline{\alpha}_1}{\alpha_1-\overline{\alpha}_1}i = \dfrac{\alpha_2+\overline{\alpha}_2}{\alpha_2-\overline{\alpha}_2}i$, so $\alpha_2\overline{\alpha}_1 = \alpha_1\overline{\alpha}_2$, and we get $\frac{\overline{\alpha}_1}{\alpha_1} = \frac{\overline{\alpha}_2}{\alpha_2}$.

(2) We have $d_1 \perp d_2$ if and only if $m_1m_2 = -1$. That is, $\alpha_2\overline{\alpha}_1 + \alpha_2\overline{\alpha}_2 = 0$, or $\dfrac{\overline{\alpha}_1}{\alpha_1} + \dfrac{\overline{\alpha}_2}{\alpha_2} = 0$.

(3) The lines d_1 and d_2 are concurrent if and only if $m_1 \neq m_2$. This condition yields $\dfrac{\overline{\alpha}_1}{\alpha_1} \neq \dfrac{\overline{\alpha}_2}{\alpha_2}$.

The results for *angular coefficient* correspond to the properties of *slope*. $\square$

The ratio $m_d = -\dfrac{\overline{\alpha}}{\alpha}$ is called the complex angular coefficient of the line d with equation

$$\overline{\alpha} \cdot \overline{z} + \alpha \cdot z + \beta = 0.$$

3.5.2 Equation of a Line Determined by Two Points

Proposition. *The equation of a line determined by the points $P_1(z_1)$ and $P_2(z_2)$ is*

$$\begin{vmatrix} z_1 & \overline{z_1} & 1 \\ z_2 & \overline{z_2} & 1 \\ z & \overline{z} & 1 \end{vmatrix} = 0.$$

Proof. The equation of a line determined by the points $P_1(x_1, \; y_1)$ and $P_2(x_2, \; y_2)$ in the Cartesian plane is

$$\begin{vmatrix} x_1 & y_1 & 1 \\ x_2 & y_2 & 1 \\ x & y & 1 \end{vmatrix} = 0.$$

Using complex numbers, we have

$$\begin{vmatrix} \frac{z_1+\overline{z_1}}{2} & \frac{z_1-\overline{z_1}}{2i} & 1 \\ \frac{z_2+\overline{z_2}}{2} & \frac{z_2-\overline{z_2}}{2i} & 1 \\ \frac{z+\overline{z}}{2} & \frac{z-\overline{z}}{2i} & 1 \end{vmatrix} = 0$$

if and only if

$$\frac{1}{4i} \begin{vmatrix} z_1 + \overline{z_1} & z_1 - \overline{z_1} & 1 \\ z_2 + \overline{z_2} & z_2 - \overline{z_2} & 1 \\ z + \overline{z} & z - \overline{z} & 1 \end{vmatrix} = 0.$$

That is,

$$\begin{vmatrix} z_1 & \overline{z_1} & 1 \\ z_2 & \overline{z_2} & 1 \\ z & \overline{z} & 1 \end{vmatrix} = 0,$$

as desired. $\square$

Remarks.

(1) The points $M_1(z_1)$, $M_2(z_2)$, $M_3(z_3)$ are collinear if and only if

$$\begin{vmatrix} z_1 & \overline{z_1} & 1 \\ z_2 & \overline{z_2} & 1 \\ z_3 & \overline{z_3} & 1 \end{vmatrix} = 0.$$

(2) The complex angular coefficient of a line determined· by the points with coordinates z_1 and z_2 is

$$m = \frac{z_2 - z_1}{\overline{z_2} - \overline{z_1}}.$$

Indeed, the equation is

$$\begin{vmatrix} z_1 & \overline{z_1} & 1 \\ z_2 & \overline{z_2} & 1 \\ z & \overline{z_3} & 1 \end{vmatrix} = 0,$$

and it is equivalent to

$$z_1\overline{z_2} + z_2\overline{z} + z\overline{z_1} - z\overline{z_2} - z_1\overline{z} - z_2\overline{z_1} = 0.$$

That is,

$$\overline{z}(z_2 - z_1) - z(\overline{z_2} - \overline{z_1}) + z_1\overline{z_2} - z_2\overline{z_1} = 0.$$

Using the definition of the complex angular coefficient, we obtain

$$m = \frac{z_2 - z_1}{\overline{z_2} - \overline{z_1}}.$$

3.5.3 The Area of a Triangle

Theorem. *The area of triangle $A_1 A_2 A_3$ whose vertices have coordinates z_1, z_2, z_3 is equal to the absolute value of the number*

$$\frac{i}{4} \begin{vmatrix} z_1 & \overline{z_1} & 1 \\ z_2 & \overline{z_2} & 1 \\ z_3 & \overline{z_3} & 1 \end{vmatrix}. \tag{1}$$

Proof. Using Cartesian coordinates, the area of a triangle with vertices $(x_1, \ y_1)$, $(x_2, \ y_2)$, $(x_3, \ y_3)$ is seen to be equal to the absolute value of the determinant

$$\Delta = \frac{1}{2} \begin{vmatrix} x_1 & y_1 & 1 \\ x_2 & y_2 & 1 \\ x_3 & y_3 & 1 \end{vmatrix}.$$

Since

$$x_k = \frac{z_k + \overline{z_k}}{2}, \quad y_k = \frac{z_k - \overline{z_k}}{2i}, \quad k = 1, 2, 3,$$

we obtain

$$\Delta = \frac{1}{8i} \begin{vmatrix} z_1 + \overline{z_1} & z_1 - \overline{z_1} & 1 \\ z_2 + \overline{z_2} & z_2 - \overline{z_2} & 1 \\ z_3 + \overline{z_3} & z_3 - \overline{z_3} & 1 \end{vmatrix} = -\frac{1}{4i} \begin{vmatrix} z_1 & \overline{z_1} & z_3 \\ z_2 & \overline{z_2} & 1 \\ z_3 & \overline{z_3} & 2 \end{vmatrix}$$

$$= \frac{i}{4} \begin{vmatrix} z_1 & \overline{z_1} & 1 \\ z_2 & \overline{z_2} & 1 \\ z_3 & \overline{z_3} & 1 \end{vmatrix},$$

as claimed.   □

It is easy to see that for a positively oriented triangle $A_1 A_2 A_3$ with vertices with coordinates z_1, z_2, z_3, the following inequality holds:

$$\frac{i}{4} \begin{vmatrix} z_1 & \overline{z_1} & 1 \\ z_2 & \overline{z_2} & 1 \\ z_3 & \overline{z_3} & 1 \end{vmatrix} > 0.$$

Corollary. *The area of a directly oriented triangle $A_1 A_2 A_3$ whose vertices have coordinates z_1, z_2, z_3 is*

$$\text{area}[A_1 A_2 A_3] = \frac{1}{2}\text{Im}(\overline{z_1}z_2 + \overline{z_2}z_3 + \overline{z_3}z_1). \tag{2}$$

Proof. The determinant in the above theorem is

$$\begin{vmatrix} z_1 & \overline{z_1} & 1 \\ z_2 & \overline{z_2} & 1 \\ z_3 & \overline{z_3} & 1 \end{vmatrix} = (z_1\overline{z_2} + z_2\overline{z_3} + z_3\overline{z_1} - \overline{z_2}z_3 - z_1\overline{z_3} - z_2\overline{z_1})$$

$$= [(z_1\overline{z_2} + z_2\overline{z_3} + z_3\overline{z_1}) - \overline{(z_1\overline{z_2} + z_2\overline{z_3} + z_3\overline{z_1})}]$$
$$= 2i\,\text{Im}(z_1\overline{z_2} + z_2\overline{z_3} + z_3\overline{z_1}) = -2i\,\text{Im}(\overline{z_1}z_2 + \overline{z_2}z_3 + \overline{z_3}z_1).$$

Replacing this value in (1), the desired formula follows.   □

We will see that formula (2) can be extended to a convex directly oriented polygon $A_1 A_2 \cdots A_n$ (see Sect. 4.3).

Problem 1. *Consider the triangle $A_1 A_2 A_3$ and the points M_1, M_2, M_3 situated on lines $A_2 A_3$, $A_1 A_3$, $A_1 A_2$, respectively. Assume that M_1, M_2, M_3 divide segments $[A_2 A_3]$, $[A_3 A_1]$, $[A_1 A_2]$ into ratios λ_1, λ_2, λ_3, respectively. Then*

$$\frac{\text{area}[M_1 M_2 M_3]}{\text{area}[A_1 A_2 A_3]} = \frac{1 - \lambda_1\lambda_2\lambda_3}{(1 - \lambda_1)(1 - \lambda_2)(1 - \lambda_3)}. \tag{3}$$

Solution. The coordinates of the points M_1, M_2, M_3 are

$$m_1 = \frac{a_2 - \lambda_1 a_3}{1 - \lambda_1}, \quad m_2 = \frac{a_3 - \lambda_2 a_1}{1 - \lambda_2}, \quad m_3 = \frac{a_1 - \lambda_3 a_2}{1 - \lambda_3}.$$

Applying formula (2), we find that

$$\text{area}[M_1 M_2 M_3] = \frac{1}{2}\text{Im}(\overline{m_1}m_2 + \overline{m_2}m_3 + \overline{m_3}m_1)$$

$$= \frac{1}{2}\text{Im}\left[\frac{(\overline{a_2} - \lambda_1\overline{a_3})(a_3 - \lambda_2 a_1)}{(1 - \lambda_1)(1 - \lambda_2)} + \frac{(\overline{a_3} - \lambda_2\overline{a_1})(a_1 - \lambda_3 a_2)}{(1 - \lambda_2)(1 - \lambda_3)}\right.$$

$$\left. + \frac{(\overline{a_1} - \lambda_3\overline{a_2})(a_2 - \lambda_1 a_3)}{(1 - \lambda_3)(1 - \lambda_1)}\right]$$

$$= \frac{1}{2}\text{Im}\left[\frac{1 - \lambda_1\lambda_2\lambda_3}{(1 - \lambda_1)(1 - \lambda_2)(1 - \lambda_3)}(\overline{a_1}a_2 + \overline{a_2}a_3 + \overline{a_3}a_1)\right]$$

$$= \frac{1 - \lambda_1\lambda_2\lambda_3}{(1 - \lambda_1)(1 - \lambda_2)(1 - \lambda_3)}\text{area}[A_1 A_2 A_3].$$

Remark. From formula (3), we derive the well-known theorem of Menelaus: *The points M_1, M_2, M_3 are collinear if and only if $\lambda_1\lambda_2\lambda_3 = 1$, i.e.,*

$$\frac{M_1 A_2}{M_1 A_3} \cdot \frac{M_2 A_3}{M_2 A_1} \cdot \frac{M_3 A_1}{M_3 A_2} = 1.$$

Problem 2. *Let a, b, c be the coordinates of the vertices A, B, C of a triangle. It is known that $|a| = |b| = |c| = 1$ and that there exists $\alpha \in \left(0, \frac{\pi}{2}\right)$ such that $a + b\cos\alpha + c\sin\alpha = 0$. Prove that*

$$1 < \text{area}[ABC] \le \frac{1 + \sqrt{2}}{2}.$$

(Romanian Mathematical Olympiad—Final Round, 2003)

Solution. Observe that

$$1 = |a|^2 = |b\cos\alpha + c\sin\alpha|^2$$

$$= (b\cos\alpha + c\sin\alpha)(\overline{b}\cos\alpha + \overline{c}\sin\alpha)$$

$$= |b|^2\cos^2\alpha + |c|^2\sin^2\alpha + (b\overline{c} + \overline{b}c)\sin\alpha\cos\alpha$$

$$= 1 + \frac{b^2 + c^2}{bc}\cos\alpha\sin\alpha.$$

It follows that $b^2 + c^2 = 0$; hence $b = \pm ic$. Applying formula (2), we obtain

$$\text{area}[ABC] = \frac{1}{2}|\text{Im}(\overline{a}b + \overline{b}c + \overline{c}a)|$$

$$= \frac{1}{2}|\text{Im}[(-\overline{b}\cos\alpha - \overline{c}\sin\alpha)b + \overline{b}c - \overline{c}(b\cos\alpha + c\sin\alpha)]|$$

$$= \frac{1}{2}|\text{Im}(-\cos\alpha - \sin\alpha - b\overline{c}\sin\alpha - b\overline{c}\cos\alpha + \overline{b}c)|$$

$$= \frac{1}{2}|\mathrm{Im}[\bar{b}c - (\sin\alpha + \cos\alpha)b\bar{c}]| = \frac{1}{2}|\mathrm{Im}[(1 + \sin\alpha + \cos\alpha)\bar{b}c]|$$

$$= \frac{1}{2}(1 + \sin\alpha + \cos\alpha)|\mathrm{Im}(\bar{b}c)| = \frac{1}{2}(1 + \sin\alpha + \cos\alpha)|\mathrm{Im}(\pm ic\bar{c})|$$

$$= \frac{1}{2}(1 + \sin\alpha + \cos\alpha)|\mathrm{Im}(\pm i)| = \frac{1}{2}(1 + \sin\alpha + \cos\alpha)$$

$$= \frac{1}{2}\left[1 + \sqrt{2}\left(\frac{\sqrt{2}}{2}\sin\alpha + \frac{\sqrt{2}}{2}\cos\alpha\right)\right] = \frac{1}{2}\left(1 + \sqrt{2}\sin\left(\alpha + \frac{\pi}{4}\right)\right).$$

Taking into account that $\frac{\pi}{4} < \alpha + \frac{\pi}{4} < \frac{3\pi}{4}$, we get that $\frac{\sqrt{2}}{2} < \sin\left(\alpha + \frac{\pi}{4}\right) \leq 1$, and the conclusion follows.

3.5.4 Equation of a Line Determined by a Point and a Direction

Proposition 1. *Let $d : \overline{\alpha}\bar{z} + \alpha \cdot z + \beta = 0$ be a line and let $P_0(z_0)$ be a point. The equation of the line parallel to d and passing through point P_0 is*

$$z - z_0 = -\frac{\overline{\alpha}}{\alpha}(\bar{z} - \overline{z_0}).$$

Proof. In Cartesian coordinates, the line parallel to d and passing through point $P_0(x_0, y_0)$ has the equation

$$y - y_0 = i\frac{\alpha + \overline{\alpha}}{\alpha - \overline{\alpha}}(x - x_0).$$

Using complex numbers, the equation takes the form

$$\frac{z - \bar{z}}{2i} - \frac{z_0 - \overline{z_0}}{2i} = i\frac{\alpha + \overline{\alpha}}{\alpha - \overline{\alpha}}\left(\frac{z + \bar{z}}{2} - \frac{z_0 + \overline{z_0}}{2}\right).$$

This is equivalent to $(\alpha - \overline{\alpha})(z - z_0 - \bar{z} + \overline{z_0}) = -(\alpha + \overline{\alpha})(z + \bar{z} - z_0 - \overline{z_0})$, or $\alpha(z - z_0) = -\overline{\alpha}(\bar{z} - \overline{z_0})$. We obtain $z - z_0 = -\frac{\overline{\alpha}}{\alpha}(\bar{z} - \overline{z_0})$.   □

Proposition 2. *Let $d : \overline{\alpha}\bar{z} + \alpha \cdot z + \beta = 0$ be a line and let $P_0(z_0)$ be a point. The line passing through point P_0 and perpendicular to d has the equation*

$$z - z_0 = \frac{\overline{\alpha}}{\alpha}(\bar{z} - \overline{z_0}).$$

Proof. In Cartesian coordinates, the line passing through point P_0 and perpendicular to d has the equation

$$y - y_0 = -\frac{1}{i} \cdot \frac{\alpha - \overline{\alpha}}{\alpha + \overline{\alpha}}(x - x_0).$$

Then we obtain

$$\frac{z - \overline{z}}{2i} - \frac{z_0 - \overline{z_0}}{2i} = i \cdot \frac{\alpha - \overline{\alpha}}{\alpha + \overline{\alpha}}\left(\frac{z + \overline{z}}{2} - \frac{z_0 + \overline{z_0}}{2}\right).$$

That is, $(\alpha + \overline{\alpha})(z - z_0 - \overline{z} + \overline{z_0}) = -(\alpha - \overline{\alpha})(z - z_0 + \overline{z} - \overline{z_0})$, or

$$(z - z_0)(\alpha + \overline{\alpha} + \alpha - \overline{\alpha}) = (\overline{z} - \overline{z_0})(-\alpha + \overline{\alpha} + \alpha + \overline{\alpha}).$$

We obtain

$$\alpha(z - z_0) = \overline{\alpha}(\overline{z} - \overline{z_0})$$

and therefore

$$z - z_0 = \frac{\overline{\alpha}}{\alpha}(\overline{z} - \overline{z_0}).$$

$\square$

3.5.5 The Foot of a Perpendicular from a Point to a Line

Proposition. *Let $P_0(z_0)$ be a point and let $d : \overline{\alpha}\overline{z} + \alpha z + \beta = 0$ be a line. The foot of the perpendicular from P_0 to d has the coordinate*

$$z = \frac{\alpha z_0 - \overline{\alpha z_0} - \beta}{2\alpha}.$$

Proof. The point z is the solution of the system

$$\begin{cases} \overline{\alpha} \cdot \overline{z} + \alpha \cdot z + \beta = 0, \\ \alpha(z - z_0) = \overline{\alpha}(\overline{z} - \overline{z_0}). \end{cases}$$

The first equation gives

$$\overline{z} = \frac{-\alpha z - \beta}{\overline{\alpha}}.$$

Substituting in the second equation yields

$$\alpha z - \alpha z_0 = -\alpha z - \beta - \overline{\alpha} \cdot \overline{z_0}.$$

Hence

$$z = \frac{\alpha z_0 - \overline{\alpha z_0} - \beta}{2\alpha},$$

as claimed.

$\square$

3.5.6 Distance from a Point to a Line

Proposition. *The distance from a point $P_0(z_0)$ to a line $d : \overline{\alpha} \cdot \overline{z} + \alpha \cdot z + \beta = 0$, $\alpha \in \mathbb{C}^*$, is equal to*

$$D = \frac{|\alpha z_0 + \overline{\alpha} \cdot \overline{z_0} + \beta|}{2\sqrt{\alpha \cdot \overline{\alpha}}}.$$

Proof. Using the previous result, we can write

$$D = \left| \frac{\alpha z_0 - \overline{\alpha} \cdot \overline{z_0} - \beta}{2\alpha} - z_0 \right| = \left| \frac{-\alpha z_0 - \overline{\alpha z_0} - \beta}{2\alpha} \right|$$

$$= \frac{|\alpha \cdot z_0 + \overline{\alpha z_0} + \beta|}{2|\alpha|} = \frac{|\alpha z_0 + \overline{\alpha z_0} + \beta|}{2\sqrt{\alpha \overline{\alpha}}}.$$

$\square$

3.6 The Circle

3.6.1 Equation of a Circle

Proposition. *The equation of a circle in the complex plane is*

$$z \cdot \overline{z} + \alpha \cdot z + \overline{\alpha} \cdot \overline{z} + \beta = 0,$$

where $\alpha \in \mathbb{C}$ and $\beta \in \mathbb{R}$, $\beta < |\alpha|^2$.

Proof. The equation of a circle in the Cartesian plane is

$$x^2 + y^2 + mx + ny + p = 0,$$

$m, n, p \in \mathbb{R}, p < \dfrac{m^2 + n^2}{4}$.

Setting $x = \dfrac{z + \overline{z}}{2}$ and $y = \dfrac{z - \overline{z}}{2i}$, we obtain

$$|z|^2 + m\frac{z + \overline{z}}{2} + n\frac{z - \overline{z}}{2i} + p = 0,$$

or

$$z \cdot \overline{z} + z\frac{m - ni}{2} + \overline{z}\frac{m + ni}{2} + p = 0.$$

Take $\alpha = \dfrac{m - ni}{2} \in \mathbb{C}$ and $\beta = p \in \mathbb{R}$ in the above equation, and the claim is proved. $\square$

Note that the radius of the circle is equal to

$$r = \sqrt{\frac{m^2}{4} + \frac{n^2}{4} - p} = \sqrt{\alpha\bar{\alpha} - \beta}.$$

Then the equation is equivalent to

$$(\bar{z} + a)(z + \bar{\alpha}) = r^2.$$

Setting

$$\gamma = -\bar{\alpha} = -\frac{m}{2} - \frac{n}{2}i$$

shows that the equation of the circle with center at γ and radius r is

$$(\bar{z} - \bar{\gamma})(z - \gamma) = r^2.$$

Problem. *Let z_1, z_2, z_3 be the coordinates of the vertices of triangle $A_1 A_2 A_3$. The coordinate z_0 of the circumcenter of triangle $A_1 A_2 A_3$ is*

$$z_0 = \frac{\begin{vmatrix} 1 & 1 & 1 \\ z_1 & z_2 & z_3 \\ |z_1|^2 & |z_2|^2 & |z_3|^2 \end{vmatrix}}{\begin{vmatrix} 1 & 1 & 1 \\ z_1 & z_2 & z_3 \\ \bar{z_1} & \bar{z_2} & \bar{z_3} \end{vmatrix}}. \tag{1}$$

Solution. The equation of the line passing through $P(z_0)$ that is perpendicular to the line $A_1 A_2$ can be written in the form

$$z(\bar{z_1} - \bar{z_2}) + \bar{z}(z_1 - z_2) = z_0(\bar{z_1} - \bar{z_2}) + \bar{z_0}(z_1 - z_2). \tag{2}$$

Applying this formula to the midpoints of the sides $[A_2 A_3]$, $[A_1 A_3]$ and the lines $A_2 A_3$, $A_1 A_3$, we obtain the equations

$$z(\bar{z_2} - \bar{z_3}) + \bar{z}(z_2 - z_3) = |z_2|^2 - |z_3|^2,$$
$$z(\bar{z_3} - \bar{z_1}) + \bar{z}(z_3 - z_1) = |z_3|^2 - |z_1|^2.$$

By eliminating $\bar{z}$ from these two equations, we see that that

$$z[(\bar{z_2} - \bar{z_3})(z_1 - z_3) + (\bar{z_3} - \bar{z_1})(z_2 - z_3)]$$

$$= (z_1 - z_3)(|z_2|^2 - |z_3|^2) + (z_2 - z_3)(|z_3|^2 - |z_1|^2),$$

whence

$$z \begin{vmatrix} 1 & 1 & 1 \\ z_1 & z_2 & z_3 \\ \bar{z_1} & \bar{z_2} & \bar{z_3} \end{vmatrix} = \begin{vmatrix} 1 & 1 & 1 \\ z_1 & z_2 & z_3 \\ |z_1|^2 & |z_2|^2 & |z_3|^2 \end{vmatrix},$$

and the desired formula follows.

Remark. We can write this formula in the following equivalent form:

$$z_0 = \frac{z_1\overline{z_1}(z_2 - z_3) + z_2\overline{z_2}(z_3 - z_1) + z_3\overline{z_3}(z_1 - z_2)}{\begin{vmatrix} 1 & 1 & 1 \\ \overline{z_1} & \overline{z_2} & \overline{z_3} \\ z_1 & z_2 & z_3 \end{vmatrix}}. \tag{3}$$

3.6.2 The Power of a Point with Respect to a Circle

Proposition. *Consider a point $P_0(z_0)$ and a circle with equation*

$$z\,\overline{z} + \alpha \cdot z + \overline{\alpha} \cdot \overline{z} + \beta = 0,$$

for $\alpha \in \mathbb{C}$ and $\beta \in \mathbb{R}$.

The power of P_0 with respect to the circle is

$$\rho(z_0) = z_0\,\overline{z_0} + \alpha z_0 + \overline{\alpha} \cdot \overline{z_0} + \beta.$$

Proof. Let $O(-\overline{\alpha})$ be the center of the circle. The power of P_0 with respect to the circle of radius r is defined by $\rho(z_0) = OP_0^2 - r^2$. In this case, we obtain

$$\rho(z_0) = OP_0^2 - r^2 = |z_0 + \overline{\alpha}|^2 - r^2 = z_0\overline{z_0} + \alpha z_0 + \overline{\alpha z_0} + \alpha\overline{\alpha} - \alpha\overline{\alpha} + \beta$$
$$= z_0\overline{z_0} + \alpha z_0 + \overline{\alpha} \cdot \overline{z_0} + \beta,$$

as claimed.   □

Given two circles with equations

$$z \cdot \overline{z} + \alpha_1 \cdot z + \overline{\alpha_1} \cdot \overline{z} + \beta_1 = 0 \text{ and } z\overline{z} + \alpha_2 \cdot z + \overline{\alpha_2} \cdot \overline{z} + \beta_2 = 0,$$

where $\alpha_1,\ \alpha_2 \in \mathbb{C}$, $\beta_1,\ \beta_2 \in \mathbb{R}$, their *radical axis* is the locus of points having equal powers with respect to the circles. If $P(z)$ is a point of this locus, then

$$z \cdot \overline{z} + \alpha_1 z + \overline{\alpha_1} \cdot \overline{z} + \beta_1 = z \cdot \overline{z} + \alpha_2 z + \overline{\alpha_2} \cdot z + \beta_2,$$

or equivalently, $(\alpha_1 - \alpha_2)z + (\overline{\alpha_1} - \overline{\alpha_2})\overline{z} + \beta_1 - \beta_2 = 0$, which is the equation of a line.

3.6.3 Angle Between Two Circles

The angle between two intersecting circles with equations

$$z \cdot \overline{z} + \alpha_1 \cdot z + \overline{\alpha_1} \cdot \overline{z} + \beta_1 = 0$$

and

$$z \cdot \overline{z} + \alpha_2 \cdot z + \overline{\alpha_2} \cdot \overline{z} + \beta_2 = 0, \quad \alpha_1, \alpha_2 \in \mathbb{C}, \quad \beta_1, \beta_2 \in \mathbb{R},$$

is the angle θ determined by the tangents to the circles at a common point.

Proposition. *The following formula holds (Fig. 3.15):*

$$\cos\theta = \left| \frac{\beta_1 + \beta_2 - (\alpha_1\overline{\alpha_2} + \overline{\alpha_1}\alpha_2)}{2r_1 r_2} \right|.$$

Proof. Let T be a common point and let $O_1(-\overline{\alpha_1})$, $O_2(-\overline{\alpha_2})$ be the centers of the circles.

The angle θ is equal to $\widehat{O_1 T O_2}$ or $\pi - \widehat{O_1 T O_2}$; hence

$$\cos\theta = |\cos\widehat{O_1 T O_2}| = \frac{|r_1^2 + r_2^2 - O_1 O_2^2|}{2r_1 r_2}$$

$$= \frac{|\alpha_1\overline{\alpha_1} - \beta_1 + \alpha_2\overline{\alpha_2} - \beta_2 - |\overline{\alpha_1} - \overline{\alpha_2}|^2|}{2r_1 r_2}$$

$$= \frac{|\alpha_1\overline{\alpha_1} + \alpha_2\overline{\alpha_2} - \beta_1 - \beta_2 - \overline{\alpha_1}\alpha_1 - \alpha_2\overline{\alpha_2} + \overline{\alpha_1}\alpha_2 + \alpha_1\overline{\alpha_2}|}{2r_1 r_2}$$

$$= \frac{|\beta_1 + \beta_2 - (\alpha_1\overline{\alpha_2} + \overline{\alpha_1}\alpha_2)|}{2r_1 r_2},$$

as claimed. $\square$

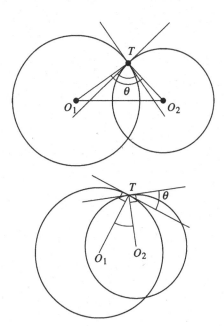

Figure 3.15.

Note that the circles are orthogonal if and only if

$$\beta_1 + \beta_2 = \alpha_1\overline{\alpha_2} + \overline{\alpha_1}\alpha_2.$$

Problem. *Let a, b be real numbers such that $|b| \leq 2a^2$. Prove that the set of points with coordinates z such that*

$$|z^2 - a^2| = |2az + b|$$

is the union of two orthogonal circles.

Solution. The relation

$$|z^2 - a^2| = |2az + b|$$

is equivalent to

$$|z^2 - a^2|^2 = |2az + b|^2,$$

i.e.,

$$(z^2 - a^2)(\overline{z}^2 - \overline{a}^2) = (2az + b)(2a\overline{z} + b).$$

We can rewrite the last relation as

$$|z|^4 - a^2(z^2 + \overline{z}^2) + a^4 = 4a^2|z|^2 + 2ab(z + \overline{z}) + b^2,$$

i.e.,

$$|z|^4 - a^2[(z + \overline{z})^2 - 2|z|^2] + a^4 = 4a^2|z|^2 + 2ab(z + \overline{z}) + b^2.$$

Hence

$$|z|^4 - 2a^2|z|^2 + a^4 = a^2(z + \overline{z})^2 + 2ab(z + \overline{z}) + b^2,$$

i.e.,

$$(|z|^2 - a^2)^2 = (a(z + \overline{z}) + b)^2.$$

It follows that

$$z \cdot \overline{z} - a^2 = a(z + \overline{z}) + b \text{ or} z \cdot \overline{z} - a^2 = -a(z + \overline{z}) - b.$$

This is equivalent to

$$(z - a)(\overline{z} - a) = 2a^2 + b \text{ or } (z + a)(\overline{z} + a) = 2a^2 - b.$$

Finally,

$$|z - a|^2 = 2a^2 + b \text{ or} |z + a|^2 = 2a^2 - b. \tag{1}$$

Since $|b| \leq 2a^2$, it follows that $2a^2 + b \geq 0$ and $2a^2 - b \geq 0$. Hence the relations (1) are equivalent to

$$|z - a| = \sqrt{2a^2 + b} \text{ or} |z + a| = \sqrt{2a^2 - b}.$$

Therefore, the points with coordinates z that satisfy $|z^2 - a^2| = |2az + b|$ lie on two circles with centers C_1 and C_2 whose coordinates are a and $-a$, and with radii $R_1 = \sqrt{2a^2 + b}$ and $R_2 = \sqrt{2a^2 - b}$. Furthermore, using the Pythagorean theorem, we have

$$C_1C_2^2 = 4a^2 = (\sqrt{2a^2 + b})^2 + (\sqrt{2a^2 - b})^2 = R_1^2 + R_2^2.$$

Hence the circles are orthogonal, as claimed.

Chapter 4
More on Complex Numbers and Geometry

4.1 The Real Product of Two Complex Numbers

The concept of the scalar product of two vectors is well known. In what follows, we will introduce this concept for complex numbers. We will see that the use of this product simplifies the solution to many problems considerably.

Let a and b be two complex numbers.

Definition. Given complex numbers a and b, we call the number given by

$$a \cdot b = \frac{1}{2}(\bar{a}b + a\bar{b})$$

the *real product* of the two numbers. It is easy to see that

$$\overline{a \cdot b} = \frac{1}{2}(a\bar{b} + \bar{a}b) = a \cdot b;$$

hence $a \cdot b$ is a real number, which justifies the name of this product.

Let $A(a), B(b)$ be points in the complex plane, and let $\theta = (\widehat{\overrightarrow{OA}, \overrightarrow{OB}})$ be the angle between the vectors $\overrightarrow{OA}, \overrightarrow{OB}$. The following formula holds:

$$a \cdot b = |a||b| \cos\theta = \overrightarrow{OA} \cdot \overrightarrow{OB}.$$

Indeed, considering the polar form of a and b, we have

$$a = |a|(\cos t_1 + i\sin t_1), \quad b = |b|(\cos t_2 + i\sin t_2),$$

and

$$a \cdot b = \frac{1}{2}(\bar{a}b + a\bar{b}) = \frac{1}{2}|a||b|[\cos(t_1 - t_2) - i\sin(t_1 - t_2) + \cos(t_1 - t_2) + i\sin(t_1 - t_2)]$$

$$= |a||b|\cos(t_1 - t_2) = |a||b|\cos\theta = \overrightarrow{OA} \cdot \overrightarrow{OB}.$$

The following properties are easy to verify.

T. Andreescu and D. Andrica, *Complex Numbers from A to ... Z,*
DOI 10.1007/978-0-8176-8415-0_4, © Springer Science+Business Media New York 2014

Proposition 1. *For all complex numbers* a, b, c, z, *the following relations hold:*

(1) $a \cdot a = |a|^2$.
(2) $a \cdot b = b \cdot a$ *(the real product is commutative).*
(3) $a \cdot (b + c) = a \cdot b + a \cdot c$ *(the real product is distributive with respect to addition).*
(4) $(\alpha a) \cdot b = \alpha(a \cdot b) = a \cdot (\alpha b)$ *for all* $\alpha \in \mathbb{R}$.
(5) $a \cdot b = 0$ *if and only if* $OA \perp OB$, *where* A *has coordinate* a *and* B *has coordinate* b.
(6) $(az) \cdot (bz) = |z|^2(a \cdot b)$.

Remark. Suppose that A and B are points with coordinates a and b. Then the real product $a \cdot b$ is equal to the power of the origin with respect to the circle of diameter AB.

Indeed, let $M\left(\dfrac{a + b}{2}\right)$ be the midpoint of $[AB]$, hence the center of this circle, and let $r = \dfrac{1}{2}AB = \dfrac{1}{2}|a - b|$ be the radius of this circle. The power of the origin with respect to the circle is

$$OM^2 - r^2 = \left|\frac{a + b}{2}\right|^2 - \left|\frac{a - b}{2}\right|^2$$

$$= \frac{(a + b)(\overline{a} + \overline{b})}{4} - \frac{(a - b)(\overline{a} - \overline{b})}{4} = \frac{a\overline{b} + b\overline{a}}{2} = a \cdot b,$$

as claimed.

Proposition 2. *Suppose that* $A(a)$, $B(b)$, $C(c)$, *and* $D(d)$ *are four distinct points. The following statements are equivalent:*

(1) $AB \perp CD$;
(2) $(b - a) \cdot (d - c) = 0$;
(3) $\dfrac{b - a}{d - c} \in i\mathbb{R}^*$ *(or equivalently,* $\text{Re}\left(\dfrac{b - a}{d - c}\right) = 0$).

Proof. Take points $M(b - a)$ and $N(d - c)$ such that $OABM$ and $OCDN$ are parallelograms. Then we have $AB \perp CD$ if and only if $OM \perp ON$. That is, $m \cdot n = (b - a) \cdot (d - c) = 0$, using property (5) of the real product.

The equivalence (2) $\Leftrightarrow$ (3) follows immediately from the definition of the real product. $\square$

Proposition 3. *The circumcenter of triangle* ABC *is at the origin of the complex plane. If* a, b, c *are the coordinates of vertices* A, B, C, *then the orthocenter* H *has the coordinate* $h = a + b + c$.

Proof. Using the real product of the complex numbers, the equations of the altitudes AA', BB', CC' of the triangle are

$$AA' : (z-a)\cdot(b-c) = 0, \quad BB' : (z-b)\cdot(c-a) = 0, \quad CC' : (z-c)\cdot(a-b) = 0.$$

We will show that the point with coordinate $h = a+b+c$ lies on all three altitudes. Indeed, we have $(h-a)\cdot(b-c) = 0$ if and only if $(b+c)\cdot(b-c) = 0$. The last relation is equivalent to $b\cdot b - c\cdot c = 0$, or $|b|^2 = |c|^2$. Similarly, $H \in BB'$ and $H \in CC'$, and we are done.   □

Remark. If the numbers a, b, c, o, h are the coordinates of the vertices of triangle ABC, the circumcenter O, and the orthocenter H of the triangle, then $h = a+b+c-2o$.

Indeed, if we take A' diametrically opposite A in the circumcircle of triangle ABC, then the quadrilateral $HBA'C$ is a parallelogram. If $\{M\} = HA' \cap BC$, then

$$z_M = \frac{b+c}{2} = \frac{z_H + z_{A'}}{2} = \frac{z_H + 2o - a}{2}, \quad \text{i.e.,} \quad z_H = a+b+c-2o.$$

Problem 1. *Let $ABCD$ be a convex quadrilateral. Prove that*

$$AB^2 + CD^2 = AD^2 + BC^2$$

if and only if $AC \perp BD$.

Solution. Using the properties of the real product of complex numbers, we have

$$AB^2 + CD^2 = BC^2 + DA^2$$

if and only if

$$(b-a)\cdot(b-a) + (d-c)\cdot(d-c) = (c-b)\cdot(c-b) + (a-d)\cdot(a-d).$$

That is,

$$a\cdot b + c\cdot d = b\cdot c + d\cdot a,$$

and finally,

$$(c-a)\cdot(d-b) = 0,$$

or equivalently, $AC \perp BD$, as required.

Problem 2. *Let M, N, P, Q, R, S be the midpoints of the sides AB, BC, CD, DE, EF, FA of a hexagon. Prove that*

$$RN^2 = MQ^2 + PS^2$$

if and only if $MQ \perp PS$.

(Romanian Mathematical Olympiad—Final Round, 1994)

Solution. Let a, b, c, d, e, f be the coordinates of the vertices of the hexagon (Fig. 4.1). The points M, N, P, Q, R, S have coordinates

$$m = \frac{a+b}{2}, \; n = \frac{b+c}{2}, \; p = \frac{c+d}{2},$$

$$q = \frac{d+e}{2}, \; r = \frac{e+f}{2}, \; s = \frac{f+a}{2},$$

respectively.

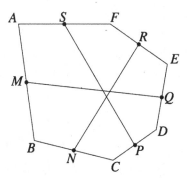

Figure 4.1.

Using the properties of the real product of complex numbers, we have

$$RN^2 = MQ^2 + PS^2$$

if and only if

$$(e+f-b-c)\cdot(e+f-b-c)=(d+e-a-b)\cdot(d+e-a-b)+(f+a-c-d)\cdot(f+a-c-d).$$

That is,

$$(d + e - a - b) \cdot (f + a - c - d) = 0;$$

hence $MQ \perp PS$, as claimed.

Problem 3. *Let $A_1 A_2 \cdots A_n$ be a regular polygon inscribed in a circle with center O and radius R. Prove that for all points M in the plane, the following relation holds:*

$$\sum_{k=1}^{n} MA_k^2 = n(OM^2 + R^2).$$

Solution. Consider the complex plane with the origin at point O, with the x-axis containing the point A_1, and let $R\varepsilon_k$ be the coordinate of vertex A_k, where ε_k are the nth-roots of unity, $k = 1, \ldots, n$. Let m be the coordinate of M.

Using the properties of the real product of the complex numbers, we have

$$\sum_{k=1}^{n} MA_k^2 = \sum_{k=1}^{n} (m - R\varepsilon_k) \cdot (m - R\varepsilon_k)$$

$$= \sum_{k=1}^{n} (m \cdot m - 2R\varepsilon_k \cdot m + R^2 \varepsilon_k \cdot \varepsilon_k)$$

$$= n|m|^2 - 2R \left(\sum_{k=1}^{n} \varepsilon_k \right) \cdot m + R^2 \sum_{k=1}^{n} |\varepsilon_k|^2$$

$$= n \cdot OM^2 + nR^2 = n(OM^2 + R^2),$$

since $\sum_{k=1}^{n} \varepsilon_k = 0$.

Remark. If M lies on the circumcircle of the polygon, then

$$\sum_{k=1}^{n} MA_k^2 = 2nR^2.$$

Problem 4. *Let O be the circumcenter of the triangle ABC, let D be the midpoint of the segment AB, and let E is the centroid of triangle ACD. Prove that lines CD and OE are perpendicular if and only if $AB = AC$.*

(Balkan Mathematical Olympiad, 1985)

Solution. Let O be the origin of the complex plane and let a, b, c, d, e be the coordinates of points A, B, C, D, E, respectively. Then

$$d = \frac{a+b}{2} \text{ and } e = \frac{a+c+d}{3} = \frac{3a+b+2c}{6}$$

Using the real product of complex numbers, if R is the circumradius of triangle ABC, then

$$a \cdot a = b \cdot b = c \cdot c = R^2.$$

Lines CD and OE are perpendicular if and only if $(d - c) \cdot e = 0$, that is,

$$(a + b - 2c) \cdot (3a + b + 2c) = 0.$$

The last relation is equivalent to

$$3a \cdot a + a \cdot b + 2a \cdot c + 3a \cdot b + b \cdot b + 2b \cdot c - 6a \cdot c - 2b \cdot c - 4c \cdot c = 0,$$

that is,

$$a \cdot b = a \cdot c. \tag{1}$$

On the other hand, $AB = AC$ is equivalent to

$$|b - a|^2 = |c - a|^2.$$

That is,

$$(b - a) \cdot (b - a) = (c - a) \cdot (c - a),$$

or

$$b \cdot b - 2a \cdot b + a \cdot a = c \cdot c - 2a \cdot c + a \cdot a,$$

whence

$$a \cdot b = a \cdot c. \tag{2}$$

The relations (1) and (2) show that $CD \perp OE$ if and only if $AB = AC$.

Problem 5. *Let a, b, c be distinct complex numbers such that $|a| = |b| = |c|$ and $|b + c - a| = |a|$. Prove that $b + c = 0$.*

Solution. Let A, B, C be the geometric images of the complex numbers a, b, c, respectively. Choose the circumcenter of triangle ABC as the origin of the complex plane and denote by R the circumradius of triangle ABC. Then

$$a\bar{a} = b\bar{b} = c\bar{c} = R^2,$$

and using the real product of the complex numbers, we have

$$|b + c - a| = |a| \text{ if and only if } |b + c - a|^2 = |a|^2.$$

That is,

$$(b + c - a) \cdot (b + c - a) = |a|^2,$$

i.e.,

$$|a|^2 + |b|^2 + |c|^2 + 2b \cdot c - 2a \cdot c - 2a \cdot b = |a|^2.$$

We obtain

$$2(R^2 + b \cdot c - a \cdot c - a \cdot b) = 0,$$

i.e.,

$$a \cdot a + b \cdot c - a \cdot c - a \cdot b = 0.$$

It follows that $(a - b) \cdot (a - c) = 0$, and hence $AB \perp AC$, i.e., $\widehat{BAC} = 90°$. Therefore, $[BC]$ is the diameter of the circumcircle of triangle ABC, so $b + c = 0$.

Problem 6. *Let E, F, G, H be the midpoints of sides AB, BC, CD, DA of the convex quadrilateral $ABCD$. Prove that lines AB and CD are perpendicular if and only if*

$$BC^2 + AD^2 = 2(EG^2 + FH^2).$$

Solution. Denote by the corresponding lowercase letter the coordinate of a point denoted by an uppercase letter. Then

$$e = \frac{a+b}{2}, \ f = \frac{b+c}{2}, \ g = \frac{c+d}{2}, \ h = \frac{d+a}{2}.$$

Using the real product of the complex numbers, the relation

$$BC^2 + AD^2 = 2(EG^2 + FH^2)$$

becomes

$$(c - b) \cdot (c - b) + (d - a) \cdot (d - a) = \frac{1}{2}(c + d - a - b) \cdot (c + d - a - b)$$
$$+ \frac{1}{2}(a + d - b - c) \cdot (a + d - b - c).$$

This is equivalent to

$$c \cdot c + b \cdot b + d \cdot d + a \cdot a - 2b \cdot c - 2a \cdot d$$
$$= a \cdot a + b \cdot b + c \cdot c + d \cdot d - 2a \cdot c - 2b \cdot d,$$

or

$$a \cdot d + b \cdot c = a \cdot c + b \cdot d.$$

The last relation shows that $(a - b) \cdot (d - c) = 0$ if and only if $AB \perp CD$, as desired.

Problem 7. *Let G be the centroid of triangle ABC and let A_1, B_1, C_1 be the midpoints of sides BC, CA, AB, respectively. Prove that*

$$MA^2 + MB^2 + MC^2 + 9MG^2 = 4(MA_1^2 + MB_1^2 + MC_1^2)$$

for all points M in the plane.

Solution. Denote by the corresponding lowercase letter the coordinate of a point denoted by an uppercase letter. Then

$$g = \frac{a+b+c}{3}, \ a_1 = \frac{b+c}{2}, \ b_1 = \frac{c+a}{2}, \ c_1 = \frac{a+b}{2}.$$

Using the real product of the complex numbers, we have

$$MA^2 + MB^2 + MC^2 + 9MG^2$$

$$= (m - a) \cdot (m - a) + (m - b) \cdot (m - b) + (m - c) \cdot (m - c)$$

$$+ 9 \left(m - \frac{a+b+c}{3} \right) \cdot \left(m - \frac{a+b+c}{3} \right)$$

$$= 12|m|^2 - 8(a + b + c) \cdot m + 2(|a|^2 + |b|^2 + |c|^2) + 2a \cdot b + 2b \cdot c + 2c \cdot a.$$

On the other hand,

$$4(MA_1^2 + MB_1^2 + MC_1^2)$$

$$= 4\left[\left(m - \frac{b+c}{2}\right) \cdot \left(m - \frac{b+c}{2}\right) + \left(m - \frac{c+a}{2}\right)\right.$$

$$\left. \cdot \left(m - \frac{c+a}{2}\right) + \left(m - \frac{a+b}{2}\right) \cdot \left(m - \frac{a+b}{2}\right)\right]$$

$$= 12|m|^2 - 8(a+b+c) \cdot m + 2(|a|^2 + |b|^2 + |c|^2) + 2a \cdot b + 2b \cdot c + 2c \cdot a,$$

so we are done.

Remark. The following generalization can be proved similarly.

Let $A_1A_2 \cdots A_n$ be a polygon with centroid G and let A_{ij} be the midpoint of the segment $[A_iA_j]$, $i < j$, i, $j \in \{1, 2, \ldots, n\}$.

Then

$$(n-2)\sum_{k=1}^{n} MA_k^2 + n^2 MG^2 = 4\sum_{i<j} MA_{ij}^2,$$

for all points M in the plane. A nice generalization is given in Theorem 3 in Sect. 4.11.

4.2 The Complex Product of Two Complex Numbers

The cross product of two vectors is a central concept in vector algebra, with numerous applications in various branches of mathematics and science. In what follows, we adapt this product to complex numbers. The reader will see that this new interpretation has multiple advantages in solving problems involving area or collinearity.

Let a and b be two complex numbers.

Definition. The complex number

$$a \times b = \frac{1}{2}(\bar{a}b - a\bar{b})$$

is called the *complex product* of the numbers a and b.

Note that

$$a \times b + \overline{a \times b} = \frac{1}{2}(\bar{a}b - a\bar{b}) + \frac{1}{2}(a\bar{b} - \bar{a}b) = 0,$$

so $\mathrm{Re}(a \times b) = 0$, which justifies the definition of this product.

Let $A(a), B(b)$ be points in the complex plane, and let $\theta = (\overrightarrow{OA}, \overrightarrow{OB})$ be the angle between the vectors $\overrightarrow{OA}$, $\overrightarrow{OB}$. The following formula holds:

$$a \times b = \varepsilon i |a||b| \sin \theta,$$

where
$$\varepsilon = \begin{cases} -1, \text{ if triangle } OAB \text{ is positively oriented}; \\ +1, \text{ if triangle } OAB \text{ is negatively oriented}. \end{cases}$$

Indeed, if $a = |a|(\cos t_1 + i \sin t_1)$ and $b = |b|(\cos t_2 + i \sin t_2)$, then

$$a \times b = i|a||b| \sin(-t_1 + t_2) = \varepsilon i|a||b| \sin\theta.$$

The connection between the real product and the complex product is given by the following Lagrange-type formula:

$$|a \cdot b|^2 + |a \times b|^2 = |a|^2|b|^2.$$

The following properties are easy to verify:

Proposition 1. *Suppose that a, b, c are complex numbers. Then:*

(1) $a \times b = 0$ if and only if $a = 0$ or $b = 0$ or $a = \lambda b$, where λ is a real number.

(2) $a \times b = -b \times a$ (the complex product is anticommutative).

(3) $a \times (b+c) = a \times b + a \times c$ (the complex product is distributive with respect to addition).

(4) $\alpha(a \times b) = (\alpha a) \times b = a \times (\alpha b)$, for all real numbers α.

(5) If $A(a)$ and $B(b)$ are distinct points other than the origin, then $a \times b = 0$ if and only if O, A, B are collinear.

Remarks.

(a) Suppose $A(a)$ and $B(b)$ are distinct points in the complex plane different from the origin (Fig. 4.2).
The complex product of the numbers a and b has the following useful geometric interpretation:

$$a \times b = \begin{cases} 2i. \text{ area } [AOB], \quad \text{if triangle } OAB \text{ is positively oriented}; \\ -2i. \text{ area } [AOB], \text{ if triangle } OAB \text{ is negatively oriented}. \end{cases}$$

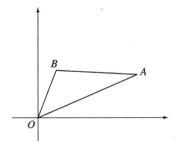

Figure 4.2.

Indeed, if triangle OAB is positively (directly) oriented, then

$$2i \cdot \text{area } [OAB] = i \cdot OA \cdot OB \cdot \sin(\widehat{AOB})$$

$$= i|a| \cdot |b| \cdot \sin\left(\arg\frac{b}{a}\right) = i \cdot |a| \cdot |b| \cdot \text{Im}\left(\frac{b}{a}\right) \cdot \frac{|a|}{|b|}$$

$$= \frac{1}{2}|a|^2 \left(\frac{b}{a} - \frac{\bar{b}}{\bar{a}}\right) = \frac{1}{2}(\bar{a}b - a\bar{b}) = a \times b.$$

In the other case, note that triangle OBA is positively oriented; hence

$$2i \cdot \text{area}[OBA] = b \times a = -a \times b.$$

(b) Suppose $A(a)$, $B(b)$, $C(c)$ are three points in the complex plane. The complex product allows us to obtain the following useful formula for the area of the triangle ABC:

$$\text{area } [ABC] = \begin{cases} \dfrac{1}{2i}(a \times b + b \times c + c \times a) \\ \text{if triangle } ABC \text{ is positively oriened;} \\ \\ -\dfrac{1}{2i}(a \times b + b \times c + c \times a) \\ \text{if triangle } ABC \text{ is negatively oriented.} \end{cases}$$

Moreover, simple algebraic manipulation shows that

$$\text{area } [ABC] = \frac{1}{2}\text{Im}(\bar{a}b + \bar{b}c + \bar{c}a)$$

if triangle ABC is directly (positively) oriented.

To prove the above formula, translate points A, B, C by the vector $-c$. The images of A, B, C are the points A', B', O with coordinates $a - c$, $b - c$, 0, respectively. Triangles ABC and $A'B'O$ are congruent with the same orientation. If ABC is positively oriented, then

$$\text{area } [ABC] = \text{area } [OA'B'] = \frac{1}{2i}((a - c) \times (b - c))$$

$$= \frac{1}{2i}((a - c) \times b - (a - c) \times c) = \frac{1}{2i}(c \times (a - c) - b \times (a - c))$$

$$= \frac{1}{2i}(c \times a - c \times c - b \times a + b \times c) = \frac{1}{2i}(a \times b + b \times c + c \times a),$$

as claimed.

The other situation can be handled similarly.

Proposition 2. *Suppose $A(a)$, $B(b)$, and $C(c)$ are distinct points. The following statements are equivalent:*

(1) Points A, B, C are collinear.
(2) $(b - a) \times (c - a) = 0$.
(3) $a \times b + b \times c + c \times a = 0$.

Proof. Points A, B, C are collinear if and only if area $[ABC] = 0$, i.e., $a \times b + b \times c + c \times a = 0$. The last equation can be written in the form $(b - a) \times (c - a) = 0$. □

Proposition 3. *Let $A(a)$, $B(b)$, $C(c)$, $D(d)$ be four points, no three of which are collinear. Then $AB \| CD$ if and only if $(b - a) \times (d - c) = 0$.*

Proof. Choose the points $M(m)$ and $N(n)$ such that $OABM$ and $OCDN$ are parallelograms; then $m = b - a$ and $n = d - c$.

Lines AB and CD are parallel if and only if points O, M, N are collinear. Using property 5, this is equivalent to $0 = m \times n = (b - a) \times (d - c)$. □

Problem 1. *Points D and E lie on sides AB and AC of the triangle ABC such that*

$$\frac{AD}{AB} = \frac{AE}{AC} = \frac{3}{4}.$$

Consider points E' and D' on the rays $(BE$ and $(CD$ such that $EE' = 3BE$ and $DD' = 3CD$. Prove the following:

(1) points D', A, E' are collinear.
(2) $AD' = AE'$.

Solution. The points D, E, D', E' have coordinates: $d = \dfrac{a + 3b}{4}$, $e = \dfrac{a + 3c}{4}$,

$$e' = 4e - 3b = a + 3c - 3b, \text{ and } d' = 4d - 3c = a + 3b - 3c,$$

respectively.

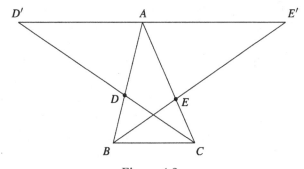

Figure 4.3.

(1) Since

$$(a - d') \times (e' - d') = (3c - 3b) \times (6c - 6b) = 18(c - b) \times (c - b) = 0,$$

it follows from Proposition 2 in Sect. 4.2 that the points D', A, E' are collinear (Fig. 4.3).

(2) Note that

$$\frac{AD'}{D'E'} = \left| \frac{a - d'}{e' - d'} \right| = \frac{1}{2},$$

so A is the midpoint of segment $D'E'$.

Problem 2. *Let $ABCDE$ be a convex pentagon and let M, N, P, Q, X, Y be the midpoints of the segments BC, CD, DE, EA, MP, NQ, respectively. Prove that $XY \| AB$.*

Solution. Let a, b, c, d, e be the coordinates of vertices A, B, C, D, E, respectively (Fig. 4.4).

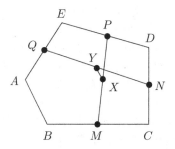

Figure 4.4.

Points M, N, P, Q, X, Y have coordinates

$$m = \frac{b + c}{2}, \ n = \frac{c + d}{2}, \ p = \frac{d + e}{2},$$

$$q = \frac{e + a}{2}, \ x = \frac{b + c + d + e}{4}, \ y = \frac{c + d + e + a}{4},$$

respectively. Then

$$\frac{y - x}{b - a} = \frac{\dfrac{a - b}{4}}{b - a} = -\frac{1}{4} \in \mathbb{R},$$

whence

$$(y - x) \times (b - a) = -\frac{1}{4}(b - a) \times (b - a) = 0.$$

From Proposition 3 in Sect. 4.2, it follows that $XY \| AB$.

4.3 The Area of a Convex Polygon

We say that the convex polygon $A_1 A_2 \cdots A_n$ is *directly* (or *positively*) *oriented* if for every point M situated in the interior of the polygon, the triangles $M A_k A_{k+1}$, $k = 1, 2, \ldots, n$, are directly oriented, where $A_{n+1} = A_1$.

Theorem. *Consider a directly oriented convex polygon $A_1 A_2 \cdots A_n$ with vertices with coordinates $a_1, a_2, \ldots, a_n$. Then*

$$\text{area } [A_1 A_2 \cdots A_n] = \frac{1}{2} \text{Im}(\overline{a_1} a_2 + \overline{a_2} a_3 + \cdots + \overline{a_{n-1}} a_n + \overline{a_n} a_1).$$

Proof. We use induction on n. The base case $n = 3$ was proved above using the complex product. Suppose that the claim holds for $n = k$, and note that

$$\text{area } [A_1 A_2 \cdots A_k A_{k+1}] = \text{area } [A_1 A_2 \cdots A_k] + \text{area } [A_k A_{k+1} A_1]$$

$$= \frac{1}{2} \text{Im}(\overline{a}_1 a_2 + \overline{a_2} a_3 + \cdots + \overline{a_{k-1}} a_k + \overline{a_k 1} a) + \frac{1}{2} \text{Im}(\overline{a_k} a_{k+1} + \overline{a_{k+1}} a_1 + \overline{a_1} a_k)$$

$$= \frac{1}{2} \text{Im}(\overline{a}_1 a_2 + \overline{a_2} a_3 + \cdots + \overline{a_{k-1}} a_k + \overline{a_k} a_{k+1} + \overline{a_{k+1}} a_1)$$

$$+ \frac{1}{2} \text{Im}(\overline{a_k} a_1 + \overline{a_1} a_k) = \frac{1}{2} \text{Im}(\overline{a_1} a_2 + \overline{a_2} a_3 + \cdots + \overline{a_k} a_{k+1} + \overline{a_{k+1}} a_1),$$

since

$$\text{Im}(\overline{a_k} a_1 + \overline{a_1} a_k) = 0.$$

Alternative proof. Choose a point M in the interior of the polygon. Applying the formula (2) in Sect. 3.5.3, we have

$$\text{area } [A_1 A_2 \cdots A_n] = \sum_{k=1}^{n} \text{area } [M A_k A_{k+1}]$$

$$= \frac{1}{2} \sum_{k=1}^{n} \text{Im}(\overline{z} a_k + \overline{a_k} a_{k+1} + \overline{a_{k+1}} z)$$

$$= \frac{1}{2} \sum_{k=1}^{n} \text{Im}(\overline{a_k} a_{k+1}) + \frac{1}{2} \sum_{k=1}^{n} \text{Im}(\overline{z} a_k + \overline{a_{k+1}} z)$$

$$= \frac{1}{2} \text{Im} \left(\sum_{k=1}^{n} \overline{a_k} a_{k+1} \right) + \frac{1}{2} \text{Im} \left(\overline{z} \sum_{k=1}^{n} a_k + z \sum_{j=1}^{n} \overline{a_j} \right) = \frac{1}{2} \text{Im} \left(\sum_{k=1}^{n} \overline{a_k} a_{k+1} \right),$$

since for any complex numbers z, w the relation $\text{Im}(\overline{z} w + z \overline{w}) = 0$ holds. $\square$

Remark. From the above formula, it follows that the points $A_1(a_1)$, $A_2(a_2), \ldots, A_n(a_n)$ as in the theorem are collinear if and only if

$$\text{Im}(\overline{a_1} a_2 + \overline{a_2} a_3 + \cdots + \overline{a_{n-1}} a_n + \overline{a_n} a_1) = 0.$$

For this result, the hypotheses in the theorem are essential, as we can see from the following counterexample.

Counterexample The points with respective complex coordinates $a_1 = 0$, $a_2 = 1$, $a_3 = i$, $a_4 = 1 + i$ are not collinear, but we have $\mathrm{Im}(\overline{a_1}a_2 + \overline{a_2}a_3 + \overline{a_3}a_4 + \overline{a_4}a_1) = \mathrm{Im}(-1) = 0$.

Problem 1. *Let $P_0P_1\cdots P_{n-1}$ be the polygon whose vertices have coordinates $1,\ \varepsilon,\ldots,\ \varepsilon^{n-1}$, and let $Q_0Q_1\cdots Q_{n-1}$ be the polygon whose vertices have coordinates $1, 1+\varepsilon,\ \ldots,\ 1+\varepsilon+\cdots+\varepsilon^{n-1}$, where $\varepsilon = \cos\dfrac{2\pi}{n} + i\sin\dfrac{2\pi}{n}$. Find the ratio of the areas of these polygons.*

Solution. Consider $a_k = 1 + \varepsilon + \cdots + \varepsilon^k$, $k = 0, 1, \ldots, n-1$, and observe that

$$\text{area}\,[Q_0Q_1\cdots Q_{n-1}] = \frac{1}{2}\mathrm{Im}\left(\sum_{k=0}^{n-1}\overline{a_k}a_{k+1}\right)$$

$$= \frac{1}{2}\mathrm{Im}\left(\sum_{k=0}^{n-1}\frac{(\overline{\varepsilon})^{k+1}-1}{\overline{\varepsilon}-1}\cdot\frac{\varepsilon^{k+2}-1}{\varepsilon-1}\right)$$

$$= \frac{1}{2|\varepsilon-1|^2}\mathrm{Im}\left[\sum_{k=0}^{n-1}(\varepsilon-(\overline{\varepsilon})^{k+1}-\varepsilon^{k+2}+1)\right]$$

$$= \frac{1}{2|\varepsilon-1|^2}\mathrm{Im}(n\varepsilon+n) = \frac{1}{2|\varepsilon-1|^2}n\sin\frac{2\pi}{n}$$

$$= \frac{n}{8\sin^2\frac{\pi}{n}}2\sin\frac{\pi}{n}\cos\frac{\pi}{n} = \frac{n}{4}\cotan\frac{\pi}{n},$$

since

$$\sum_{k=0}^{n-1}\overline{\varepsilon}^{k+1} = 0 \text{ and } \sum_{k=0}^{n-1}\varepsilon^{k+2} = 0.$$

On the other hand, it is clear that

$$\text{area}\,[P_0P_1\cdots P_{n-1}] = n\,\text{area}\,[P_0OP_1] = \frac{n}{2}\sin\frac{2\pi}{n} = n\sin\frac{\pi}{n}\cos\frac{\pi}{n}.$$

We obtain

$$\frac{\text{area}[P_0P_1\cdots P_{n-1}]}{\text{area}[Q_0Q_1\cdots Q_{n-1}]} = \frac{n\sin\frac{\pi}{n}\cos\frac{\pi}{n}}{\frac{n}{4}\cotan\frac{\pi}{n}} = 4\sin^2\frac{\pi}{n}. \tag{1}$$

Remark. We have $Q_kQ_{k+1} = |a_{k+1}-a_k| = |\varepsilon^{k+1}| = 1$ and $P_kP_{k+1} = |\varepsilon^{k+1}-\varepsilon^k| = |\varepsilon^k(\varepsilon-1)| = |\varepsilon^k||1-\varepsilon| = |1-\varepsilon| = 2\sin\frac{\pi}{n}$, $k = 0,1,\ldots,n-1$. It follows that

$$\frac{P_k P_{k+1}}{Q_k Q_{k+1}} = 2 \sin \frac{\pi}{n}, \ k = 0, 1, \ \ldots, \ n-1.$$

That is, the polygons $P_0 P_1 \cdots P_{n-1}$ and $Q_0 Q_1 \cdots Q_{n-1}$ are similar, and the result in (1) follows.

Problem 2: *Let $A_1 A_2 \cdots A_n (n \geq 5)$ be a convex polygon and let B_k be the midpoint of the segment $[A_k A_{k+1}]$, $k = 1, 2, \ \ldots, \ n$, where $A_{n+1} = A_1$. Then the following inequality holds:*

$$\text{area } [B_1 B_2 \cdots B_n] \geq \frac{1}{2} \text{ area } [A_1 A_2 \cdots A_n].$$

Solution. Let a_k and b_k be the coordinates of points A_k and B_k, $k = 1, 2, \ \ldots, \ n$. It is clear that the polygon $B_1 B_2 \cdots B_n$ is convex, and if we assume that $A_1 A_2 \cdots A_n$ is positively oriented, then $B_1 B_2 \cdots B_n$ also has this property. Choose as the origin O of the complex plane a point situated in the interior of polygon $A_1 A_2 \cdots A_n$.

We have $b_k = \frac{1}{2}(a_k + a_{k+1})$, $k = 1, 2, \ \ldots, \ n$, and

$$\text{area } [B_1 B_2 \cdots B_n] = \frac{1}{2} \text{Im}\left(\sum_{k=1}^{n} \overline{b_k} b_{k+1}\right) = \frac{1}{8} \text{Im} \sum_{k=1}^{n} (\overline{a_k} + \overline{a_{k+1}})(a_{k+1} + a_{k+2})$$

$$= \frac{1}{8} \text{Im}\left(\sum_{k=1}^{n} \overline{a_k} a_{k+1}\right) + \frac{1}{8} \text{Im}\left(\sum_{k=1}^{n} \overline{a_{k+1}} a_{k+2}\right) + \frac{1}{8} \text{Im}\left(\sum_{k=1}^{n} \overline{a_k} a_{k+2}\right)$$

$$= \frac{1}{2} \text{area } [A_1 A_2 \cdots A_n] + \frac{1}{8} \text{Im}\left(\sum_{k=1}^{n} \overline{a_k} a_{k+2}\right)$$

$$= \frac{1}{2} \text{area } [A_1 A_2 \cdots A_n] + \frac{1}{8} \sum_{k=1}^{n} \text{Im}(\overline{a_k} a_{k+2})$$

$$= \frac{1}{2} \text{area } [A_1 A_2 \cdots A_n] + \frac{1}{8} \sum_{k=1}^{n} OA_k \cdot OA_{k+2} \sin A_k \widehat{O} A_{k+2}$$

$$\geq \frac{1}{2} \text{area } [A_1 A_2 \cdots A_n],$$

where we have used the relations

$$\text{Im}\left(\sum_{k=1}^{n} \overline{a_k} a_{k+1}\right) = \text{Im}\left(\sum_{k=1}^{n} \overline{a_{k+1}} a_{k+2}\right) = 2 \text{ area } [A_1 A_2 \cdots A_n]$$

and $\sin A_k \widehat{O} A_{k+2} \geq 0$, $k = 1, 2, \ \ldots, \ n$, where $A_{n+2} = A_2$.

4.4 Intersecting Cevians and Some Important Points in a Triangle

Proposition. *Consider the points A', B', C' on the sides BC, CA, AB of the triangle ABC such that AA', BB', CC' intersect at point Q and let*

$$\frac{BA'}{A'C} = \frac{p}{n}, \frac{CB'}{B'A} = \frac{m}{p}, \frac{AC'}{C'B} = \frac{n}{m}.$$

If a, b, c are the coordinates of points A, B, C, respectively, then the coordinate of point Q is

$$q = \frac{ma + nb + pc}{m + n + p}.$$

Proof. The coordinates of A', B', C' are $a' = \dfrac{nb + pc}{n + p}$, $b' = \dfrac{ma + pc}{m + p}$, and $c' = \dfrac{ma + nb}{m + n}$, respectively. Let Q be the point with coordinate $q = \frac{ma+nb+pc}{m+n+p}$. We prove that AA', BB', CC' meet at Q.

The points A, Q, A' are collinear if and only if $(q - a) \times (a' - a) = 0$. This is equivalent to

$$\left(\frac{ma + nb + pc}{m + n + p} - a \right) \times \left(\frac{nb + pc}{n + p} - a \right) = 0,$$

or $(nb + pc - (n+p)a) \times (nb + pc - (n+p)a) = 0$, which is clear by definition of the complex product.

Likewise, Q lies on lines BB' and CC', so the proof is complete.   □

Some Important Points in a Triangle

(1) If $Q = G$, the centroid of the triangle ABC, we have $m = n = p$. Then we obtain again that the coordinate of G is

$$z_G = \frac{a + b + c}{3}.$$

(2) Suppose that the lengths of the sides of triangle ABC are $BC = \alpha$, $CA = \beta$, $AB = \gamma$. If $Q = I$, the incenter of triangle ABC, then using a known result concerning the angle bisector, it follows that $m = \alpha$, $n = \beta$, $p = \gamma$. Therefore, the coordinate of I is

$$z_I = \frac{\alpha a + \beta b + \gamma c}{\alpha + \beta + \gamma} = \frac{1}{2s}[(\alpha a + \beta b + \gamma c)],$$

where $s = \frac{1}{2}(\alpha + \beta + \gamma)$.

(3) If $Q = H$, the orthocenter of the triangle ABC, we easily obtain the relations

$$\frac{BA'}{A'C} = \frac{\tan C}{\tan B}, \quad \frac{CB'}{B'A} = \frac{\tan A}{\tan C}, \quad \frac{AC'}{C'B} = \frac{\tan B}{\tan A}.$$

It follows that $m = \tan A$, $n = \tan B$, $p = \tan C$, and the coordinate of H is given by

$$z_H = \frac{(\tan A)a + (\tan B)b + (\tan C)c}{\tan A + \tan B + \tan C}.$$

Remark. The above formula can also be extended to the limiting case in which the triangle ABC is a right triangle. Indeed, assume that $A \to \dfrac{\pi}{2}$. Then $\tan A \to \pm\infty$ and $\dfrac{(\tan B)b + (\tan C)c}{\tan A} \to 0$, $\dfrac{\tan B + \tan C}{\tan A} \to 0$. In this case, $z_H = a$, i.e., the orthocenter of triangle ABC is the vertex A.

(4) The Gergonne[1] point J is the intersection of the cevians AA', BB', CC', where A', B', C' are the points of tangency of the incircle to the sides BC, CA, AB, respectively. Then

$$\frac{BA'}{A'C} = \frac{\frac{1}{s-\gamma}}{\frac{1}{s-\beta}}, \quad \frac{CB'}{B'A} = \frac{\frac{1}{s-\alpha}}{\frac{1}{s-\gamma}}, \quad \frac{AC'}{C'B} = \frac{\frac{1}{s-\beta}}{\frac{1}{s-\alpha}},$$

and the coordinate z_J is obtained from the same proposition, where

$$z_J = \frac{r_\alpha a + r_\beta b + r_\gamma c}{r_\alpha + r_\beta + r_\gamma}.$$

Here r_α, r_β, r_γ denote the radii of the three excircles of triangle. It is not difficult to show that the following formulas hold:

$$r_\alpha = \frac{K}{s-\alpha}, \quad r_\beta = \frac{K}{s-\beta}, \quad r_\gamma = \frac{K}{s-\gamma},$$

where $K = \text{area } [ABC]$ and $s = \frac{1}{2}(\alpha + \beta + \gamma)$.

(5) The Lemoine[2] point K is the intersection of the symmedians of the triangle (the symmedian is the reflection of the bisector across the median). Using the notation from the proposition, we obtain

$$\frac{BA'}{A'C} = \frac{\gamma^2}{\beta^2}, \quad \frac{CB'}{B'A} = \frac{\alpha^2}{\gamma^2}, \quad \frac{AC'}{C'B} = \frac{\beta^2}{\alpha^2}.$$

---

[1] Joseph Diaz Gergonne (1771–1859), French mathematician, founded the journal *Annales de Mathématiques Pures et Appliquées* in 1810.

[2] Émile Michel Hyacinthe Lemoine (1840–1912), French mathematician, made important contributions to geometry.

It follows that

$$z_K = \frac{\alpha^2 a + \beta^2 b + \gamma^2 c}{\alpha^2 + \beta^2 + \gamma^2}.$$

(6) The Nagel[3] point N is the intersection of the cevian AA', BB', CC', where A', B', C' are the points of tangency of the excircles with respective sides BC, CA, AB. Then

$$\frac{BA'}{A'C} = \frac{s - \gamma}{s - \beta}, \quad \frac{CB'}{B'A} = \frac{s - \alpha}{s - \gamma}, \quad \frac{AC'}{C'B} = \frac{s - \beta}{s - \alpha},$$

and the proposition mentioned above gives the coordinate z_N of the Nagel point N:

$$z_N = \frac{(s - \alpha)a + (s - \beta)b + (s - \gamma)c}{(s - \alpha) + (s - \beta) + (s - \gamma)} = \frac{1}{s}[(s - \alpha)a + (s - \beta)b + (s - \gamma)c]$$

$$= \left(1 - \frac{\alpha}{s}\right)a + \left(1 - \frac{\beta}{s}\right)b + \left(1 - \frac{\gamma}{s}\right)c.$$

Problem. *Let α, β, γ be the lengths of sides BC, CA, AB of triangle ABC and suppose $\alpha < \beta < \gamma$. If points O, I, H are the circumcenter, the incenter, and the orthocenter of triangle ABC, respectively, prove that*

$$\text{area } [OIH] = \frac{1}{8r}(\alpha - \beta)(\beta - \gamma)(\gamma - \alpha),$$

where r is the inradius of ABC.

Solution. Consider triangle ABC, directly oriented in the complex plane centered at point O.

Using the complex product and the coordinates of I and H, we have

$$\text{area } [OIH] = \frac{1}{2i}(z_I \times z_H) = \frac{1}{2i}\left[\frac{\alpha a + \beta b + \gamma c}{\alpha + \beta + \gamma} \times (a + b + c)\right]$$

$$= \frac{1}{4si}[(\alpha - \beta)a \times b + (\beta - \gamma)b \times c + (\gamma - \alpha)c \times a]$$

$$= \frac{1}{2s}[(\alpha - \beta) \cdot \text{area } [OAB] + (\beta - \gamma) \cdot \text{area } [OBC] + (\gamma - \alpha) \cdot \text{area } [OCA]]$$

$$= \frac{1}{2s}\left[(\alpha - \beta)\frac{R^2 \sin 2C}{2} + (\beta - \gamma)\frac{R^2 \sin 2A}{2} + (\gamma - \alpha)\frac{R^2 \sin 2B}{2}\right]$$

---

[3] Christian Heinrich von Nagel (1803–1882), German mathematician. His contributions to triangle geometry were included in the book *The Development of Modern Triangle Geometry* [21].

$$= \frac{R^2}{4s}[(\alpha - \beta)\sin 2C + (\beta - \gamma)\sin 2A + (\gamma - \alpha)\sin 2B]$$

$$= \frac{1}{8r}(\alpha - \beta)(\beta - \gamma)(\gamma - \alpha),$$

as desired.

4.5 The Nine-Point Circle of Euler

Given a triangle ABC, choose its circumcenter O to be the origin of the complex plane and let a, b, c be the coordinates of the vertices A, B, C. We have seen in Sect. 4.1, Proposition 3, that the coordinate of the orthocenter H is $z_H = a + b + c$.

Let us denote by A_1, B_1, C_1 the midpoints of sides BC, CA, AB; by A', B', C' the feet of the altitudes; and by A'', B'', C'' the midpoints of segments AH, BH, CH, respectively (Fig. 4.5).

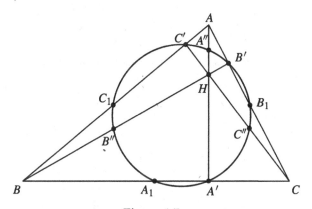

Figure 4.5.

It is clear that for the points A_1, B_1, C_1, A'', B'', C'', we have the following coordinates:

$$z_{A_1} = \frac{1}{2}(b + c), \ z_{B_1} = \frac{1}{2}(c + a), \ z_{C_1} = \frac{1}{2}(a + b),$$

$$z_{A''} = a + \frac{1}{2}(b + c), \ z_{B''} = b + \frac{1}{2}(c + a), \ z_{C''} = c + \frac{1}{2}(a + b).$$

It is not so easy to find the coordinates of A', B', C'.

Proposition. *Consider the point $X(x)$ on the circumcircle of triangle ABC. Let P be the projection of X onto line BC. Then the coordinate of P is given by*

$$p = \frac{1}{2}\left(x - \frac{bc}{R^2}\overline{x} + b + c\right),$$

where R is the circumradius of triangle ABC.

Proof. Using the complex product and the real product, we can write the equations of lines BC and XP as follows:

$$BC : (z - b) \times (c - b) = 0,$$
$$XP : (z - x) \cdot (c - b) = 0.$$

The coordinate p of P satisfies both equations; hence we have

$$(p - b) \times (c - b) = 0 \text{ and } (p - x) \cdot (c - b) = 0.$$

These equations are equivalent to

$$(p - b)(\overline{c} - \overline{b}) - (\overline{p} - \overline{b})(c - b) = 0$$

and

$$(p - x)(\overline{c} - \overline{b}) + (\overline{p} - \overline{x})(c - b) = 0.$$

Adding the above relations, we obtain

$$(2p - b - x)(\overline{c} - \overline{b}) + (\overline{b} - \overline{x})(c - b) = 0.$$

It follows that

$$p = \frac{1}{2}\left[b + x + \frac{c-b}{\overline{c} - \overline{b}}(\overline{x} - \overline{b})\right] = \frac{1}{2}\left[b + x + \frac{c-b}{\dfrac{R^2}{c} - \dfrac{R^2}{b}}(\overline{x} - \overline{b})\right]$$

$$= \frac{1}{2}\left[b + x - \frac{bc}{R^2}(\overline{x} - \overline{b})\right] = \frac{1}{2}\left(x - \frac{bc}{R^2}\overline{x} + b + c\right). \qquad \Box$$

From the above proposition, we see that the coordinates of A', B', C' are

$$z_{A'} = \frac{1}{2}\left(a + b + c - \frac{bc\overline{a}}{R^2}\right),$$

$$z_{B'} = \frac{1}{2}\left(a + b + c - \frac{ca\overline{b}}{R^2}\right),$$

$$z_{C'} = \frac{1}{2}\left(a + b + c - \frac{ab\overline{c}}{R^2}\right).$$

Theorem 1 (The nine-point circle). *In every triangle ABC, the points A_1, B_1, C_1, A', B', C', A'', B'', C'' are all on the same circle, whose center is at the midpoint of the segment OH and whose radius is one-half the circumradius.*

Proof. Denote by O_9 the midpoint of the segment OH. Using our initial assumption, it follows that $z_{O9} = \frac{1}{2}(a + b + c)$. Also, we have $|a| = |b| = |c| = R$, where R is the circumradius of triangle ABC.

Observe that $O_9A_1 = |z_{A_1} - z_{O_9}| = \frac{1}{2}|a| = \frac{1}{2}R$, and also $O_9B_1 = O_9C_1 = \frac{1}{2}R$.

We can write $O_9A'' = |z_{A''} - z_{O_9}| = \frac{1}{2}|a| = \frac{1}{2}R$, and also $O_9B'' = O_9C'' = \frac{1}{2}R$.

The distance O_9A' is also not difficult to compute:

$$O_9A' = |z_{A'} - z_{O_9}| = \left| \frac{1}{2}\left(a + b + c - \frac{bc\bar{a}}{R^2} \right) - \frac{1}{2}(a+b+c) \right|$$

$$= \frac{1}{2R^2}|bc\bar{a}| = \frac{1}{2R^2}|\bar{a}||b||c| = \frac{R^3}{2R^2} = \frac{1}{2}R.$$

Similarly, we get $O_9B' = O_9C' = \frac{1}{2}R$. Therefore, $O_9A_1 = O_9B_1 = O_9C_1 = O_9A' = O_9B' = O_9C' = O_9A'' = O_9B'' = O_9C'' = \frac{1}{2}R$, and the desired property follows. $\square$

Theorem 2.

(1) (Euler[4] line of a triangle.) In any triangle ABC the points O, G, H are collinear.

(2) (Nagel line of a triangle.) In any triangle ABC the points I, G, N are collinear.

Proof.

(1) If the circumcenter O is the origin of the complex plane, we have $z_O = 0$, $z_G = \frac{1}{3}(a + b + c)$, $z_H = a + b + c$. Hence these points are collinear by Proposition 2 in Sect. 3.2 or 4.2.

(2) We have $z_I = \frac{\alpha}{2s}a + \frac{\beta}{2s}b + \frac{\gamma}{2s}c$, $z_G = \frac{1}{3}(a+b+c)$, and $z_N = \left(1 - \frac{\alpha}{s}\right)a + \left(1 - \frac{\beta}{s}\right)b + \left(1 - \frac{\gamma}{s}\right)c$, and we can write $z_N = 3z_G - 2z_I$.

Applying the result mentioned above and properties of the complex product, we obtain $(z_G - z_I) \times (z_N - z_I) = (z_G - z_I) \times [3(z_G - z_I)] = 0$; hence the points I, G, N are collinear. $\square$

---

[4] Leonhard Euler (1707–1783), one of the most important mathematicians of all time, created much of modern calculus and contributed significantly to almost every existing branch of pure mathematics, adding proofs and arranging the whole in a consistent form. Euler wrote an immense number of memoirs on a great variety of mathematical subjects. We recommend William Dunham's book *Euler: The Master of Us All* [33] for more details concerning Euler's contributions to mathematics.

Remark. Note that $NG = 2GI$, and hence the triangles OGI and HGN are similar. It follows that the lines OI and NH are parallel, and we have the basic configuration of triangle ABC shown in Fig. 4.6.

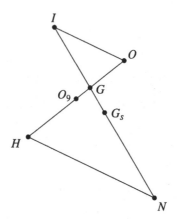

Figure 4.6.

If G_s is the midpoint of segment $[IN]$, then its coordinate is

$$z_{G_s} = \frac{1}{2}(z_I + z_N) = \frac{(\beta + \gamma)}{4s}a + \frac{(\gamma + \alpha)}{4s}b + \frac{(\alpha + \beta)}{4s}c.$$

The point G_s is called the *Spiecker point* of triangle ABC, and it is easy to verify that it is the incenter of the medial triangle $A_1B_1C_1$.

Problem 1. *Consider a point M on the circumcircle of triangle ABC. Prove that the nine-point centers of triangles MBC, MCA, MAB are the vertices of a triangle similar to triangle ABC.*

Solution. Let A', B', C' be the nine-point centers of the triangles MBC, MCD, MAB, respectively. Take the origin of the complex plane to be at the circumcenter of triangle ABC. Denote by the corresponding lowercase letter the coordinate of the point denoted by an uppercase letter. Then

$$a' = \frac{m+b+c}{2}, \ b' = \frac{m+c+a}{2}, \ c' = \frac{m+a+b}{2},$$

since M lies on the circumcircle of triangle ABC. Then

$$\frac{b'-a'}{c'-a'} = \frac{a-b}{a-c} = \frac{b-a}{c-a},$$

and hence triangles $A'B'C'$ and ABC are similar.

Problem 2. *Show that triangle ABC is a right triangle if and only if its circumcircle and its nine-point circle are tangent.*

Solution. Take the origin of the complex plane to be at the circumcenter O of triangle ABC, and denote by a, b, c the coordinates of vertices A, B, C, respectively. Then the circumcircle of triangle ABC is tangent to the nine-point circle of triangle ABC if and only if $OO_9 = \dfrac{R}{2}$. This is equivalent to $OO_9^2 = \dfrac{R^2}{4}$, that is, $|a + b + c|^2 = R^2$.

Using properties of the real product, we have

$$|a + b + c|^2 = (a + b + c) \cdot (a + b + c) = |a|^2 + |b|^2 + |c|^2 + 2(a \cdot b + b \cdot c + c \cdot a)$$

$$= 3R^2 + 2(a \cdot b + b \cdot c + c \cdot a) = 3R^2 + (2R^2 - \alpha^2 + 2R^2 - \beta^2 + 2R^2 - \gamma^2)$$

$$= 9R^2 - (\alpha^2 + \beta^2 + \gamma^2),$$

where α, β, γ are the lengths of the sides of triangle ABC. We have used the formulas $a \cdot b = R^2 - \dfrac{\gamma^2}{2}$, $b \cdot c = R^2 - \dfrac{\alpha^2}{2}$, $c \cdot a = R^2 - \dfrac{\beta^2}{2}$, which can be easily derived from the definition of the real product of complex numbers (see also the lemma in Sect. 4.6.2).

Therefore, $\alpha^2 + \beta^2 + \gamma^2 = 8R^2$, which is the same as $\sin^2 A + \sin^2 B + \sin^2 C = 2$. We can write the last relation as $1 - \cos 2A + 1 - \cos 2B + 1 - \cos 2C = 4$. This is equivalent to $2 \cos(A + B) \cos(A - B) + 2 \cos^2 C = 0$, i.e., $4 \cos A \cos B \cos C = 0$, and the desired conclusion follows.

Problem 3. *Let ABCD be a cyclic quadrilateral and let E_a, E_b, E_c, E_d be the nine-point centers of triangles BCD, CDA, DAB, ABC, respectively. Prove that the lines AE_a, BE_b, CE_c, DE_d are concurrent.*

Solution. Take the origin of the complex plane to be the center O of the circumcircle of $ABCD$. Then the coordinates of the nine-point centers are

$$e_a = \frac{1}{2}(b + c + d), \ e_b = \frac{1}{2}(c + d + a), \ e_c = \frac{1}{2}(d + a + b), \ e_d = \frac{1}{2}(a + b + c).$$

We have $AE_a : z = ka + (1 - k)e_a$, $k \in \mathbb{R}$, and the analogous equations for the lines BE_b, CE_c, DE_d. Observe that the point with coordinate $\frac{1}{3}(a + b + c + d)$ lies on all four lines $\left(k = \dfrac{1}{3}\right)$, and we are done.

4.6 Some Important Distances in a Triangle

4.6.1 Fundamental Invariants of a Triangle

Consider the triangle ABC with sides α, β, γ; semiperimeter

$$s = \frac{1}{2}(\alpha + \beta + \gamma);$$

inradius r; and circumradius R. The numbers s, r, R are called the *fundamental invariants* of triangle ABC.

Theorem. *The sides α, β, γ are the roots of the cubic equation*

$$t^3 - 2st^2 + (s^2 + r^2 + 4Rr)t - 4sRr = 0.$$

Proof. Let us prove that α satisfies the equation. We have

$$\alpha = 2R\sin A = 4R\sin\frac{A}{2}\cos\frac{A}{2} \text{ and } s - \alpha = r\cot\frac{A}{2} = r\frac{\cos\dfrac{A}{2}}{\sin\dfrac{A}{2}},$$

whence

$$\cos^2\frac{A}{2} = \frac{\alpha(s - \alpha)}{4Rr} \text{ and } \sin^2\frac{A}{2} = \frac{\alpha r}{4R(s - \alpha)}.$$

From the formula $\cos^2\dfrac{A}{2} + \sin^2\dfrac{A}{2} = 1$, it follows that

$$\frac{\alpha(s - \alpha)}{4Rr} + \frac{\alpha r}{4R(s - \alpha)} = 1.$$

That is, $\alpha^3 - 2s\alpha^2 + (s^2 + r^2 + 4Rr)\alpha - 4sRr = 0$. We can show analogously that β and γ are roots of the above equation. $\square$

From the above theorem, using the relations between the roots and the coefficients, it follows that

$$\alpha + \beta + \gamma = 2s,$$

$$\alpha\beta + \beta\gamma + \gamma\alpha = s^2 + r^2 + 4Rr,$$

$$\alpha\beta\gamma = 4sRr.$$

Corollary. *The following formulas hold in every triangle ABC:*

$$\alpha^2 + \beta^2 + \gamma^2 = 2(s^2 - r^2 - 4Rr),$$

$$\alpha^3 + \beta^3 + \gamma^3 = 2s(s^2 - 3r^2 - 6Rr).$$

Proof. We have

$$\alpha^2 + \beta^2 + \gamma^2 = (\alpha+\beta+\gamma)^2 - 2(\alpha\beta+\beta\gamma+\gamma\alpha) = 4s^2 - 2(s^2 + r^2 + 4Rr)$$
$$= 2s^2 - 2r^2 - 8Rr = 2(s^2 - r^2 - 4Rr).$$

In order to prove the second identity, we can write

$$\alpha^3 + \beta^3 + \gamma^3 = (\alpha+\beta+\gamma)(\alpha^2+\beta^2+\gamma^2 - \alpha\beta - \beta\gamma - \gamma\alpha) + 3\alpha\beta\gamma$$

$$= 2s(2s^2 - 2r^2 - 8Rr - s^2 - r^2 - 4Rr) + 12sRr = 2s(s^2 - 3r^2 - 6Rr). \quad \square$$

4.6.2 The Distance OI

Assume that the circumcenter O of the triangle ABC is the origin of the complex plane, and let a, b, c be the coordinates of the vertices A, B, C, respectively.

Lemma. *The real products $a \cdot b$, $b \cdot c$, $c \cdot a$ are given by*

$$a \cdot b = R^2 - \frac{\gamma^2}{2}, \quad b \cdot c = R^2 - \frac{\alpha^2}{2}, \quad c \cdot a = R^2 - \frac{\beta^2}{2}.$$

Proof. Using the properties of the real product, we have

$$\gamma^2 = |a-b|^2 = (a-b)\cdot(a-b) = a\cdot a - 2a\cdot b + b\cdot b = |a|^2 - 2a\cdot b + |b|^2 = 2R^2 - 2a\cdot b,$$

and the first formula follows. $\square$

In order to simplify the formulas, we will use the symbol $\sum_{\text{cyc}}$, called the *cyclic sum*:

$$\sum_{\text{cyc}} f(x_1, x_2, x_3) = f(x_1, x_2, x_3) + f(x_2, x_3, x_1) + f(x_3, x_1, x_2),$$

where the sum is taken over all cyclic permutations of the variables.

Theorem (Euler). *The following formula holds:*

$$OI^2 = R^2 - 2Rr.$$

Proof. The coordinate of the incenter is given by

$$z_I = \frac{\alpha}{2s}a + \frac{\beta}{2s}b + \frac{\gamma}{2s}c,$$

so we can write

$$OI^2 = |z_I|^2 = \left(\frac{\alpha}{2s}a + \frac{\beta}{2s}b + \frac{\gamma}{2s}c\right) \cdot \left(\frac{\alpha}{2s}a + \frac{\beta}{2s}b + \frac{\gamma}{2s}c\right)$$

$$= \frac{1}{4s^2}(\alpha^2 + \beta^2 + \gamma^2)R^2 + 2\frac{1}{4s^2}\sum_{cyc}(\alpha\beta)a \cdot b.$$

Using the above lemma, we find that

$$OI^2 = \frac{1}{4s^2}(\alpha^2 + \beta^2 + \gamma^2)R^2 + \frac{2}{4s^2}\sum_{cyc}\alpha\beta\left(R^2 - \frac{\gamma^2}{2}\right)$$

$$= \frac{1}{4s^2}(\alpha + \beta + \gamma)^2 R^2 - \frac{1}{4s^2}\sum_{cyc}\alpha\beta\gamma^2 = R^2 - \frac{1}{4s^2}\alpha\beta\gamma(\alpha + \beta + \gamma)$$

$$= R^2 - \frac{1}{2s}\alpha\beta\gamma = R^2 - 2\frac{\alpha\beta\gamma}{4K} \cdot \frac{K}{s} = R^2 - 2\mathrm{Rr},$$

where the well-known formulas

$$R = \frac{\alpha\beta\gamma}{4K}, \quad r = \frac{K}{s},$$

are used. Here K is the area of triangle ABC.   □

Corollary (Euler's inequality). *In every triangle ABC, the following inequality holds:*

$$R \geq 2r.$$

We have equality if and only if triangle ABC is equilateral.

Proof. From the above theorem. we have $OI^2 = R(R-2r) \geq 0$, hence $R \geq 2r$. The equality $R - 2r = 0$ holds if and only if $OI^2 = 0$, i.e., $O = I$. Therefore, triangle ABC is equilateral.   □

4.6.3 The Distance ON

Theorem 1. *If N is the Nagel point of triangle ABC, then*

$$ON = R - 2r.$$

Proof. The coordinate of the Nagel point of the triangle is given by

$$z_N = \left(1 - \frac{\alpha}{s}\right)a + \left(1 - \frac{\beta}{s}\right)b + \left(1 - \frac{\gamma}{s}\right)c.$$

Therefore,

$$ON^2 = |z_N|^2 = z_N \cdot z_N = R^2 \sum_{\text{cyc}} \left(1 - \frac{\alpha}{s}\right)^2 + 2 \sum_{\text{cyc}} \left(1 - \frac{\alpha}{s}\right)\left(1 - \frac{\beta}{s}\right) a \cdot b$$

$$= R^2 \sum_{\text{cyc}} \left(1 - \frac{\alpha}{s}\right)^2 + 2 \sum_{\text{cyc}} \left(1 - \frac{\alpha}{s}\right)\left(1 - \frac{\beta}{s}\right)\left(R^2 - \frac{\gamma^2}{2}\right)$$

$$= R^2 \left(3 - \frac{\alpha + \beta + \gamma}{s}\right)^2 - \sum_{\text{cyc}} \left(1 - \frac{\alpha}{s}\right)\left(1 - \frac{\beta}{s}\right)\gamma^2$$

$$= R^2 - \sum_{\text{cyc}} \left(1 - \frac{\alpha}{s}\right)\left(1 - \frac{\beta}{s}\right)\gamma^2 = R^2 - E.$$

To calculate E, we note that

$$E = \sum_{\text{cyc}} \left(1 - \frac{\alpha + \beta}{s} + \frac{\alpha\beta}{s^2}\right)\gamma^2 = \sum_{\text{cyc}} \gamma^2 - \frac{1}{s}\sum_{\text{cyc}}(\alpha + \beta)\gamma^2 + \frac{1}{s^2}\sum_{\text{cyc}} \alpha\beta\gamma^2$$

$$= \sum_{\text{cyc}} \gamma^2 - \frac{1}{s}\sum_{\text{cyc}}(2s - \gamma)\gamma^2 + \frac{2\alpha\beta\gamma}{s} = -\sum_{\text{cyc}} \alpha^2 + \frac{1}{s}\sum_{\text{cyc}} \alpha^3 + 8\frac{\alpha\beta\gamma}{4K}\cdot\frac{K}{s}$$

$$= -\sum_{\text{cyc}} \alpha^2 + \frac{1}{s}\sum_{\text{cyc}} \alpha^3 + 8Rr.$$

Applying the formula in the corollary of Sect. 4.6.1, we conclude that

$$E = -2(s^2 - r^2 - 4Rr) + 2(s^2 - 3r^2 - 6Rr) + 8Rr = -4r^2 + 4Rr.$$

Hence $ON^2 = R^2 - E = R^2 - 4Rr + 4r^2 = (R - 2r)^2$, and the desired formula is proved by Euler's inequality. $\square$

Theorem 2 (Feuerbach[5]). *In any triangle the incircle and the nine-point circle of Euler are tangent.*

Proof. Using the configuration in Sect. 4.5 we observe that

$$\frac{1}{2} = \frac{GI}{GN} = \frac{GO_9}{GO}.$$

Therefore, triangles GIO_9 and GNO are similar. It follows that the lines IO_9 and ON are parallel and $IO_9 = \frac{1}{2}ON$. Applying Theorem 1 in Sect. 4.6.3, we get $IO_9 = \frac{1}{2}(R - 2r) = \frac{R}{2} - r = R_9 - r$, and hence the incircle is tangent to the nine-point circle. $\square$

---

[5] Karl Wilhelm Feuerbach (1800–1834), German geometer, published the result of Theorem 2 in 1822.

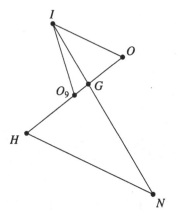

Figure 4.7.

The point of tangency of these two circles is denoted by φ and is called the *Feuerbach point* of the triangle (Fig. 4.7).

4.6.4 The Distance OH

Theorem. *If H is the orthocenter of triangle ABC, then*

$$OH^2 = 9R^2 + 2r^2 + 8Rr - 2s^2.$$

Proof. Assuming that the circumcenter O is the origin of the complex plane, the coordinate of H is

$$z_H = a + b + c.$$

Using the real product, we can write

$$OH^2 = |z_H|^2 = z_H \cdot z_H = (a + b + c) \cdot (a + b + c)$$
$$= \sum_{\text{cyc}} |a|^2 + 2 \sum_{\text{cyc}} a \cdot b = 3R^2 + 2 \sum_{\text{cyc}} a \cdot b.$$

Applying the formulas in the lemma and then the first formula in Corollary 4.6.1, we obtain

$$OH^2 = 3R^2 + 2 \sum_{\text{cyc}} \left(R^2 - \frac{\gamma^2}{2} \right) = 9R^2 - (\alpha^2 + \beta^2 + \gamma^2)$$

$$= 9R^2 - 2(s^2 - r^2 - 4Rr) = 9R^2 + 2r^2 + 8Rr - 2s^2. \qquad \square$$

Corollary 1. *The following formulas hold:*

(1) $OG^2 = R^2 + \dfrac{2}{9}r^2 + \dfrac{8}{9}Rr - \dfrac{2}{9}s^2;$

(2) $OO_9^2 = \dfrac{9}{4}R^2 + \dfrac{1}{2}r^2 + 2Rr - \dfrac{1}{2}s^2.$

Corollary 2. *In every triangle ABC, the inequality*

$$\alpha^2 + \beta^2 + \gamma^2 \leq 9R^2$$

is valid. Equality holds if and only if the triangle is equilateral.

4.6.5 Blundon's Inequalities

Given a triangle ABC, denote by O its circumcenter, I the incenter, G the centroid, N the Nagel point, s the semiperimeter, R the circumradius, and r the inradius. In what follows, we present a geometric proof to the so-called fundamental triangle inequality. This relation contains, in fact, two inequalities, and it was first proved by E. Rouché in 1851, answering a question of Ramus concerning necessary and sufficient conditions for three positive real numbers s, R, r to be the semiperimeter, circumradius, and inradius of a triangle. The standard simple proof was first given by W.J. Blundon, and it is based on the following algebraic property of the roots of a cubic equation: The roots x_1, x_2, x_3 of the equation

$$x^3 + a_1 x^2 + a_2 x + a_3 = 0$$

are the side lengths of a (nondegenerate) triangle if and only if the following three conditions are satisfied:

(i) $18a_1 a_2 a_3 + a_1^2 a_2^2 - 27a_3^2 - 4a_2^3 - 4a_1^3 a_3 > 0;$
(ii) $-a_1 > 0, \ a_2 > 0, \ -a_3 > 0;$
(iii) $a_1^3 - 4a_1 a_4 + 8a_3 > 0.$

The following result contains a simple geometric proof of the fundamental inequality of a triangle, as presented in the article [15].

Theorem 1. *Assume that the triangle ABC is not equilateral. The following relation holds:*

$$\cos \widehat{ION} = \frac{2R^2 + 10Rr - r^2 - s^2}{2(R - 2r)\sqrt{R^2 - 2Rr}}.$$

Proof. It is known (see Theorem 2 in Sect. 4.5) that the points N, G, and I are collinear on a line called Nagel's line of the triangle, and we have $NI = 3GI$. If we use Stewart's theorem in the triangle ION, then we get

$$ON^2 \cdot GI + OI^2 \cdot NG - OG^2 \cdot NI = GI \cdot GN \cdot NI,$$

and it follows that

$$ON^2 \cdot GI + OI^2 \cdot 2GI - OG^2 - 3GI = 6GI^3.$$

This relation is equivalent to

$$ON^2 + 2OI^2 - 3OG^2 = 6GI^2.$$

Now, using formulas for ON, OI, and OG, we obtain

$$GI^2 = \frac{1}{6}\left(\frac{a^2+b^2+c^2}{3} - 8Rr + 4r^2\right) = \frac{1}{6}\left(\frac{2(s^2-r^2-4Rr)}{3} - 8Rr + 4r^2\right).$$

So we get

$$NI^2 = 9GI^2 = 5r^2 + s^2 - 16Rr.$$

We use the law of cosines in the triangle ION to obtain

$$\cos \widehat{ION} = \frac{ON^2 + OI^2 - NI^2}{2ON \cdot OI}$$

$$= \frac{(R-2r)^2 + (R^2-2Rr) - (5r^2+s^2-16Rr)}{2(R-2r)\sqrt{R^2-2Rr}} = \frac{2R^2+10Rr-r^2-s^2}{2(R-2r)\sqrt{R^2-2Rr}},$$

and we are done.

If the triangle ABC is equilateral, then the points I, O, N coincide, i.e., triangle ION degenerates to a single point. In this case, we extend the formula by $\cos \widehat{ION} = 1$. $\qquad\square$

Theorem 2 (Blundon's inequalities). *A necessary and sufficient condition for the existence of a triangle with elements s, R, and r is*

$$2R^2 + 10Rr - r^2 - 2(R-2r)\sqrt{R^2-2Rr}$$

$$\leq s^2 \leq 2R^2 + 10Rr - r^2 + 2(R^2-2r)\sqrt{R^2-2Rr}.$$

Proof. If we have $R = 2r$, then the triangle must be equilateral, and we are done. If we assume that $R - 2r \neq 0$, then the desired inequalities are direct consequences of the fact that $-1 \leq \cos \widehat{ION} \leq 1$. $\qquad\square$

Equilateral triangles give the trivial situation in which we have equality. Suppose that we are not working with equilateral triangles, i.e., we have $R - 2r \neq 0$. Denote by $\mathcal{T}(R,r)$ the family of all triangles with circumradius R and inradius r. Blundon's inequalities give, in terms of R and r, the exact interval for the semiperimeter s of triangles in the family $\mathcal{T}(R,r)$. We have

$$s_{\min}^2 = 2R^2 + 10Rr - r^2 - 2(R-2r)\sqrt{R^2-2Rr}$$

and

$$s_{\max}^2 = 2R^2 + 10Rr - r^2 + 2(R - 2r)\sqrt{R^2 - 2Rr}.$$

If we fix the circumcenter O and the incenter I such that $OI = \sqrt{R^2 - 2Rr}$, then the triangle in the family $\mathcal{T}(R, r)$ with minimal semiperimeter corresponds to the case $\cos \widehat{ION} = 1$ of equality, i.e., points I, O, N are collinear, and I and N belong to the same ray with the origin O. Taking into account the well-known property that points O, G, H belong to Euler's line of the triangle, we see that O, I, G must be collinear, and hence in this case, triangle ABC is isosceles. In Fig. 4.8, this triangle is denoted by $A_{\min} B_{\min} C_{\min}$. Also, the triangle in the family $\mathcal{T}(R, r)$ with maximal semiperimeter corresponds to the case of equality $\cos \widehat{ION} = -1$, i.e., points I, O, N are collinear, and O is situated between I and N. Using again the Euler line of the triangle, we see that triangle ABC is isosceles. In Fig. 4.8, this triangle is denoted by $A_{\max} B_{\max} C_{\max}$.

Note that we have $B_{\min} C_{\min} > B_{\max} C_{\max}$. The triangles in the family $\mathcal{T}(R, r)$ are "between" these two extremal triangles (see Fig. 4.8). According to Poncelet's closure theorem, they are inscribed in the circle $\mathcal{C}(O; R)$, and their sides are externally tangent to the circle $\mathcal{C}(I; r)$.

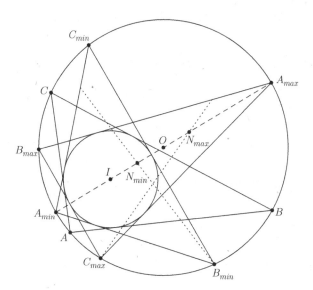

Figure 4.8.

4.7 Distance Between Two Points in the Plane of a Triangle

4.7.1 Barycentric Coordinates

Consider a triangle ABC and let α, β, γ be the lengths of sides BC, CA, AB, respectively.

Proposition. *Let a, b, c be the coordinates of vertices A, B, C and let P be a point in the plane of the triangle. If z_P is the coordinate of P, then there exist unique real numbers μ_a, μ_b, μ_c such that*

$$z_P = \mu_a a + \mu_b b + \mu_c c \text{ and } \mu_a + \mu_b + \mu_c = 1.$$

Proof. Assume that P is in the interior of triangle ABC and consider the point A' such that $AP \cap BC = \{A'\}$. Let $k_1 = \dfrac{PA}{PA'}$, $k_2 = \dfrac{A'B}{A'C}$, and observe that

$$z_P = \frac{a + k_1 z_{A'}}{1 + k_1}, \quad z_{A'} = \frac{b + k_2 c}{1 + k_2}.$$

Hence in this case, we can write

$$z_P = \frac{1}{1 + k_1}a + \frac{k_1}{(1 + k_1)(1 + k_2)}b + \frac{k_1 k_2}{(1 + k_1)(1 + k_2)}c.$$

Moreover, if we consider

$$\mu_a = \frac{1}{1 + k_1}, \quad \mu_b = \frac{k_1}{(1 + k_1)(1 + k_2)}, \quad \mu_c = \frac{k_1 k_2}{(1 + k_1)(1 + k_2)},$$

we have

$$\mu_a + \mu_b + \mu_c = \frac{1}{1 + k_1} + \frac{k_1}{(1 + k_1)(1 + k_2)} + \frac{k_1 k_2}{(1 + k_1)(1 + k_2)}$$
$$= \frac{1 + k_1 + k_2 + k_1 k_2}{(1 + k_1)(1 + k_2)} = 1.$$

We proceed in an analogous way when the point P is situated in the exterior of triangle ABC.

If the point P is situated on the support line of a side of triangle ABC (i.e., the line determined by two vertices), then

$$z_P = \frac{1}{1 + k}b + \frac{k}{1 + k}c = 0 \cdot a + \frac{1}{1 + k}b + \frac{k}{1 + k}c,$$

where $k = \dfrac{PB}{PC}$.   □

The real numbers μ_a, μ_b, μ_c are called the *absolute barycentric coordinates* of P with respect to triangle ABC.

The signs of the numbers μ_a, μ_b, μ_c depend on the regions of the plane in which the point P is situated. Triangle ABC determines seven such regions (Fig. 4.9).

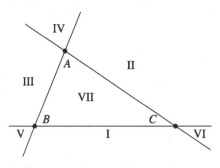

Figure 4.9.

In the following table, we give the signs of μ_a, μ_b, μ_c:

	I	II	III	IV	V	VI	VII
μ_a	−	+	+	+	−	−	+
μ_b	+	−	+	−	+	−	+
μ_c	+	+	−	−	−	+	+

4.7.2 Distance Between Two Points in Barycentric Coordinates

In what follows, in order to simplify the formulas, we will use again the cyclic sum symbol defined above, $\sum_{\text{cyc}} f(x_1, x_2, \ldots, x_n)$. The most important example for our purposes is

$$\sum_{\text{cyc}} f(x_1, x_2, x_3) = f(x_1, x_2, x_3) + f(x_2, x_3, x_1) + f(x_3, x_1, x_2).$$

Theorem 1. *In the plane of triangle ABC, consider the points P_1 and P_2 with coordinates z_{P_1} and z_{P_2}, respectively. If $z_{P_k} = \alpha_k a + \beta_k b + \gamma_k c$, where α_k, β_k, γ_k are real numbers such that $\alpha_k + \beta_k + \gamma_k = 1$, $k = 1, 2$, then*

$$P_1 P_2^2 = -\sum_{\text{cyc}} (\alpha_2 - \alpha_1)(\beta_2 - \beta_1)\gamma^2.$$

Proof. Choose the origin of the complex plane to be located at the circumcenter O of the triangle ABC. Using properties of the real product, we have

$$P_1 P_2^2 = |z_{P_2} - z_{P_1}|^2 = |(\alpha_2 - \alpha_1)a + (\beta_2 - \beta_1)b + (\gamma_2 - \gamma_1)c|^2$$

$$= \sum_{\text{cyc}} (\alpha_2 - \alpha_1)^2 a \cdot a + 2 \sum_{\text{cyc}} (\alpha_2 - \alpha_1)(\beta_2 - \beta_1) a \cdot b$$

$$= \sum_{\text{cyc}} (\alpha_2 - \alpha_1)^2 R^2 + 2 \sum_{\text{cyc}} (\alpha_2 - \alpha_1)(\beta_2 - \beta_1) \left(R^2 - \frac{\gamma^2}{2} \right)$$

$$= R^2 (\alpha_2 + \beta_2 + \gamma_2 - \alpha_1 - \beta_1 - \gamma_1)^2 - \sum_{\text{cyc}} (\alpha_2 - \alpha_1)(\beta_2 - \beta_1) \gamma^2$$

$$= - \sum_{\text{cyc}} (\alpha_2 - \alpha_1)(\beta_2 - \beta_1) \gamma^2,$$

since $\alpha_1 + \beta_1 + \gamma_1 = \alpha_2 + \beta_2 + \gamma_2 = 1$. $\qquad\square$

Theorem 2. *The points A_1, A_2, B_1, B_2, C_1, C_2 are situated on the sides BC, CA, AB of triangle ABC such that lines AA_1, BB_1, CC_1 meet at point P_1, and lines AA_2, BB_2, CC_2 meet at point P_2. If*

$$\frac{BA_k}{A_k C} = \frac{p_k}{n_k}, \ \frac{CB_k}{B_k A} = \frac{m_k}{p_k}, \ \frac{AC_k}{C_k B} = \frac{n_k}{m_k}, \ k = 1, 2,$$

where m_k, n_k, p_k are nonzero real numbers, $k = 1, 2$, and $S_k = m_k + n_k + p_k$, $k = 1, 2$, then

$$P_1 P_2^2 = \frac{1}{S_1^2 S_2^2} \left[S_1 S_2 \sum_{\text{cyc}} (n_1 p_2 + p_1 n_2) a^2 - S_1^2 \sum_{\text{cyc}} n_2 p_2 a^2 - S_2^2 \sum_{\text{cyc}} n_1 p_1 a^2 \right].$$

Proof. The coordinates of points P_1 and P_2 are

$$z_{P_k} = \frac{m_k a + n_k b + p_k c}{m_k + n_k + p_k}, \ k = 1, 2.$$

It follows that in this case, the absolute barycentric coordinates of points P_1 and P_2 are given by

$$\alpha_k = \frac{m_k}{m_k + n_k + p_k} = \frac{m_k}{S_k}, \ \beta_k = \frac{n_k}{m_k + n_k + p_k} = \frac{n_k}{S_k},$$

$$\gamma_k = \frac{p_k}{m_k + n_k + p_k} = \frac{p_k}{S_k}, \ k = 1, 2.$$

Substituting in the formula in Theorem 1 in Sect. 4.7.2, we obtain

$$P_1 P_2^2 = -\sum_{\text{cyc}} \left(\frac{n_2}{S_2} - \frac{n_1}{S_1} \right) \left(\frac{p_2}{S_2} - \frac{p_1}{S_1} \right) \alpha^2$$

$$= -\frac{1}{S_1^2 S_2^2} \sum_{\text{cyc}} (S_1 n_2 - S_2 n_1)(S_1 p_2 - S_2 p_1) \alpha^2$$

$$= -\frac{1}{S_1^2 S_2^2} \sum_{\text{cyc}} \left[S_1^2 n_2 p_2 + S_2^2 n_1 p_1 - S_1 S_2 (n_1 p_2 + n_2 p_1) \right] \alpha^2$$

$$= \frac{1}{S_1^2 S_2^2} \left[S_1 S_2 \sum_{\text{cyc}} (n_1 p_2 + n_2 p_1) \alpha^2 - S_1^2 \sum_{\text{cyc}} n_2 p_2 \alpha^2 - S_2^2 \sum_{\text{cyc}} n_1 p_1 \alpha^2 \right],$$

and the desired formula follows. $\qquad\qquad\qquad\qquad\qquad\qquad\qquad\qquad$ □

Corollary 1. *For real numbers* $\alpha_k, \beta_k, \gamma_k$ *with* $\alpha_k + \beta_k + \gamma_k = 1,\ k = 1, 2,$ *the following inequality holds:*

$$\sum_{\text{cyc}} (\alpha_2 - \alpha_1)(\beta_2 - \beta_1)\gamma^2 \leq 0,$$

with equality if and only if $\alpha_1 = \alpha_2,\ \beta_1 = \beta_2,\ \gamma_1 = \gamma_2.$

Corollary 2. *For nonzero real numbers* $m_k,\ n_k,\ p_k,\ k = 1, 2,$ *with* $S_k = m_k + n_k + p_k,\ k = 1, 2,$ *the lengths of sides* $\alpha,\ \beta,\ \gamma$ *of triangle* ABC *satisfy the inequality*

$$\sum_{\text{cyc}} (n_1 p_2 + p_1 n_2)^2 \geq \frac{S_1}{S_2} \sum_{\text{cyc}} n_2 p_2 \alpha^2 + \frac{S_2}{S_1} \sum_{\text{cyc}} n_1 p_1 \alpha^2,$$

with equality if and only if $\dfrac{p_1}{n_1} = \dfrac{p_2}{n_2},\ \dfrac{m_1}{p_1} = \dfrac{m_2}{p_2},\ \dfrac{n_1}{m_1} = \dfrac{n_2}{m_2}.$

Applications

(1) Let us use the formula in Theorem 2 in Sect. 4.7.2 to compute the distance GI, used in Sect. 4.6.5, where G is the centroid and I is the incenter of the triangle.

We have $m_1 = n_1 = p_1 = 1$ and $m_2 = \alpha,\ n_2 = \beta,\ p_2 = \gamma$; hence

$$S_1 = \sum_{\text{cyc}} m_1 = 3; \quad S_2 = \sum_{\text{cyc}} m_2 = \alpha + \beta + \gamma = 2s;$$

$$\sum_{\text{cyc}} (n_1 p_2 + n_2 p_1) \alpha^2 = (\beta + \gamma)\alpha^2 + (\gamma + \alpha)\beta^2 + (\alpha + \beta)\gamma^2$$

$$= (\alpha + \beta + \gamma)(\alpha\beta + \beta\gamma + \gamma\alpha) - 3\alpha\beta\gamma = 2s(s^2 + r^2 + 4rR) - 12sRr$$

$$= 2s^3 + 2sr^2 - 4sRr.$$

On the other hand,

$$\sum_{\text{cyc}} n_2 p_2 \alpha^2 = \alpha^2 \beta\gamma + \beta^2 \gamma\alpha + \gamma^2 \alpha\beta = \alpha\beta\gamma(\alpha + \beta + \gamma) = 8s^2 Rr$$

and

$$\sum_{\text{cyc}} n_1 p_1 \alpha^2 = \alpha^2 + \beta^2 + \gamma^2 = 2s^2 - 2r^2 - 8Rr.$$

Then

$$GI^2 = \frac{1}{9}(s^2 + 5r^2 - 16Rr).$$

(2) Let us prove that in every triangle ABC with sides α, β, γ, the following inequality holds:

$$\sum_{\text{cyc}} (2\alpha - \beta - \gamma)(2\beta - \alpha - \gamma)\gamma^2 \leq 0.$$

In the inequality in Corollary 1 in Sect. 4.7.2, we consider the points $P_1 = G$ and $P_2 = I$. Then $\alpha_1 = \beta_1 = \gamma_1 = \frac{1}{3}$ and $\alpha_2 = \frac{\alpha}{2s}$, $\beta_2 = \frac{\beta}{2s}$, $\gamma_2 = \frac{\gamma}{2s}$, and the above inequality follows. We have equality if and only if $P_1 = P_2$, that is, $G = I$, so the triangle is equilateral.

4.8 The Area of a Triangle in Barycentric Coordinates

Consider the triangle ABC with a, b, c the respective coordinates of its vertices. Let α, β, γ be the lengths of sides BC, CA, and AB.

Theorem. Let $P_j(z_{p_j})$, $j = 1, 2, 3$, be three points in the plane of triangle ABC with $z_{P_j} = \alpha_j a + \beta_j b + \gamma_j c$, where α_j, β_j, γ_j are the barycentric coordinates of P_j. If the triangles ABC and $P_1 P_2 P_3$ have the same orientation, then

$$\frac{\text{area}[P_1 P_2 P_3]}{\text{area}[ABC]} = \begin{vmatrix} \alpha_1 & \beta_1 & \gamma_1 \\ \alpha_2 & \beta_2 & \gamma_2 \\ \alpha_3 & \beta_3 & \gamma_3 \end{vmatrix}.$$

Proof. Suppose that the triangles ABC and $P_1 P_2 P_3$ are positively oriented. If O denotes the origin of the complex plane, then using the complex product, we can write

$$2i \, \text{area}[P_1 O P_2] = z_{P_1} \times z_{P_2} = (\alpha_1 a + \beta_1 b + \gamma_1 c) \times (\alpha_2 a + \beta_2 b + \gamma_2 c)$$

$$= (\alpha_1 \beta_2 - \alpha_2 \beta_1) a \times b + (\beta_1 \gamma_2 - \beta_2 \gamma_1) b \times c + (\gamma_1 \alpha_1 - \gamma_2 \alpha_1) c \times a$$

$$= \begin{vmatrix} a \times b & b \times c & c \times a \\ \gamma_1 & \gamma_1 & \beta_1 \\ \gamma_2 & \alpha_2 & \beta_2 \end{vmatrix} = \begin{vmatrix} a \times b & b \times c & 2i \text{ area } [ABC] \\ \gamma_1 & \alpha_1 & 1 \\ \gamma_2 & \alpha_2 & 1 \end{vmatrix}.$$

Analogously, we obtain

$$2i \text{ area } [P_2OP_3] = \begin{vmatrix} a \times b & b \times c & 2i \text{ area } [ABC] \\ \gamma_2 & \alpha_2 & 1 \\ \gamma_3 & \alpha_3 & 1 \end{vmatrix},$$

$$2i \text{ area } [P_3OP_1] = \begin{vmatrix} a \times b & b \times c & 2i \text{ area } [ABC] \\ \gamma_3 & \alpha_3 & 1 \\ \gamma_1 & \alpha_1 & 1 \end{vmatrix}.$$

Assuming that the origin O is situated in the interior of triangle $P_1P_2P_3$, it follows that

$$\text{area}[P_1P_2P_3] = \text{area } [P_1OP_2] + \text{area } [P_2OP_3] + \text{area } [P_3OP_1]$$

$$= \frac{1}{2i}(\alpha_1 - \alpha_2 + \alpha_2 - \alpha_3 + \alpha_3 - \alpha_1)a \times b - \frac{1}{2i}(\gamma_1 - \gamma_2 + \gamma_2 - \gamma_3 + \gamma_3 - \gamma_1)b \times c$$

$$+ (\gamma_1\alpha_2 - \gamma_2\alpha_1 + \gamma_2\alpha_3 - \gamma_3\alpha_2 + \gamma_3\alpha_1 - \gamma_1\alpha_3)\text{area } [ABC]$$

$$= (\gamma_1\alpha_2 - \gamma_2\alpha_1 + \gamma_2\alpha_3 - \gamma_3\alpha_2 + \gamma_3\alpha_1 - \gamma_1\alpha_3) \text{ area } [ABC]$$

$$= \text{area } [ABC] \begin{vmatrix} 1 & \gamma_1 & \alpha_1 \\ 1 & \gamma_2 & \alpha_2 \\ 1 & \gamma_3 & \alpha_3 \end{vmatrix} = \text{area}[ABC] \begin{vmatrix} \alpha_1 & \beta_1 & \gamma_1 \\ \alpha_2 & \beta_2 & \gamma_2 \\ \alpha_3 & \beta_3 & \gamma_3 \end{vmatrix},$$

and the desired formula is obtained.   □

Corollary 1. *Consider the triangle ABC and the points A_1, B_1, C_1 situated on the respective lines BC, CA, AB (Fig. 4.10) such that*

$$\frac{A_1B}{A_1C} = k_1, \quad \frac{B_1C}{B_1A} = k_2, \quad \frac{C_1A}{C_1B} = k_3.$$

If $AA_1 \cap BB_1 = \{P_1\}$, $BB_1 \cap CC_1 = \{P_2\}$, and $CC_1 \cap AA_1 = \{P_3\}$, then

$$\frac{\text{area}[P_1P_2P_3]}{\text{area}[ABC]} = \frac{(1 - k_1k_2k_3)^2}{(1 + k_1 + k_1k_2)(1 + k_2 + k_2k_3)(1 + k_3 + k_3k_1)}.$$

Proof. Applying the well-known Menelaus's theorem to triangle AA_1B, we find that

$$\frac{C_1A}{C_1B} \cdot \frac{CB}{CA_1} \cdot \frac{P_3A_1}{P_3A} = 1.$$

Hence

$$\frac{P_3A}{P_3A_1} = \frac{C_1A}{C_1B} \cdot \frac{CB}{CA_1} = k_3(1 + k_1).$$

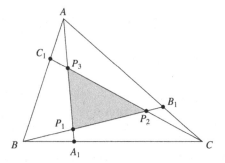

Figure 4.10.

The coordinate of P_3 is given by

$$z_{P_3} = \frac{a + k_3(1+k_1)z_{A_1}}{1 + k_3(1+k_1)} = \frac{a + k_3(1+k_1)\frac{b+k_1c}{1+k_1}}{1 + k_3 + k_3k_1} = \frac{a + k_3b + k_3k_1c}{1 + k_3 + k_3k_1}.$$

In an analogous way, we find that

$$z_{P_1} = \frac{k_1k_2a + b + k_1c}{1 + k_1 + k_1k_2} \text{ and } z_{P_2} = \frac{k_2a + k_2k_3b + c}{1 + k_2 + k_2k_3}.$$

The triangles ABC and $P_1P_2P_3$ have the same orientation; hence by applying the formula in the above theorem, we find that

$$\frac{\text{area}[P_1P_2P_3]}{\text{area}[ABC]} = \frac{1}{(1+k_1+k_1k_2)(1+k_2+k_2k_3)(1+k_3+k_3k_1)} \begin{vmatrix} k_1k_2 & 1 & k_1 \\ k_2 & k_2k_3 & 1 \\ 1 & k_3 & k_3k_1 \end{vmatrix}$$

$$= \frac{(1 - k_1k_2k_3)^2}{(1 + k_1 + k_1k_2)(1 + k_2 + k_2k_3)(1 + k_3 + k_3k_1)}. \qquad \square$$

Remark. When $k_1 = k_2 = k_3 = k$, from Corollary 1 in Sect. 4.8, we obtain Problem 3 in Sect. 4.9.2 from the 23rd Putnam Mathematical Competition.

Let A_j, B_j, C_j be points on the lines BC, CA, AB, respectively, such that

$$\frac{BA_j}{A_jC} = \frac{p_j}{n_j}, \frac{CB_j}{B_jA} = \frac{m_j}{p_j}, \frac{AC_j}{C_jB} = \frac{n_j}{m_j}, j = 1, 2, 3.$$

Corollary 2. *If P_j is the intersection point of lines AA_j, BB_j, CC_j, $j = 1, 2, 3$, and the triangles ABC, $P_1P_2P_3$ have the same orientation, then*

$$\frac{\text{area}[P_1P_2P_3]}{\text{area}[ABC]} = \frac{1}{S_1S_2S_3} \begin{vmatrix} m_1 & n_1 & p_1 \\ m_2 & n_2 & p_2 \\ m_3 & n_3 & p_3 \end{vmatrix},$$

where $S_j = m_j + n_j + p_j$, $j = 1, 2, 3$.

Proof. In terms of the coordinates of the triangle, the coordinates of the points P_j are

$$z_{P_j} = \frac{m_j a + n_j b + p_j c}{m_j + n_j + p_j} = \frac{1}{S_j}(m_j a + n_j b + p_j c), \; j = 1, 2, 3.$$

The formula above follows directly from the above theorem. $\square$

Corollary 3. *In triangle ABC, let us consider the cevians AA', BB', and CC' such that*

$$\frac{A'B}{AC} = m, \; \frac{B'C}{B'A} = n, \; \frac{C'A}{C'B} = p.$$

Then the following formula holds:

$$\frac{\text{area}[A'B'C']}{\text{area}[ABC]} = \frac{1 + mnp}{(1+m)(1+n)(1+p)}.$$

Proof. Observe that the coordinates of A', B', C' are given by

$$z_{A'} = \frac{1}{1+m}b + \frac{m}{1+m}c, \; z_{B'} = \frac{1}{1+n}c + \frac{n}{1+n}a, \; z_{C'} = \frac{1}{1+p}a + \frac{p}{1+p}b.$$

Applying the formula in Corollary 2 in Sect. 4.8, we obtain

$$\frac{\text{area}[A'B'C']}{\text{area}[ABC]} = \frac{1}{(1+m)(1+n)(1+p)} \begin{vmatrix} 0 & 1 & m \\ n & 0 & 1 \\ 1 & p & 0 \end{vmatrix}$$

$$= \frac{1 + mnp}{(1+m)(1+n)(1+p)}. \qquad \square$$

Applications

(1) (Steinhaus)[6] Let A_j, B_j, C_j be points on lines BC, CA, AB, respectively, $j = 1, 2, 3$. Assume that

$$\frac{BA_1}{A_1C} = \frac{2}{4}, \; \frac{CB_1}{B_1A} = \frac{1}{2}, \; \frac{AC_1}{C_1B} = \frac{4}{1};$$

$$\frac{BA_2}{A_2C} = \frac{4}{1}, \; \frac{CB_2}{B_2A} = \frac{2}{4}, \; \frac{AC_2}{C_2B} = \frac{1}{2};$$

$$\frac{BA_3}{A_3C} = \frac{1}{2}, \; \frac{CB_3}{B_3A} = \frac{4}{1}, \; \frac{AC_3}{C_3B} = \frac{2}{4}.$$

If P_j is the intersection point of lines AA_j, BB_j, CC_j, $j = 1, 2, 3$, and triangles ABC, $P_1P_2P_3$ are of the same orientation, then from Corollary 3 above, we obtain

---

[6] Hugo Dyonizy Steinhaus (1887–1972), Polish mathematician, made important contributions to functional analysis and other branches of modern mathematics.

$$\frac{\text{area}[P_1 P_2 P_3]}{\text{area}[ABC]} = \frac{1}{7 \cdot 7 \cdot 7} \begin{vmatrix} 1 & 4 & 2 \\ 2 & 1 & 4 \\ 4 & 2 & 1 \end{vmatrix} = \frac{49}{7^3} = \frac{1}{7}.$$

(2) If the cevians AA', BB', CC' are concurrent at point P, let us denote by K_P the area of triangle $A'B'C'$. We can use the formula in Corollary 3 above to compute the areas of some triangles determined by the feet of the cevians of some notable points in a triangle.

(i) If I is the incenter of triangle ABC, we have

$$K_I = \frac{1 + \dfrac{\gamma}{\beta} \cdot \dfrac{\beta}{\alpha} \cdot \dfrac{\alpha}{\gamma}}{\left(1 + \dfrac{\gamma}{\beta}\right)\left(1 + \dfrac{\beta}{\alpha}\right)\left(1 + \dfrac{\alpha}{\gamma}\right)}\text{area}[ABC]$$

$$= \frac{2\alpha\beta\gamma}{(\alpha + \beta)(\beta + \gamma)(\gamma + \alpha)}\text{area}[ABC] = \frac{2\alpha\beta\gamma sr}{(\alpha + \beta)(\beta + \gamma)(\gamma + \alpha)}.$$

(ii) For the orthocenter H of the acute triangle ABC, we obtain

$$K_H = \frac{1 + \dfrac{\tan C}{\tan B} \cdot \dfrac{\tan B}{\tan A} \cdot \dfrac{\tan A}{\tan C}}{\left(1 + \dfrac{\tan C}{\tan B}\right)\left(1 + \dfrac{\tan B}{\tan A}\right)\left(1 + \dfrac{\tan A}{\tan C}\right)}\text{area}[ABC]$$

$$= (2\cos A \cos B \cos C)\text{area}[ABC] = (2\cos A \cos B \cos C)sr.$$

(iii) For the Nagel point of triangle ABC, we can write

$$K_N = \frac{1 + \dfrac{s - \gamma}{s - \beta} \cdot \dfrac{s - \alpha}{s - \gamma} \cdot \dfrac{s - \beta}{s - \alpha}}{\left(1 + \dfrac{s - \gamma}{s - \beta}\right)\left(1 + \dfrac{s - \alpha}{s - \gamma}\right)\left(1 + \dfrac{s - \beta}{s - \alpha}\right)}\text{area}[ABC]$$

$$= \frac{2(s - \alpha)(s - \beta)(s - \gamma)}{\alpha\beta\gamma}\text{area}[ABC] = \frac{4\text{area}^2[ABC]}{2s\alpha\beta\gamma}\text{area}[ABC]$$

$$= \frac{r}{2R}\text{area}[ABC] = \frac{sr^2}{2R}.$$

If we proceed in the same way for the Gergonne point J, we obtain the relation

$$K_J = \frac{r}{2R}\text{area}[ABC] = \frac{sr^2}{2R}.$$

Remark. Two cevians AA' and AA'' are *isotomic* if the points A' and A'' are symmetric with respect to the midpoint of the segment BC. Assuming that

$$\frac{A'B}{A'C} = m, \quad \frac{B'C}{B'A} = n, \quad \frac{C'A}{C'B} = p,$$

then for the corresponding isotomic cevians, we have

$$\frac{A''B}{A''C} = \frac{1}{m}, \quad \frac{B''C}{B''A} = \frac{1}{n}, \quad \frac{C''A}{C''B} = \frac{1}{p}.$$

Applying the formula in Corollary 3 above yields that

$$\frac{\text{area}[A'B'C']}{\text{area}[ABC]} = \frac{1 + mnp}{(1+m)(1+n)(1+p)}$$

$$= \frac{1 + \dfrac{1}{mnp}}{\left(1 + \dfrac{1}{m}\right)\left(1 + \dfrac{1}{n}\right)\left(1 + \dfrac{1}{p}\right)} = \frac{\text{area}[A''B''C'']}{\text{area}[ABC]}.$$

Therefore, area $[A'B'C'] = \text{area}[A''B''C'']$. A special case of this relation is $K_N = K_J$, since the points N and J are isotomic (i.e., these points are intersections of isotomic cevians).

(3) Consider the excenters I_α, I_β, I_γ of triangle ABC. It is not difficult to see that the coordinates of these points are

$$z_{I_\alpha} = -\frac{\alpha}{2(s-\alpha)}a + \frac{\beta}{2(s-\beta)}b + \frac{\gamma}{2(s-\gamma)}c,$$

$$z_{I_\beta} = \frac{\alpha}{2(s-\alpha)}a - \frac{\beta}{2(s-\beta)}b + \frac{\gamma}{2(s-\gamma)}c,$$

$$z_{I_\gamma} = \frac{\alpha}{2(s-\alpha)}a + \frac{\beta}{2(s-\beta)}b - \frac{\gamma}{2(s-\gamma)}c.$$

From the formula in the theorem above, it follows that

$$\text{area}[I_\alpha I_\beta I_\gamma] = \begin{vmatrix} -\frac{\alpha}{2(s-\alpha)} & \frac{\beta}{2(s-\beta)} & \frac{\gamma}{2(s-\gamma)} \\ \frac{\alpha}{2(s-\alpha)} & -\frac{\beta}{2(s-\beta)} & \frac{\gamma}{2(s-\gamma)} \\ \frac{\alpha}{2(s-\alpha)} & \frac{\beta}{2(s-\beta)} & -\frac{\gamma}{2(s-\gamma)} \end{vmatrix} \text{area}[ABC]$$

$$= \frac{\alpha\beta\gamma}{8(s-\alpha)(s-\beta)(s-\gamma)} \begin{vmatrix} -1 & 1 & 1 \\ 1 & -1 & 1 \\ 1 & 1 & -1 \end{vmatrix} \text{area}[ABC]$$

$$= \frac{s\alpha\beta\gamma\,\text{area}[ABC]}{2s(s-\alpha)(s-\beta)(s-\gamma)} = \frac{s\alpha\beta\gamma\,\text{area}[ABC]}{2\text{area}^2[ABC]} = \frac{2s\alpha\beta\gamma}{4\text{area}[ABC]} = 2sR.$$

(4) (Nagel line) Using the formula in the theorem above, we give a different proof for the so-called Nagel line: the points I, G, N are collinear. We have seen that the coordinates of these points are

$$z_I = \frac{\alpha}{2s}a + \frac{\beta}{2s}b + \frac{\gamma}{2s}c,$$

$$z_G = \frac{1}{3}a + \frac{1}{3}b + \frac{1}{3}c,$$

$$z_N = \left(1 - \frac{\alpha}{s}\right)a + \left(1 - \frac{\beta}{s}\right)b + \left(1 - \frac{\gamma}{s}\right)c.$$

Then

$$\text{area}[IGN] = \begin{vmatrix} \frac{\alpha}{2s} & \frac{\beta}{2s} & \frac{\gamma}{2s} \\ \frac{1}{3} & \frac{1}{3} & \frac{1}{3} \\ 1 - \frac{\alpha}{s} & 1 - \frac{\beta}{s} & 1 - \frac{\gamma}{s} \end{vmatrix} \cdot \text{area}[ABC] = 0,$$

and hence the points I, G, N are collinear.

4.9 Orthopolar Triangles

4.9.1 The Simson–Wallace Line and the Pedal Triangle

Consider the triangle ABC, and let M be a point situated in the plane of the triangle. Let P, Q, R be the projections of M onto lines BC, CA, AB, respectively.

Theorem 1 (The Simson line[7]). *The points P, Q, R are collinear if and only if M is on the circumcircle of triangle ABC.*

Proof. We will give a standard geometric argument.

Suppose that M lies on the circumcircle of triangle ABC. Without loss of generality, we may assume that M is on the arc $\overset{\frown}{BC}$. In order to prove the collinearity of R, P, Q, it suffices to show that the angles $\widehat{BPR}$ and $\widehat{CPQ}$ are congruent. The quadrilaterals $PRBM$ and $PCQM$ are cyclic (since $\widehat{BRM} \equiv \widehat{BPM}$ and $\widehat{MPC} + \widehat{MQC} = 180°$); hence we have $\widehat{BPR} \equiv \widehat{BMR}$ and $\widehat{CPQ} \equiv \widehat{CMQ}$. But $\widehat{BMR} = 90° - \widehat{ABM} = 90° - \widehat{MCQ}$, since the quadrilateral $ABMC$ is cyclic, too. Finally, we obtain $\widehat{BMR} = 90° - \widehat{MCQ} = \widehat{CMQ}$, so the angles $\widehat{BPR}$ and $\widehat{CPQ}$ are congruent (Fig. 4.11).

To prove the converse, we note that if the points P, Q, R are collinear, then the angles $\widehat{BPR}$ and $\widehat{CPQ}$ are congruent; hence $\widehat{ABM} + \widehat{ACM} = 180°$, i.e., the quadrilateral $ABMC$ is cyclic. Therefore, the point M is situated on the circumcircle of triangles ABC. □

---

[7] Robert Simson (1687–1768), Scottish mathematician. This line was attributed to Simson by Poncelet, but it is now generally known as the Simson–Wallace line, since it does not actually appear in any work of Simson. William Wallace (1768–1843) was also a Scottish mathematician, who possibly published the theorem above concerning the Simson line in 1799.

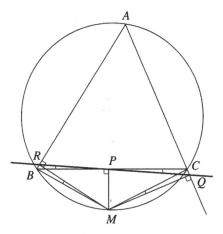

Figure 4.11.

When M lies on the circumcircle of triangle ABC, the line in the above theorem is called the *Simson–Wallace line of M* with respect to triangle ABC.

We continue with a nice generalization of the property contained in Theorem 1 above. For an arbitrary point X in the plane of triangle ABC, consider its projections P, Q, and R on the lines BC, CA and AB, respectively.

The triangle PQR is called the *pedal triangle* of point X with respect to the triangle ABC. Let us choose the circumcenter O of triangle ABC as the origin of the complex plane.

Theorem 2. *The area of the pedal triangle of X with respect to the triangle ABC is given by*

$$\text{area}[PQR] = \frac{\text{area}[ABC]}{4R^2}\big||x|^2 - R^2\big|, \tag{1}$$

where R is the circumradius of triangle ABC.

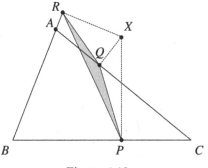

Figure 4.12.

Proof. Applying the formula in the proposition of Sect. 4.5, we obtain the coordinates p, q, r of the points P, Q, R, respectively (Fig. 4.12):

$$p = \frac{1}{2}\left(x - \frac{bc}{R^2}\overline{x} + b + c\right),$$

$$q = \frac{1}{2}\left(x - \frac{ca}{R^2}\overline{x} + c + a\right),$$

$$r = \frac{1}{2}\left(x - \frac{ab}{R^2}\overline{x} + a + b\right).$$

Taking into account the formula in Sect. 3.5.3, we have

$$\text{area}[PQR] = \left|\frac{i}{4}\begin{vmatrix} p & \overline{p} & 1 \\ q & \overline{q} & 1 \\ r & \overline{r} & 1 \end{vmatrix}\right| = \left|\frac{i}{4}\begin{vmatrix} q - p & \overline{q} - \overline{p} \\ r - p & \overline{r} - \overline{p} \end{vmatrix}\right|.$$

For the coordinates p, q, r, we obtain

$$\overline{p} = \frac{1}{2}\left(\overline{x} - \frac{\overline{bc}}{R^2}x + \overline{b} + \overline{c}\right),$$

$$\overline{q} = \frac{1}{2}\left(\overline{x} - \frac{\overline{ca}}{R^2}x + \overline{c} + \overline{a}\right),$$

$$\overline{r} = \frac{1}{2}\left(\overline{x} - \frac{\overline{ab}}{R^2}x + \overline{a} + \overline{b}\right).$$

It follows that

$$q - p = \frac{1}{2}(a - b)\left(1 - \frac{c\overline{x}}{R^2}\right) \text{ and } r - p = \frac{1}{2}(a - c)\left(1 - \frac{b\overline{x}}{R^2}\right), \qquad (2)$$

$$\overline{q} - \overline{p} = \frac{1}{2abc}(a - b)(x - c)R^2 \text{ and } \overline{r} - \overline{p} = \frac{1}{2abc}(a - c)(x - b)R^2.$$

Therefore,

$$\text{area}[PQR] = \left|\frac{i}{4}\begin{vmatrix} q - p & \overline{q} - \overline{p} \\ r - p & \overline{r} - \overline{p} \end{vmatrix}\right| = \left|\frac{i(a - b)(a - c)}{16abc}\begin{vmatrix} 1 - \frac{c\overline{x}}{R^2} & (x - c)R^2 \\ 1 - \frac{b\overline{x}}{R^2} & (x - b)R^2 \end{vmatrix}\right|$$

$$= \left|\frac{i(a - b)(a - c)}{16abc}\begin{vmatrix} R^2 - c\overline{x} & x - c \\ R^2 - b\overline{x} & x - b \end{vmatrix}\right| = \left|\frac{i(a - b)(a - c)}{16abc}\begin{vmatrix} (b - c)\overline{x} & b - c \\ R^2 - b\overline{x} & x - b \end{vmatrix}\right|$$

$$= \left|\frac{i(a - b)(b - c)(a - c)}{16abc}\begin{vmatrix} \overline{x} & 1 \\ R^2 - b\overline{x} & x - b \end{vmatrix}\right| = \left|\frac{i(a - b)(b - c)(a - c)}{16abc}(x\overline{x} - R^2)\right|.$$

We find that

$$\text{area}[PQR] = \frac{|a - b||b - c||c - a|}{16|a||b||c|} \left| |x|^2 - R^2 \right| = \frac{\alpha\beta\gamma}{16R^3} \left| |x|^2 - R^2 \right|$$

$$= \frac{\text{area}[ABC]}{4R^2} \left| |x|^2 - R^2 \right|,$$

where α, β, γ are the side lengths of triangle ABC. □

Remarks.

(1) The formula in Theorem 2 above contains the Simson–Wallace line property. Indeed, points P, Q, R are collinear if and only if area $[PQR] = 0$, that is, $|x\bar{x} - R^2| = 0$, i.e., $x\bar{x} = R^2$. It follows that $|x| = R$, so X lies on the circumcircle of triangle ABC.

(2) If X lies on a circle of radius R_1 and center O (the circumcenter of triangle ABC), then $x\bar{x} = R_1^2$, and from Theorem 2 above, we obtain

$$\text{area}[PQR] = \frac{\text{area}[ABC]}{4R^2} |R_1^2 - R^2|.$$

It follows that the area of triangle PQR does not depend on the point X.

The converse is also true. The locus of all points X in the plane of triangle ABC such that area $[PQR] = k$ (constant) is defined by

$$\left| |x|^2 - R^2 \right| = \frac{4R^2 k}{\text{area}[ABC]}.$$

This is equivalent to

$$|x|^2 = R^2 \pm \frac{4R^2 k}{\text{area}[ABC]} = R^2 \left(1 \pm \frac{4k}{\text{area}[ABC]} \right).$$

If $k > \frac{1}{4}\text{area}[ABC]$, then the locus is a circle with center O and radius

$$R_1 = R\sqrt{1 + \frac{4k}{\text{area}[ABC]}}.$$

If $k \leq \frac{1}{4}\text{area}[ABC]$, then the locus consists with two circles of center O and radii $R\sqrt{1 \pm \frac{4k}{\text{area}[ABC]}}$, one of which degenerates to O when $k = \frac{1}{4}\text{area}[ABC]$.

Theorem 3. *For every point X in the plane of triangle ABC, we can construct a triangle with sides $AX \cdot BC$, $BX \cdot CA$, $CX \cdot AB$. This triangle is then similar to the pedal triangle of point X with respect to the triangle ABC.*

Proof. Let PQR be the pedal triangle of X with respect to triangle ABC. From formula (2), we obtain

$$q - p = \frac{1}{2}(a - b)(x - c)\frac{R^2 - c\bar{x}}{R^2(x - c)}. \tag{3}$$

Taking moduli in (3), we obtain

$$|q - p| = \frac{1}{2R^2}|a - b||x - c|\left|\frac{R^2 - c\bar{x}}{x - c}\right|. \tag{4}$$

On the other hand,

$$\left|\frac{R^2 - c\bar{x}}{x - c}\right|^2 = \frac{R^2 - c\bar{x}}{x - c} \cdot \frac{R^2 - \bar{c}x}{\bar{x} - \bar{c}} = \frac{R^2 - c\bar{x}}{x - c} \cdot \frac{R^2 - \bar{c}x}{\bar{x} - \dfrac{R^2}{c}}$$

$$= \frac{R^2 - c\bar{x}}{x - c} \cdot \frac{R^2(c - x)}{c\bar{x} - R^2} = R^2,$$

whence from (4), we derive the relation

$$|q - p| = \frac{1}{2R}|a - b||x - c|. \tag{5}$$

Therefore,

$$\frac{PQ}{CX \cdot AB} = \frac{QR}{AX \cdot BC} = \frac{RP}{BX \cdot CA} = \frac{1}{2R}, \tag{6}$$

and the conclusion follows. □

Corollary 1. *In the plane of triangle ABC, consider the point X and denote by $A'B'C'$ the triangle with sides $AX \cdot BC$, $BX \cdot CA$, $CX \cdot AB$. Then*

$$\text{area}[A'B'C'] = \text{area}[ABC]\big||x|^2 - R^2\big|. \tag{7}$$

Proof. From formula (6), it follows that area $[A'B'C'] = 4R^2$ area $[PQR]$, where PQR is the pedal triangle of X with respect to triangle ABC. Replacing this result in (1), we obtain the desired formula. □

Corollary 2 (Ptolemy's inequality). *The following inequality holds for every quadrilateral $ABCD$:*

$$AC \cdot BD \le AB \cdot CD + BC \cdot AD. \tag{8}$$

Corollary 3 (Ptolemy's theorem). *The convex quadrilateral $ABCD$ is cyclic if and only if*

$$AC \cdot BD = AB \cdot CD + BC \cdot AD. \tag{9}$$

Proof. If the relation (9) holds, then triangle $A'B'C'$ in Corollary 1 above is degenerate; i.e., area $[A'B'C'] = 0$. From formula (7), it follows that $d \cdot \overline{d} = R^2$, where d is the coordinate of D and R is the circumradius of triangle ABC. Hence the point D lies on the circumcircle of triangle ABC.

If quadrilateral $ABCD$ is cyclic, then the pedal triangle of point D with respect to triangle ABC is degenerate. From (6), we obtain the relation (9).   □

Corollary 4 (Pompeiu's theorem[8]). *For every point X in the plane of the equilateral triangle ABC, three segments with lengths XA, XB, XC can be taken as the sides of a triangle.*

Proof. In Theorem 3 above, we have $BC = CA = AB$, and the desired conclusion follows.   □

The triangle in Corollary 4 above is called the *Pompeiu triangle* of X with respect to the equilateral triangle ABC. This triangle is degenerate if and only if X lies on the circumcircle of ABC. Using the second part of Theorem 3, we find that Pompeiu's triangle of the point X is similar to the pedal triangle of X with respect to triangle ABC and

$$\frac{CX}{PQ} = \frac{AX}{QR} = \frac{BX}{RP} = \frac{2R}{\alpha} = \frac{2\sqrt{3}}{3}. \tag{10}$$

Problem 1. *Let A, B, and C be equidistant points on the circumference of a circle of unit radius centered at O, and let X be any point in the circle's interior. Let d_A, d_B, d_C be the distances from X to A, B, C, respectively. Show that there is a triangle with sides d_A, d_B, d_C, and that the area of this triangle depends only on the distance from X to O.*

(2003 Putnam Mathematical Competition)

Solution. The first assertion is just the property contained in Corollary 4 above. Taking into account the relations (10), we see that the area of Pompeiu's triangle of point X is $\frac{4}{3}$area$[PQR]$. From Theorem 2 above, we get that area $[PQR]$ depends only on the distance from X to O, as desired.

Problem 2. *Let X be a point in the plane of the equilateral triangle ABC such that X does not lie on the circumcircle of triangleABC, and let $XA = u$, $XB = v$, $XC = w$. Express the side length α of triangle ABC in terms of real numbers u, v, w.*

(1978 GDR Mathematical Olympiad)

---

[8] Dimitrie Pompeiu (1873–1954), Romanian mathematician, made important contributions in the fields of mathematical analysis, functions of a complex variable, and rational mechanics. He was a Ph.D student of Henri Poincaré.

Solution. The segments $[XA]$, $[XB]$, $[XC]$ are the sides of Pompeiu's triangle of point X with respect to equilateral triangle ABC. Denote this triangle by $A'B'C'$. From relations (10) and from Theorem 2 in Sect. 4.9.1 it follows that

$$\text{area}[A'B'C'] = \left(\frac{2\sqrt{3}}{3}\right)^2 \text{area}[PQR] = \frac{1}{3R^2}\text{area}[ABC]|x\cdot\overline{x} - R^2|$$

$$= \frac{1}{3R^2}\cdot\frac{a^2\sqrt{3}}{4}||x|^2 - R^2| = \frac{\sqrt{3}}{4}|XO^2 - R^2|. \tag{1}$$

On the other hand, using the well-known formula of Heron, we obtain, after a few simple computations,

$$\text{area}[A'B'C'] = \frac{1}{4}\sqrt{(u^2+v^2+w^2)^2 - 2(u^4+v^4+w^4)}.$$

Substituting in (1), we obtain

$$|XO^2 - R^2| = \frac{1}{\sqrt{3}}\sqrt{(u^2+v^2+w^2)^2 - 2(u^4+v^4+w^4)}. \tag{11}$$

Now we consider the following two cases:

Case 1. If X lies in the interior of the circumcircle of triangle ABC, then $XO^2 < R^2$. Using the relation (see also formula (4) in Sect. 4.11)

$$XO^2 = \frac{1}{3}(u^2+v^2+w^2 - 3R^2),$$

from (11) we find that

$$2R^2 = \frac{1}{3}(u^2+v^2+w^2) + \frac{1}{\sqrt{3}}\sqrt{(u^2+v^2+w^2)^2 - 2(u^4+v^4+w^4)},$$

and hence

$$\alpha^2 = \frac{1}{2}(u^2+v^2+w^2) + \frac{\sqrt{3}}{2}\sqrt{(u^2+v^2+w^2)^2 - 2(u^4+v^4+w^4)}.$$

Case 2. If X lies in the exterior of the circumcircle of triangle ABC, then $XO^2 > R^2$, and after some similar computations we obtain

$$\alpha^2 = \frac{1}{2}(u^2+v^2+w^2) - \frac{\sqrt{3}}{2}\sqrt{(u^2+v^2+w^2)^2 - 2(u^4+v^4+w^4)}.$$

4.9.2 Necessary and Sufficient Conditions for Orthopolarity

Consider a triangle ABC and points X, Y, Z situated on its circumcircle. Triangles ABC and XYZ are called *orthopolar triangles* (or *S-triangles*)[9] if the Simson–Wallace line of point X with respect to triangle ABC is perpendicular (orthogonal) to line YZ.

Let us choose the circumcenter O of triangle ABC to lie at the origin of the complex plane. Points A, B, C, X, Y, Z have the coordinates a, b, c, x, y, z with

$$|a| = |b| = |c| = |x| = |y| = |z| = R,$$

where R is the circumradius of the triangle ABC.

Theorem. *Triangles ABC and XYZ are orthopolar triangles if and only if* $abc = xyz$.

Proof. Let P, Q, R be the feet of the orthogonal lines from the point X to the lines BC, CA, AB, respectively.

Points P, Q, R are on the same line, namely the Simson–Wallace line of point X with respect to triangle ABC.

The coordinates of P, Q, R are denoted by p, q, r, respectively. Using the formula in Proposition of Sect. 4.5, we have

$$p = \frac{1}{2}\left(x - \frac{bc}{R^2}\overline{x} + b + c\right),$$

$$q = \frac{1}{2}\left(x - \frac{ca}{R^2}\overline{x} + c + a\right),$$

$$r = \frac{1}{2}\left(x - \frac{ab}{R^2}\overline{x} + a + b\right).$$

We study two cases.

Case 1. Point X is not a vertex of triangle ABC.

Then PQ is orthogonal to YZ if and only if $(p - q) \cdot (y - z) = 0$. That is,

$$\left[(b - a)\left(1 - \frac{c\overline{x}}{R^2}\right)\right] \cdot (y - z) = 0,$$

or

$$(\overline{b} - \overline{a})(R^2 - \overline{c}x)(y - z) + (b - a)(R^2 - c\overline{x})(\overline{y} - \overline{z}) = 0.$$

We obtain

$$\left(\frac{R^2}{b} - \frac{R^2}{a}\right)\left(R^2 - \frac{R^2}{c}x\right)(y - z) + (b - a)\left(R^2 - c\frac{R^2}{x}\right)\left(\frac{R^2}{y} - \frac{R^2}{z}\right) = 0;$$

---

[9] This definition was given in 1915 by the Romanian mathematician Traian Lalescu (1882–1929). He is famous for his book *La géometrie du triangle* [43].

hence

$$\frac{1}{abc}(a-b)(c-x)(y-z) - \frac{1}{xyz}(a-b)(c-x)(y-z) = 0.$$

The last relation is equivalent to

$$(abc - xyz)\,(a-b)(c-x)(y-z) = 0,$$

and finally, we get $abc = xyz$, as desired.

Case 2.   Point X is a vertex of triangle ABC. Without loss of generality, assume that $X = B$.

Then the Simson–Wallace line of point $X = B$ is the orthogonal line from B to AC. It follows that BQ is orthogonal to YZ if and only if lines AC and YZ are parallel. This is equivalent to $ac = yz$. Because $b = x$, we obtain $abc = xyz$, as desired.

□

Remark. Due to the symmetry of the relation $abc = xyz$, we observe that the Simson–Wallace line of every vertex of triangle XYZ with respect to ABC is orthogonal to the opposite side of the triangle XYZ. Moreover, the same property holds for the vertices of triangle ABC.

Hence ABC and XYZ are orthopolar triangles if and only if XYZ and ABC are orthopolar triangles. Therefore the orthopolarity relation is symmetric.

Problem 1. *The median and the orthic triangles of a triangle ABC are orthopolar in the nine-point circle.*

Solution. Consider the origin of the complex plane at the circumcenter O of triangle ABC. Let M, N, P be the midpoints of AB, BC, CA and let A', B', C' be the feet of the altitudes of triangles ABC from A, B, C, respectively.

If m, n, p, a', b', c' are coordinates of M, N, P, A', B', C', then we have

$$m = \frac{1}{2}(a+b), \ n = \frac{1}{2}(b+c), \ p = \frac{1}{2}(c+a)$$

and

$$a' = \frac{1}{2}\left(a+b+c-\frac{bc}{R^2}\overline{a}\right) = \frac{1}{2}(a+b+c-\frac{bc}{a}),$$

$$b' = \frac{1}{2}\left(a+b+c-\frac{ca}{b}\right), \ c' = \frac{1}{2}\left(a+b+c-\frac{ab}{2}\right).$$

The nine-point center O_9 is the midpoint of the segment OH, where $H(a+b+c)$ is the orthocenter of triangle ABC. The coordinate of O_9 is $\omega = \frac{1}{2}(a+b+c)$.

Now observe that

$$(a' - w)(b' - w)(c' - w) = (m - w)(n - w)(p - w) = -\frac{1}{8}abc,$$

and the claim is proved.

Problem 2. *The altitudes of triangle ABC meet its circumcircle at points A_1, B_1, C_1, respectively. If A'_1, B'_1, C'_1 are the antipodal points of A_1, B_1, C_1 on the circumcircle ABC, then ABC and $A'_1 B'_1 C'_1$ are orthopolar triangles.*

Solution. The coordinates of A_1, B_1, C_1 are $-\dfrac{bc}{a}$, $-\dfrac{ca}{b}$, $-\dfrac{ab}{c}$, respectively. Indeed, the equation of line AH in terms of the real product is

$$AH : (z - a) \cdot (b - c) = 0.$$

It suffices to show that the point with coordinate $-\dfrac{bc}{a}$ lies both on AH and on the circumcircle of triangle ABC. First, let us note that

$$\left| -\frac{bc}{a} \right| = \frac{|b||c|}{|a|} = \frac{R \cdot R}{R} = R;$$

hence this point is situated on the circumcircle of triangle ABC. Now we shall show that the complex number $-\dfrac{bc}{a}$ satisfies the equation of the line AH. This is equivalent to

$$\left(\frac{bc}{a} + a \right) \cdot (b - c) = 0.$$

Using the definition of the real product, this reduces to

$$\left(\frac{\overline{bc}}{\overline{a}} + \overline{a} \right) (b - c) + \left(\frac{bc}{a} + a \right) (\overline{b} - \overline{c}) = 0,$$

or

$$\left(\frac{a\overline{bc}}{R^2} + \overline{a} \right) (b - c) + \left(\frac{bc}{a} + a \right) \left(\frac{R^2}{b} - \frac{R^2}{c} \right) = 0.$$

Finally, this comes down to

$$(b - c) \left(\frac{a\overline{bc}}{R^2} + \overline{a} - \frac{R^2}{a} - \frac{aR^2}{bc} \right) = 0,$$

a relation that is clearly true.

It follows that A'_1, B'_1, C'_1 have coordinates $\dfrac{bc}{a}$, $\dfrac{ca}{b}$, $\dfrac{ab}{c}$, respectively. Because

$$\frac{bc}{a} \cdot \frac{ca}{c} \cdot \frac{ab}{c} = abc,$$

we obtain that the triangles ABC and $A'_1 B'_1 C'_1$ are orthopolar (Fig. 4.13).

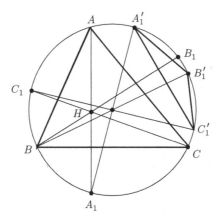

Figure 4.13.

Problem 3. *Let P and P' be distinct points on the circumcircle of triangle ABC such that lines AP and AP' are symmetric with respect to the bisector of angle $\widehat{BAC}$. Then triangles ABC and APP' are orthopolar* (Fig. 4.14).

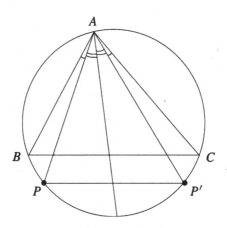

Figure 4.14.

Solution. Let us consider p and p' the coordinates of points P and P', respectively. It is clear that the lines PP' and BC are parallel. Using the complex product, it follows that $(p - p') \times (b - c) = 0$. This relation is equivalent to

$$(p - p')(\overline{b} - \overline{c}) - (\overline{p} - \overline{p'})(b - c) = 0.$$

Considering the origin of the complex plane at the circumcenter O of triangle ABC, we have

$$(p - p') \left(\frac{R^2}{b} - \frac{R^2}{c} \right) - \left(\frac{R^2}{p} - \frac{R^2}{p} \right) (b - c) = 0,$$

so

$$R^2 (p - p')(b - c) \left(\frac{1}{bc} - \frac{1}{pp'} \right) = 0.$$

Therefore, $bc = pp'$, i.e., $abc = app'$. From the theorem at the beginning of this subsection, it follows that ABC and APP' are orthopolar triangles.

4.10 Area of the Antipedal Triangle

Consider a triangle ABC and a point M. The perpendicular lines from A, B, C to MA, MB, MC, respectively, determine a triangle; we call this triangle the *antipedal* triangle of M with respect to ABC (Fig. 4.15).

Recall that M' is the *isogonal point* of M if the pairs of lines AM, AM'; BM, BM'; CM, CM' are isogonal, i.e., the following relations hold:

$$\widehat{MAC} \equiv \widehat{M'AB}, \ \widehat{MBC} \equiv \widehat{M'BA}, \ \widehat{MCA} \equiv \widehat{M'CB}.$$

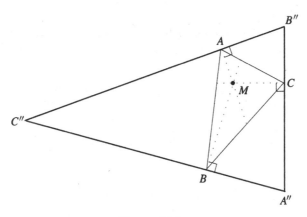

Figure 4.15.

Theorem. *Consider M a point in the plane of triangle ABC, M' the isogonal point of M, and $A''B''C''$ the antipedal triangle of M with respect to ABC. Then*

$$\frac{\text{area}[ABC]}{\text{area}[A''B''C'']} = \frac{|R^2 - OM'^2|}{4R^2} = \frac{|\rho(M')|}{4R^2},$$

where $\rho(M')$ is the power of M' with respect to the circumcircle of triangle ABC.

Proof. Consider point O the origin of the complex plane and let m, a, b, c be the coordinates of M, A, B, C. Then

$$R^2 = a\bar{a} = b\bar{b} = c\bar{c} \text{ and } \rho(M) = R^2 - m\bar{m}. \tag{1}$$

Let O_1, O_2, O_3 be the circumcenters of triangles BMC, CMA, AMB, respectively. It is easy to verify that O_1, O_2, O_3 are the midpoints of segments MA'', MB'', MC'', respectively, and so

$$\frac{\text{area}[O_1 O_2 O_3]}{\text{area}[A'' B'' C'']} = \frac{1}{4}. \tag{2}$$

The coordinate of the circumcenter of the triangle with vertices with coordinates z_1, z_2, z_3 is given by the following formula (see formula (1) in Sect. 3.6.1):

$$z_O = \frac{z_1 \bar{z_1}(z_2 - z_3) + z_2 \bar{z_2}(z_3 - z_1) + z_3 \bar{z_3}(z_1 - z_2)}{\begin{vmatrix} z_1 & \bar{z_1} & 1 \\ z_2 & \bar{z_2} & 1 \\ z_3 & \bar{z_3} & 1 \end{vmatrix}}.$$

The bisector line of the segment $[z_1, z_2]$ has the following equation in terms of the real product: $\left[z - \dfrac{1}{2}(z_1 + z_2)\right] \cdot (z_1 - z_2) = 0$. It is sufficient to check that z_O satisfies this equation, since that implies, by symmetry, that z_0 belongs to the perpendicular bisectors of segments $[z_2, z_3]$ and $[z_3, z_1]$.

The coordinate of O_1 is

$$z_{O_1} = \frac{m\bar{m}(b - c) + b\bar{b}(c - m) + c\bar{c}(m - b)}{\begin{vmatrix} m & \bar{m} & 1 \\ b & \bar{b} & 1 \\ c & \bar{c} & 1 \end{vmatrix}}$$

$$= \frac{(R^2 - m\bar{m})(c - b)}{\begin{vmatrix} m & \bar{m} & 1 \\ b & \bar{b} & 1 \\ c & \bar{c} & 1 \end{vmatrix}} = \frac{\rho(M)(c - b)}{\begin{vmatrix} m & \bar{m} & 1 \\ b & \bar{b} & 1 \\ c & \bar{c} & 1 \end{vmatrix}}.$$

Let

$$\Delta = \begin{vmatrix} a & \bar{a} & 1 \\ b & \bar{b} & 1 \\ c & \bar{c} & 1 \end{vmatrix}$$

and consider

$$\alpha = \frac{1}{\Delta} \begin{vmatrix} m & \bar{m} & 1 \\ b & \bar{b} & 1 \\ c & \bar{c} & 1 \end{vmatrix}, \quad \beta = \frac{1}{\Delta} \begin{vmatrix} m & \bar{m} & 1 \\ c & \bar{c} & 1 \\ a & \bar{a} & 1 \end{vmatrix},$$

and

$$\gamma = \frac{1}{\Delta} \begin{vmatrix} m & \overline{m} & 1 \\ a & \overline{a} & 1 \\ b & \overline{b} & 1 \end{vmatrix}.$$

With this notation we obtain

$$(\alpha a + \beta b + \gamma c) \cdot \Delta = \sum_{\text{cyc}} m(a\overline{b} - a\overline{c}) - \sum_{\text{cyc}} \overline{m}(ab - ac) + \sum_{\text{cyc}} a(b\overline{c} - \overline{b}c)$$

$$= m\Delta - \overline{m} \cdot 0 + \sum_{\text{cyc}} a\left(b\frac{R^2}{c} - \frac{R^2}{c}a\right) = m\Delta + R^2 \sum_{\text{cyc}} \left(\frac{ab}{c} - \frac{ac}{b}\right) = m\Delta,$$

and consequently,

$$\alpha a + \beta b + \gamma c = m,$$

since it is clear that $\Delta \neq 0$.

We note that α, β, γ are real numbers and $\alpha + \beta + \gamma = 1$, so α, β, γ are the barycentric coordinates of point M.

Since

$$z_{O_1} = \frac{(c - b) \cdot \rho(M)}{\alpha \cdot \Delta}, \quad z_{O_2} = \frac{(c - a) \cdot \rho(M)}{\beta \Delta}, \quad z_{O_3} = \frac{(a - b) \cdot \rho(M)}{\gamma \cdot \Delta},$$

we have

$$\frac{\text{area}[O_1 O_2 O_3]}{\text{area}[ABC]} = \left| \frac{\frac{i}{4} \begin{vmatrix} z_{O_1} & \overline{z_{O_1}} & 1 \\ z_{O_2} & \overline{z_{O_2}} & 1 \\ z_{O_3} & \overline{z_{O_3}} & 1 \end{vmatrix}}{\frac{i}{4}\Delta} \right|$$

$$= \left| \frac{1}{\Delta} \cdot \frac{\rho^2(M)}{\Delta^2} \cdot \frac{1}{\alpha\beta\gamma} \begin{vmatrix} b - c & \overline{b} - \overline{c} & \alpha \\ c - a & \overline{c} - \overline{a} & \beta \\ a - b & \overline{a} - \overline{b} & \gamma \end{vmatrix} \right|$$

$$= \left| \frac{\rho^2(M)}{\Delta^3} \cdot \frac{1}{\alpha\beta\gamma} \begin{vmatrix} c - a & \overline{c} - \overline{a} \\ a - b & \overline{a} - \overline{b} \end{vmatrix} \right|$$

$$= \left| \frac{\rho^2(M)}{\Delta^3} \cdot \frac{1}{\alpha\beta\gamma} \cdot \Delta \right| = \left| \frac{\rho^2(M)}{\Delta^2} \cdot \frac{1}{\alpha\beta\gamma} \right|. \tag{3}$$

Relations (2) and (3) imply that

$$\frac{\text{area}[ABC]}{\text{area}[A''B''C'']} = \frac{|\Delta^2 \alpha\beta\gamma|}{4\rho^2(M)}. \tag{4}$$

Because α, β, γ are the barycentric coordinates of M, it follows that

$$z_M = \alpha z_A + \beta z_B + \gamma z_C.$$

Using the real product, we find that

$$OM^2 = z_M \cdot z_M = (\alpha z_A + \beta z_B + \gamma z_C) \cdot (\alpha z_A + \beta z_B + \gamma z_C)$$

$$= (\alpha^2 + \beta^2 + \gamma^2)R^2 + 2\sum_{\mathrm{cyc}} \alpha\beta z_A \cdot z_B$$

$$= (\alpha^2 + \beta^2 + \gamma^2)R^2 + 2\sum_{\mathrm{cyc}} \alpha\beta \left(R^2 - \frac{AB^2}{2} \right)$$

$$= (\alpha + \beta + \gamma)^2 R^2 - \sum_{\mathrm{cyc}} \alpha\beta AB^2 = R^2 - \sum_{\mathrm{cyc}} \alpha\beta AB^2.$$

Therefore, the power of M' with respect to the circumcircle of triangle ABC can be expressed in the form

$$\rho(M) = R^2 - OM^2 = \sum_{\mathrm{cyc}} \alpha\beta AB^2.$$

On the other hand, if α, β, γ are the barycentric coordinates of the point M, then its isogonal point M' has barycentric coordinates given by

$$\alpha' = \frac{\beta\gamma BC^2}{\beta\gamma BC^2 + \alpha\gamma CA^2 + \alpha\beta AB^2}, \quad \beta' = \frac{\gamma\alpha CA^2}{\beta\gamma BC^2 + \alpha\gamma CA^2 + \alpha\beta AB^2},$$

$$\gamma' = \frac{\alpha\beta AB^2}{\beta\gamma BC^2 + \alpha\gamma CA^2 + \alpha\beta AB^2}.$$

Therefore,

$$\rho(M') = \sum_{\mathrm{cyc}} \alpha'\beta' AB^2$$

$$= \frac{\alpha\beta\gamma AB^2 \cdot BC^2 \cdot CA^2}{(\beta\gamma BC^2 + \alpha\gamma CA^2 + \alpha\beta AB^2)^2} = \frac{\alpha\beta\gamma AB^2 \cdot BC^2 \cdot CA^2}{\rho^2(M)}. \tag{5}$$

On the other hand, we have

$$\Delta^2 = \left| \left(\frac{4}{i} \cdot \frac{i}{4}\Delta \right)^2 \right| = \left| \frac{4}{i} \cdot \mathrm{area}[ABC] \right|^2 = \frac{AB^2 \cdot BC^2 \cdot CA^2}{R^2}. \tag{6}$$

The desired conclusion follows from the relations (4), (5), and (6).   □

Applications

(1) If M is the orthocenter H, then M' is the circumcenter O, and

$$\frac{\mathrm{area}[ABC]}{\mathrm{area}[A''B''C'']} = \frac{R^2}{4R^2} = \frac{1}{4}.$$

(2) If M is the circumcenter O, then M' is the orthocenter H, and we obtain

$$\frac{\text{area}[ABC]}{\text{area}[A''B''C'']} = \frac{|R^2 - OH^2|}{4R^2}.$$

Using the formula in the theorem of Sect. 4.6.4, it follows that

$$\frac{\text{area}[ABC]}{\text{area}[A''B''C'']} = \frac{|(2R+r)^2 - s^2|}{2R^2}.$$

(3) If M is the Lemoine point K, then M' is the centroid G, and

$$\frac{\text{area}[ABC]}{\text{area}[A''B''C'']} = \frac{|R^2 - OG^2|}{4R^2}.$$

Applying the formula in Corollary 1 in Sect. 4.6.4, then the first formula in Corollary of Sect. 4.6.1, it follows that

$$\frac{\text{area}[ABC]}{\text{area}[A''B''C'']} = \frac{2(s^2 - r^2 - 4Rr)}{36R^2} = \frac{\alpha^2 + \beta^2 + \gamma^2}{36R^2},$$

where α, β, γ are the sides of triangle ABC.
From the inequality $\alpha^2 + \beta^2 + \gamma^2 \leq 9R^2$ (Corollary 2 in Sect. 4.6.4), we obtain

$$\frac{\text{area}[ABC]}{\text{area}[A''B''C'']} \leq \frac{1}{4}.$$

(4) If M is the incenter I of triangle ABC, then $M' = I$, and using Euler's formula $OI^2 = R^2 - 2Rr$ (see the theorem of Sect. 4.6.2), we find that

$$\frac{\text{area}[ABC]}{\text{area}[A''B''C'']} = \frac{|R^2 - OI^2|}{4R^2} = \frac{2Rr}{4R^2} = \frac{r}{4R}.$$

Applying Euler's inequality $R \geq 2r$ (corollary of Sect. 4.6.2), it follows that

$$\frac{\text{area}[ABC]}{\text{area}[A''B''C'']} \leq \frac{1}{4}.$$

4.11 Lagrange's Theorem and Applications

Consider the distinct points $A_1(z_1)$, ..., $A_n(z_n)$ in the complex plane. Let m_1, ..., m_n be nonzero real numbers such that $m_1 + \cdots + m_n \neq 0$. Let $m = m_1 + \cdots + m_n$.

The point G with coordinate

$$z_G = \frac{1}{m}(m_1 z_1 + \cdots + m_n z_n)$$

is called the *barycenter of the set* $\{A_1, \ldots, A_n\}$ with respect to the weights $m_1, \ldots, m_n$.

In the case $m_1 = \cdots = m_n = 1$, the point G is the *centroid* of the set $\{A_1, \ldots, A_n\}$.

When $n = 3$ and the points A_1, A_2, A_3 are not collinear, we obtain the absolute barycentric coordinates of G with respect to the triangle $A_1 A_2 A_3$ (see Sect. 4.7.1):

$$\mu_{z_1} = \frac{m_1}{m}, \ \mu_{z_2} = \frac{m_2}{m}, \ \mu_{z_3} = \frac{m_3}{m}.$$

Theorem 1 (Lagrange[10]). *Consider the points* $A_1, \ldots, A_n$ *and the nonzero real numbers* $m_1, \ldots, m_n$ *such that* $m = m_1 + \cdots + m_n \neq 0$. *If* G *denotes the barycenter of the set* $\{A_1, \ldots, A_n\}$ *with respect to the weights* $m1, \ldots, m_n$, *then for every point* M *in the plane, the following relation holds:*

$$\sum_{j=1}^{n} m_j M A_j^2 = mMG^2 + \sum_{j=1}^{n} m_j GA_j^2. \tag{1}$$

Proof. Without loss of generality, we can assume that the barycenter G is the origin of the complex plane; that is, $z_G = 0$.

Using properties of the real product, we obtain for all $j = 1, \ldots, n$, the relations

$$MA_j^2 = |z_M - z_j|^2 = (z_M - z_j) \cdot (z_M - z_j)$$
$$= |z_M|^2 - 2z_M \cdot z_j + |z|^2,$$

i.e.,

$$MA_j^2 = |z_M|^2 - 2z_M \cdot z_j + |z_j|^2.$$

Multiplying by m_j and adding the relations obtained for $j = 1, \ldots, n$ yields

$$\sum_{j=1}^{n} m_j M A_j^2 = \sum_{j=1}^{n} m_j(|z_M|^2 - 2z_M \cdot z_j + |z_j|^2)$$

$$= m|z_M|^2 - 2z_M \cdot \left(\sum_{j=1}^{n} m_j z_j\right) + \sum_{j=1}^{n} m_j|z_j|^2$$

---

[10] Joseph Louis Lagrange (1736–1813), French mathematician, one of the greatest mathematicians of the eighteenth century. He made important contributions in all branches of mathematics, and his results have greatly influenced modern science.

$$= m|z_M|^2 - 2z_M \cdot (mz_G) + \sum_{j=1}^{n} m_j |z_j|^2$$

$$= m|z_M|^2 + \sum_{j=1}^{n} m_j |z_j|^2 = m|z_M - z_G|^2 + \sum_{j=1}^{n} m_j |z_j - z_G|^2$$

$$= mMG^2 + \sum_{j=1}^{n} m_j GA_j^2. \qquad \qquad \square$$

Corollary 1. *Consider the distinct points A_1, ..., A_n and the nonzero real numbers m_1, ..., m_n such that $m_1 + \cdots + m_n \neq 0$. The following inequality holds for every point M in the plane:*

$$\sum_{j=1}^{n} m_j MA_j^2 \geq \sum_{j=1}^{n} m_j GA_j^2, \qquad \qquad (2)$$

with equality if and only if $M = G$, the barycenter of set $\{A_1, ..., A_n\}$ with respect to the weights m_1, ..., m_n.

Proof. The inequality (2) follows directly from Lagrange's relation (1). $\quad\square$

If $m_1 = \cdots = m_n = 1$, then from Theorem 1 above, one obtains the following corollary.

Corollary 2 (Leibniz[11]). *Consider the distinct points A_1, ..., A_n and the centroid G of the set $\{A_1, ..., A_n\}$. The following relation holds for every point M in the plane:*

$$\sum_{j=1}^{n} MA_j^2 = nMG^2 + \sum_{j=1}^{n} GA_j^2. \qquad \qquad (3)$$

Remark. The relation (3) is equivalent to the following identity: For all complex numbers z, z_1, ..., z_n, we have

$$\sum_{j=1}^{n} |z - z_j|^2 = n \left| z - \frac{z_1 + \cdots + z_n}{n} \right|^2 + \sum_{j=1}^{n} \left| z_j - \frac{z_1 + \cdots + z_n}{n} \right|^2.$$

Applications. We will use formula (3) in determining some important distances in a triangle. Let us consider the triangle ABC and let us take $n = 3$ in the formula (3). We find that the following formula holds for every point M in the plane of triangle ABC:

$$MA^2 + MB^2 + MC^2 = 3MG^2 + GA^2 + GB^2 + GC^2, \qquad (4)$$

---

[11] Gottfried Wilhelm Leibniz (1646–1716) was a German philosopher, mathematician, and logician who is probably best known for having invented the differential and integral calculus independently of Sir Isaac Newton.

where G is the centroid of triangle ABC. Assume that the circumcenter O of the triangle ABC is the origin of the complex plane.

(1) In the relation (4) we choose $M = O$, and we get

$$3R^2 = 3OG^2 + GA^2 + GB^2 + GC^2.$$

Applying the well-known median formula yields

$$GA^2 + GB^2 + GC^2 = \frac{4}{9}(m_\alpha^2 + m_\beta^2 + m_\gamma^2)$$
$$= \frac{4}{9}\sum_{\text{cyc}}\frac{1}{4}[2(\beta^2 + \gamma^2) - \alpha^2] = \frac{1}{3}(\alpha^2 + \beta^2 + \gamma^2),$$

where α, β, γ are the sides of triangle ABC. We obtain

$$OG^2 = R^2 - \frac{1}{9}(\alpha^2 + \beta^2 + \gamma^2). \tag{5}$$

An equivalent form of the distance OG is given in terms of the basic invariants of a triangle in Corollary 1, Sect. 4.6.4.

(2) Using the collinearity of points O, G, H and the relation $OH = 3OG$ (see Theorem 1 in Sect. 3.1), it follows that

$$OH^2 = 9OG^2 = 9R^2 - (\alpha^2 + \beta^2 + \gamma^2). \tag{6}$$

An equivalent form for the distance OH was obtained in terms of the fundamental invariants of the triangle in the theorem of Sect. 4.6.4.

(3) In (4), consider $M = I$, the incenter of triangle ABC (Fig. 4.16). We obtain

$$IA^2 + IB^2 + IC^2 = 3IG^2 + \frac{1}{3}(\alpha^2 + \beta^2 + \gamma^2).$$

On the other hand, we have the following relations:

$$IA = \frac{r}{\sin\frac{A}{2}}, \quad IB = \frac{r}{\sin\frac{B}{2}}, \quad IC = \frac{r}{\sin\frac{C}{2}},$$

where r is the inradius of triangle ABC. It follows that

$$IG^2 = \frac{1}{3}\left[r^2\left(\frac{1}{\sin^2\frac{A}{2}} + \frac{1}{\sin^2\frac{B}{2}} + \frac{1}{\sin^2\frac{C}{2}}\right) - \frac{1}{3}(\alpha^2 + \beta^2 + \gamma^2)\right].$$

Taking into account the well-known formula

$$\sin^2\frac{A}{2} = \frac{(s-\beta)(s-\gamma)}{\beta\gamma},$$

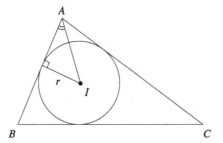

Figure 4.16.

we obtain

$$\sum_{\text{cyc}} \frac{1}{\sin^2 \frac{A}{2}} = \sum_{\text{cyc}} \frac{\beta\gamma}{(s-\beta)(s-\gamma)} = \sum_{\text{cyc}} \frac{\beta\gamma(s-\alpha)}{(s-\alpha)(s-\beta)(s-\gamma)}$$

$$= \frac{s}{K^2} \sum_{\text{cyc}} \beta\gamma(s-\alpha) = \frac{s}{K^2} \left[s \sum \beta\gamma - 3\alpha\beta\gamma \right]$$

$$= \frac{s}{K^2}[s(s^2 + r^2 + 4Rr) - 12sRr] = \frac{1}{r^2}(s^2 + r^2 - 8Rr),$$

where we have used the formulas in Sect. 4.6.1. Therefore,

$$IG^2 = \frac{1}{3}\left[s^2 + r^2 - 8Rr - \frac{1}{3}(\alpha^2 + \beta^2 + \gamma^2) \right]$$

$$= \frac{1}{3}\left[s^2 + r^2 - 8Rr - \frac{2}{3}(s^2 - r^2 - 4Rr) \right] = \frac{1}{9}(s^2 + 5r^2 - 16Rr),$$

where the first formula in Corollary 1 in this section was used. That is,

$$IG^2 = \frac{1}{9}(s^2 + 5r^2 - 16Rr), \tag{7}$$

and hence we obtain again the formula in Application 1 of Sect. 4.7.2.

Problem. *Let z_1, z_2, z_3 be distinct complex numbers having modulus R. Prove that*

$$\frac{9R^2 - |z_1 + z_2 + z_3|^2}{|z_1 - z_2| \cdot |z_2 - z_3| \cdot |z_3 - z_1|} \geq \frac{\sqrt{3}}{R}.$$

Solution. Let A, B, C be the geometric images of the complex numbers z_1, z_2, z_3 and let G be the centroid of the triangle ABC.

The coordinate of G is equal to $\dfrac{z_1 + z_2 + z_3}{3}$, and $|z_1 - z_2| = \gamma$, $|z_2 - z_3| = \alpha$, $|z_3 - z_1| = \beta$.

The inequality becomes

$$\frac{9R^2 - 9OG^2}{\alpha\beta\gamma} \geq \frac{\sqrt{3}}{R}. \tag{1}$$

Using the formula

$$OG^2 = R^2 - \frac{1}{9}(\alpha^2 + \beta^2 + \gamma^2),$$

we see that (1) is equivalent to

$$\alpha^2 + \beta^2 + \gamma^2 \geq \frac{\alpha\beta\gamma\sqrt{3}}{R} = \frac{4RK}{R}\sqrt{3} = 4K\sqrt{3}.$$

Here is a proof of this famous inequality using Heron's formula and the arithmetic–geometric mean (AM-GM) inequality:

$$K = \sqrt{s(s-\alpha)(s-\beta)(s-\gamma)} \leq \sqrt{s\frac{(s-\alpha+s-\beta+s-\gamma)^3}{27}} = \sqrt{s\frac{s^3}{27}}$$

$$= \frac{s^2}{3\sqrt{3}} = \frac{(\alpha+\beta+\gamma)^2}{12\sqrt{3}} \leq \frac{3(\alpha^2+\beta^2+\gamma^2)}{12\sqrt{3}} = \frac{\alpha^2+\beta^2+\gamma^2}{4\sqrt{3}}.$$

We now extend Leibniz's relation in Corollary 2 above. First, we need the following result.

Theorem 2. Let $n \geq 2$ be a positive integer. Consider the distinct points $A_1, \ldots, A_n$, and let G be the centroid of the set $\{A_1, \ldots, A_n\}$. Then the following formula holds for every point in the plane:

$$n^2 MG^2 = n\sum_{j=1}^{n} MA_j^2 - \sum_{1 \leq i < k \leq n} A_i A_k^2. \tag{8}$$

Proof. We assume that the barycenter G is the origin of the complex plane. Using properties of the real product, we have

$$MA_j^2 = |z_M - z_j|^2 = (z_M - z_j) \cdot (z_M - z_j) = |z_M|^2 - 2z_M \cdot z_j + |z_j|^2$$

and

$$A_i A_k^2 = |z_i - z_k|^2 = |z_i|^2 - 2z_i \cdot z_k + |z_k|^2,$$

where the complex number z_j is the coordinate of the point A_j, $j = 1, 2, \ldots, n$.

The relation (8) is equivalent to

$$n^2|z_M|^2 = n\sum_{j=1}^{n}(|z_M|^2 - 2z_M \cdot z_j + |z_j|^2) - \sum_{1 \leq i < k \leq n}|(|z_i|^2 - 2z_i \cdot z_k + |z_k|^2).$$

That is,

$$n \sum_{j=1}^{n} |z_j|^2 = 2n \sum_{j=1}^{n} z_M \cdot z_j + \sum_{1 \le i < k \le n} (|z_i|^2 - 2z_i z_k + |z_k|^2).$$

Taking into account the hypothesis that G is the origin of the complex plane, we have

$$\sum_{j=1}^{n} z_M \cdot z_j = z_M \cdot \left(\sum_{j=1}^{n} z_j \right) = n(z_M \cdot z_G) = n(z_M \cdot 0) = 0.$$

Hence, the relation (8) is equivalent to

$$\sum_{j=1}^{n} |z_j|^2 = -2 \sum_{1 \le i < k \le n} z_i \cdot z_k.$$

The last relation can be obtained as follows:

$$0 = |z_G|^2 = z_G \cdot z_G = \frac{1}{n^2} \left(\sum_{i=1}^{n} z_i \right) \cdot \left(\sum_{k=1}^{n} z_k \right)$$

$$= \frac{1}{n^2} \cdot \left(\sum_{j=1}^{n} |z_j|^2 + 2 \sum_{1 \le i < k \le n} z_i \cdot z_k \right).$$

Therefore the relation (8) is proved. □

Remark. The formula (8) is equivalent to the following identity: for all complex numbers $z, z_1, \ldots, z_n$, we have

$$\frac{1}{n} \sum_{j=1}^{n} |z - z_j|^2 - \left| z - \frac{z_1 + \cdots + z_n}{n} \right|^2 = \frac{1}{n} \sum_{1 \le i < k \le n} |z_i - z_k|^2.$$

Applications

(1) If $A_1, \ldots, A_n$ are points on the circle with center O and radius R, then if we take $M = O$ in (8), it follows that

$$\sum_{1 \le i < k \le n} A_i A_k^2 = n^2 (R^2 - OG^2).$$

If $n = 3$, we obtain the formula (5).

(2) The following inequality holds for every point M in the plane:

$$\sum_{j=1}^{n} MA_j^2 \ge \frac{1}{n} \sum_{1 \le i < k \le n} A_i A_k^2,$$

with equality if and only if $M = G$, the centroid of the set $\{A_1, \ldots, A_n\}$.

Let $n \geq 2$ be a positive integer, and let k be an integer such that $2 \leq k \leq n$. Consider the distinct points $A_1, \ldots, A_n$ and let G be the centroid of the set $\{A_1, \ldots, A_n\}$. For indices $i_1 < \cdots < i_k$, let us denote by $G_{i_1, \ldots, i_k}$ the centroid of the set $\{A_{i_1}, \ldots, A_{i_k}\}$. We have the following result:

Theorem 3. *For every point M in the plane,*

$$(n-k) \binom{n}{k} \sum_{j=1}^{n} MA_j^2 + n^2(k-1) \binom{n}{k} MG^2$$

$$= kn(n-1) \sum_{1 \leq i_1 < \cdots < i_k \leq n} MG_{i_1 \cdots i_k}^2. \tag{9}$$

Proof. It is not difficult to see that the barycenter of the set $\{G_{i_1 \cdots i_k} : 1 \leq i_1 < \cdots < i_k \leq n\}$ is G. Applying Leibniz's relation, one obtains

$$\sum_{j=1}^{n} MA_j^2 = nMG^2 + \sum_{j=1}^{n} GA_j^2, \tag{10}$$

$$\sum_{1 \leq i_1 < \cdots < i_k \leq n} MG_{i_1 \cdots i_k}^2 = \binom{n}{k} MG^2 + \sum_{1 \leq i_1 < \cdots < i_k \leq n} GG_{i_1 \cdots i_k}^2, \tag{11}$$

$$\sum_{s=1}^{k} MA_{i_s}^2 = kMG_{i_1 \cdots i_k}^2 + \sum_{s=1}^{k} G_{i_1 \cdots i_k} A_{i_s}^2. \tag{12}$$

Considering in (12) $M = G$ and adding all these relations yields

$$\sum_{1 \leq i_1 < \cdots < i_k \leq n} \sum_{s=1}^{k} GA_{i_s}^2 = k \sum_{1 \leq i_1 < \cdots < i_k \leq n} GG_{i_1 \cdots i_k}^2$$

$$+ \sum_{1 \leq i_1 < \cdots < i_k \leq n} \sum_{s=1}^{k} G_{i_1 \cdots i_k} A_{i_s}^2. \tag{13}$$

Applying formula (8) in Theorem 3 above to the sets $\{A_1, \ldots, A_n\}$ and $\{A_{i_1}, \ldots, A_{i_k}\}$, respectively, we get

$$n^2 MG^2 = n \sum_{j=1}^{n} MA_j^2 - \sum_{1 \leq i < k \leq n} A_i A_k^2, \tag{14}$$

$$k^2 MG_{i_1 \cdots i_k}^2 = k \sum_{s=1}^{k} MA_{i_s}^2 - \sum_{1 \leq p < q \leq k} A_{i_p} A_{i_q}^2. \tag{15}$$

Taking $M = G_{i_1 \cdots i_k}$ in (15) yields

$$\sum_{s=1}^{k} G_{i_1 \cdots i_k} A_{i_s}^2 = \frac{1}{k} \sum_{1 \leq p < q \leq k} A_{i_p} A_{i_q}^2. \tag{16}$$

From (16) and (13), we obtain

$$\sum_{1\le i_1<\cdots<i_k\le n}\sum_{s=1}^{k}GA_{i_s}^2 = k\sum_{1\le i_1<\cdots<i_k\le n}GG_{i_1\cdots i_k}^2$$
$$+\frac{1}{k}\sum_{1\le i_1<\cdots<i_k\le n}\sum_{1\le p<q\le n}A_{i_p}A_{i_q}^2. \tag{17}$$

If we rearrange the terms in formula (17), we get

$$\frac{\binom{k}{1}\binom{n}{k}}{\binom{n}{1}}\sum_{j=1}^{n}GA_j^2 = k\sum_{1\le i_1<\cdots<i_k\le n}GG_{i_1\cdots i_k}^2+\frac{1}{k}\frac{\binom{k}{2}\binom{n}{k}}{\binom{n}{2}}\sum_{1\le i<k\le n}A_iA_j^2.$$

$$\tag{18}$$

From relations (10), (11), (14), and (18), we readily derive formula (9). $\square$

Remark. The relation (9) is equivalent to the following identity: for all complex numbers z, z_1, ..., z_n, we have

$$(n-k)\binom{n}{k}\sum_{j=1}^{n}|z-z_j|^2 + n^2(k-1)\binom{n}{k}\left|z-\frac{z_1+\cdots+z_n}{n}\right|^2$$
$$= kn(n-1)\sum_{1\le i_1<\cdots<i_k\le n}\left|z-\frac{z_{i_1}+\cdots+z_{i_k}}{k}\right|^2.$$

Applications

(1) In the case $k=2$, from (9) we obtain that the following relation holds for every point M in the plane:

$$(n-2)\sum_{j=1}^{n}MA_j^2 + n^2MG^2 = 4\sum_{1\le i_1<i_2\le n}MG_{i_1i_2}^2.$$

In this case, $G_{i_1i_2}$ is the midpoint of the segment $[A_{i_1}A_{i_2}]$.
(2) If $k=3$, from (9) we get that the relation

$$(n-3)(n-2)\sum_{j=1}^{n}MA_j^2 + 2n^2(n-2)MG^2 = 18\sum_{1\le i_1<i_2<i_3\le n}MG_{i_1i_2i_3}^2$$

holds for every point M in the plane. Here the point $G_{i_1i_2i_3}$ is the centroid of triangle $A_{i_1}A_{i_2}A_{i_3}$.

4.12 Euler's Center of an Inscribed Polygon

Consider a polygon $A_1 A_2 \cdots A_n$ inscribed in a circle centered at the origin of the complex plane and let a_1, a_2, $\ldots$, a_n be the coordinates of its vertices.

By definition, the point E with coordinate

$$z_E = \frac{a_1 + a_2 + \cdots + a_n}{2}$$

is called *Euler's center* of the polygon $A_1 A_2 \cdots A_n$. In the case $n = 3$, it is clear that E is equal to O_9, the center of Euler's nine-point circle.

Remarks.

(a) Let $G(z_G)$ and $H(z_H)$ be the centroid and orthocenter of the inscribed polygon $A_1 A_2 \cdots A_n$. Then

$$z_E = \frac{n z_G}{2} = \frac{z_H}{2} \text{ and } OE = \frac{nOG}{2} = \frac{OH}{2}.$$

Recall that the orthocenter of the polygon $A_1 A_2 \cdots A_n$ is the point H with coordinate $z_H = a_1 + a_2 + \cdots + a_n$.

(b) For $n = 4$, point E is also called *Mathot's point* of the inscribed quadrilateral $A_1 A_2 A_3 A_4$.

Proposition. *In the above notation, the following relation holds:*

$$\sum_{i=1}^{n} EA_i^2 = nR^2 + (n-4)EO^2. \tag{1}$$

Proof. Using the identity (8) in Theorem 4, Sect. 2.17 for $M = E$ and $M = O$, namely

$$n^2 \cdot MG^2 = n \sum_{i=1}^{n} MA_i^2 - \sum_{1 \le i < j \le n} A_i A_j^2,$$

we obtain

$$n^2 \cdot EG^2 = n \sum_{i=1}^{n} EA_i^2 - \sum_{1 \le i < j \le n} A_i A_j^2 \tag{2}$$

and

$$n^2 \cdot OG^2 = nR^2 - \sum_{1 \le i < j \le n} A_i A_j^2. \tag{3}$$

Setting $s = \sum_{i=1}^{n} a_i$, we have

$$EG = |z_E - z_G| = \left| \frac{s}{2} - \frac{s}{n} \right| = \left| \frac{s}{2} \right| \cdot \frac{n-2}{n} = \frac{n-2}{n} \cdot OE. \tag{4}$$

From the relations (2), (3), and (4), we derive that

$$n \sum_{i=1}^{n} EA_i^2 = n^2 \cdot EG^2 - n^2 \cdot OG^2 + n^2 R^2$$

$$= (n-2)^2 OE^2 - 4OE^2 + n^2 R^2 = n(n-4) \cdot EO^2 + n^2 R^2,$$

or equivalently,

$$\sum_{i=1}^{n} EA_i^2 = nR^2 + (n-4)EO^2,$$

as desired. $\square$

Applications

(1) For $n = 3$, from relation (1), we obtain

$$O_9 A_1^2 + O_9 A_2^2 + O_9 A_3^2 = 3R^2 - OO_9^2. \tag{5}$$

Using the formula in Corollary 1 in Sect. 4.6.4, we can express the right-hand side in (5) in terms of the fundamental invariants of triangle $A_1 A_2 A_3$:

$$O_9 A_1^2 + O_9 A_2^2 + O_9 A_3^2 = \frac{3}{4}R^2 - \frac{1}{2}r^2 - 2Rr + \frac{1}{2}s^2. \tag{6}$$

From formula (5), it follows that the following inequality holds for every triangle $A_1 A_2 A_3$:

$$O_9 A_1^2 + O_9 A_2^2 + O_9 A_3^2 \leq 3R^2, \tag{7}$$

with equality if and only if the triangle is equilateral.

(2) For $n = 4$, we obtain the interesting relation

$$\sum_{i=1}^{4} EA_i^2 = 4R^2. \tag{8}$$

The point E is the unique point in the plane of the quadrilateral $A_1 A_2 A_3 A_4$ satisfying relation (8).

(3) For $n > 4$, from relation (1), the inequality

$$\sum_{i=1}^{n} EA_i^2 \geq nR^2 \tag{9}$$

follows. Equality holds only in the polygon $A_1 A_2 \cdots A_n$ with the property $E = O$.

(4) The Cauchy–Schwarz inequality and inequality (7) give

$$\left(\sum_{i=1}^{3} R \cdot O_9 A_i\right)^2 \leq (3R^2) \sum_{i=1}^{3} O_9 A_i^2 \leq 9R^2.$$

This is equivalent to

$$O_9 A_1 + O_9 A_2 + O_9 A_3 \leq 3R. \tag{10}$$

(5) Using the same inequality and the relation (8), we have

$$\left(R\sum_{i=1}^{4} EA_i\right)^2 \leq 4R^2 \cdot \sum_{i=1}^{4} EA_i = 16R^4,$$

or equivalently,

$$\sum_{i=1}^{4} EA_i \leq 4R. \tag{11}$$

(6) Using the relation

$$2EA_i = 2|e - a_i| = 2\left|\frac{s}{2} - a_i\right| = |s - 2a_i|,$$

the inequalities (4), (5) become respectively

$$\sum_{\text{cyc}} |-a_1 + a_2 + a_3| \leq 6R$$

and

$$\sum_{\text{cyc}} |-a_1 + a_2 + a_3 + a_4| \leq 8R.$$

The above inequalities hold for all complex numbers of the same modulus R.

4.13 Some Geometric Transformations of the Complex Plane

4.13.1 Translation

Let z_0 be a fixed complex number and let t_{z_0} be the mapping defined by

$$t_{z_0} : \mathbb{C} \to \mathbb{C}, \ t_{z_0}(z) = z + z_0.$$

The mapping t_{z_0} is called the *translation* of the complex plane by complex number z_0.

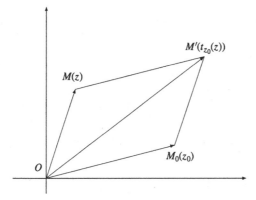

Figure 4.17.

Taking into account the geometric interpretation of the addition of two complex numbers (see Sect. 1.2.3), we have Fig. 4.17, giving the geometric image of $t_{z_0}(z)$.

In Fig. 4.17, $OM_0M'M$ is a parallelogram and OM' is one of its diagonals. Therefore, the mapping t_{z_0} corresponds in the complex plane $\mathbb{C}$ to the translation $t_{\overrightarrow{OM_0}}$ by the vector $\overrightarrow{OM_0}$ in the case of the Euclidean plane.

It is clear that the composition of two translations t_{z_1} and t_{z_2} satisfies the relation

$$t_{z_1} \circ t_{z_2} = t_{z_1+z_2}.$$

It is also clear that the set $\mathcal{T}$ of all translations of the complex plane is a group with respect to the composition of mappings. The group $(\mathcal{T}, \circ)$ is abelian, and its unit is $t_O = 1_{\mathbb{C}}$, translation by the complex number 0.

4.13.2 Reflection in the Real Axis

Consider the mapping $s : \mathbb{C} \to \mathbb{C}$, $s(z) = \overline{z}$. If M is the point with coordinate z, then the point $M'(s(z))$ is obtained by reflecting M across the real axis (see Fig. 4.18). The mapping s is called the *reflection in the real axis*. It is clear that $s \circ s = 1_{\mathbb{C}}$.

4.13.3 Reflection in a Point

Consider the mapping $s_0 : \mathbb{C} \to \mathbb{C}$, $s_0(z) = -z$. Since $s_0(z)+z = 0$, the origin O is the midpoint of the segment $[M(z)M'(z)]$; hence M' is the reflection of point M across O (Fig. 4.19).

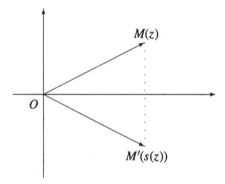

Figure 4.18.

The mapping s_0 is called the *reflection in the origin*.

Consider a fixed complex number z_0 and the mapping

$$s_{z_0} : \mathbb{C} \to \mathbb{C}, \ s_{z_0}(z) = 2z_0 - z.$$

If z_0, z, $s_{z_0}(z)$ are the coordinates of points M_0, M, M', then M_0 is the midpoint of the segment $[MM']$. Hence M' is the reflection of M in M_0 (Fig. 4.20).

The mapping s_{z_0} is called the *reflection in the point* $M_0(z_0)$. It is clear that the following relation holds: $s_{z_0} \circ s_{z_0} = 1_{\mathbb{C}}$.

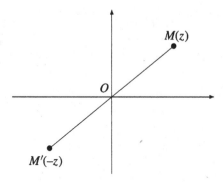

Figure 4.19.

4.13.4 Rotation

Let $a = \cos t_0 + i \sin t_0$ be a complex number having modulus 1 and let r_a be the mapping given by $r_a : \mathbb{C} \to \mathbb{C}, \ r_a(z) = az$. If $z = \rho(\cos t + i \sin t)$, then

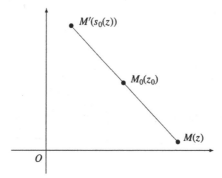

Figure 4.20.

$$r_a(z) = az = \rho[\cos(t + t_0) + i\sin(t + t_0)],$$

and hence $M'(r_a(z))$ is obtained by rotating point $M(z)$ about the origin through the angle t_0 (Fig. 4.21).

The mapping r_a is called the *rotation* with center O and angle $t_0 = \arg a$.

4.13.5 Isometric Transformation of the Complex Plane

A mapping $f : \mathbb{C} \to \mathbb{C}$ is called an *isometry* if it preserves distance, i.e., for all z_1, $z_2 \in \mathbb{C}$, $|f(z_1) - f(z_2)| = |z_1 - z_2|$.

Theorem 1. *Translations, reflections (in the real axis or in a point), and rotations about center O are isometries of the complex plane.*

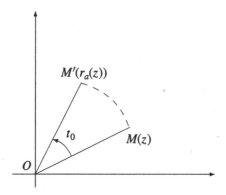

Figure 4.21.

Proof. For the translation t_{z_0}, we have

$$|t_{z_0}(z_1) - t_{z_0}(z_2)| = |(z_1 + z_0) - (z_2 + z_0)| = |z_1 - z_2|.$$

For the reflection s across the real axis, we obtain

$$|s(z_1) - s(z_2)| = |\overline{z_1} - \overline{z_2}| = |\overline{z_1 - z_2}| = |z_1 - z_2|,$$

and the same goes for the reflection in a point. Finally, if r_a is a rotation, then

$$|r_a(z_1) - r_a(z_2)| = |az_1 - az_2| = |a||z_1 - z_2| = |z_1 - z_2|, \text{ since } |a| = 1. \quad \square$$

We can easily check that the composition of two isometries is also an isometry. The set $\text{Iso}(\mathbb{C})$ of all isometries of the complex plane is a group with respect to the composition of mappings, and $(\mathcal{T}, \circ)$ is a subgroup of that group.

Problem. *Let $A_1A_2A_3A_4$ be a cyclic quadrilateral inscribed in a circle with center O, and let H_1, H_2, H_3, H_4 be the orthocenters of triangles $A_2A_3A_4$, $A_1A_3A_4$, $A_1A_2A_4$, $A_1A_2A_3$, respectively.*
Prove that quadrilaterals $A_1A_2A_3A_4$ and $H_1H_2H_3H_4$ are congruent.

(Balkan Mathematical Olympiad, 1984)

Solution. Consider the complex plane with origin at the circumcenter, and denote by the corresponding lowercase letter the coordinates of a point denoted by an uppercase letter.

If $s = a_1 + a_2 + a_3 + a_4$, then $h_1 = a_2 + a_3 + a_4 = s - a_1$, $h_2 = s - a_2$, $h_3 = s - a_3$, $h_4 = s - a_4$. Hence the quadrilateral $H_1H_2H_3H_4$ is the reflection of quadrilateral $A_1A_2A_3A_4$ across the point with coordinate $\frac{s}{2}$.

The following result describes all isometries of the complex plane.

Theorem 2. *Every isometry of the complex plane is a mapping $f : \mathbb{C} \to \mathbb{C}$ with $f(z) = az + b$ or $f(z) = a\overline{z} + b$, where a, $b \in \mathbb{C}$ and $|a| = 1$.*

Proof. Let $b = f(0)$, $c = f(1)$, and $a = c - b$. Then

$$|a| = |c - b| = |f(1) - f(0)| = |1 - 0| = 1.$$

Consider the mapping $g : \mathbb{C} \to \mathbb{C}$, given by $g(z) = az + b$. It is not difficult to prove that g is an isometry, with $g(0) = b = f(0)$ and $g(1) = a + b = c = f(1)$. Hence $h = g^{-1}$ is an isometry, with 0 and 1 as fixed points. By definition, it follows that every real number is a fixed point of h, and hence $h = 1_{\mathbb{C}}$ or $h = s$, the reflection in the real axis. Hence $g = f$ or $g = f \circ s$, and the proof is complete. $\square$

The above result shows that every isometry of the complex plane is the composition of a rotation and a translation or the composition of a rotation with a reflection in the origin O and a translation.

4.13.6 Morley's Theorem

In 1899, Frank Morley, then professor of mathematics at Haverford College, came across a result so surprising that it entered mathematical folklore under the name "Morley's Miracle." Morley's marvelous theorem states that *the three points of intersection of the adjacent trisectors of the angles of any triangle form an equilateral triangle.*

The theorem was mistakenly attributed to Napoleon Bonaparte, who made some contributions to geometry.

There are various proofs of this nice result, such as those by J. Conway, D.J. Newman, L. Bankoff, and N. Dergiades.

Here we present a new proof published in 1998, by Alain Connes. His proof is derived from the following result:

Theorem 1 (Alain Connes). *Consider the transformations* $f_i : \mathbb{C} \to \mathbb{C}$, $f_i(z) = a_i z + b_i$, $i = 1, 2, 3$, *of the complex plane, where all coefficients* a_i *are different from zero. Assume that the mappings* $f_1 \circ f_2$, $f_2 \circ f_3$, $f_3 \circ f_1$, *and* $f_1 \circ f_2 \circ f_3$ *are not translations, equivalently, that* $a_1 a_2$, $a_2 a_3, a_3 a_1$, $a_1 a_2 a_3 \in \mathbb{C} \backslash \{1\}$. *Then the following statements are equivalent:*

(1) $f_1^3 \circ f_2^3 \circ f_3^3 = 1_{\mathbb{C}}$.
(2) $j^3 = 1$ *and* $\alpha + j\beta + j^2\gamma = 0$, *where* $j = a_1 a_2 a_3 \neq 1$ *and* α, β, γ *are the respective unique fixed points of the mappings* $f_1 \circ f_2$, $f_2 \circ f_3$, $f_3 \circ f_1$.

Proof. Note that $(f_1 \circ f_2)(z) = a_1 a_2 z + a_1 b_2 + b_1$, $a_1 a_2 \neq 1$,

$$(f_2 \circ f_3)(z) = a_2 a_3 z + a_2 b_3 + b_2, \ a_2 a_3 \neq 1,$$
$$(f_3 \circ f_1)(z) = a_3 a_1 z + a_3 b_1 + b_3, \ a_3 a_1 \neq 1,$$
$$\text{Fix } (f_1 \circ f_2) = \left\{ \frac{a_1 b_2 + b_1}{1 - a_1 a_2} \right\} = \left\{ \frac{a_1 a_3 b_2 + a_3 b_1}{a_3 - j} =: \alpha \right\},$$
$$\text{Fix } (f_2 \circ f_3) = \left\{ \frac{a_2 b_3 + b_2}{1 - a_2 a_3} \right\} = \left\{ \frac{a_1 a_2 b_3 + a_1 b_2}{a_1 - j} =: \beta \right\},$$
$$\text{Fix } (f_3 \circ f_1) = \left\{ \frac{a_3 b_1 + b_3}{1 - a_3 a_1} \right\} = \left\{ \frac{a_2 a_3 b_1 + a_2 b_3}{a_2 - j} =: \gamma \right\},$$

where Fix(f) denotes the set of fixed points of the mapping f.

For the cubes of f_1, f_2, f_3, we have the formulas

$$f_1^3(z) = a_1^3 z + b_1(a_1^2 + a_1 + 1),$$
$$f_2^3(z) = a_2^3 z + b_2(a_2^2 + a_2 + 1),$$
$$f_3^3(z) = a_3^3 + b_3(a_3^2 + a_3 + 1),$$

whence

$$(f_1^3 \circ f_2^3 \circ f_3^3)(z) = a_1^3 a_2^3 a_3^3 z + a_1^3 a_2^3 b_3(a_3^2 + a_3 + 1)$$
$$+ a_1^3 b_2(a_2^2 + a_2 + 1) + b_1(a_1^2 + a_1 + 1).$$

Therefore, $f_1^3 \circ f_2^3 \circ f_3^3 = id_C$ if and only if $a_1^3 a_2^3 a_3^3 = 1$ and

$$a_1^3 a_2^3 b_3 (a_3^2 + a_3 + 1) + a_1^3 b_2 (a_2^2 + a_2 + 1) + b_1 (a_1^2 + a_1 + 1) = 0.$$

To prove the equivalence of statements (1) and (2) we have to show that $\alpha + j\beta + j^2\gamma$ is different from the free term of $f_1^3 \circ f_2^3 \circ f_3^3$ by a multiplicative constant. Indeed, using the relation $j^3 = 1$ and implicitly $j^2 + j + 1 = 0$, we have successively

$$\alpha + j\beta + j^2\gamma = \alpha + j\beta + (-1-j)\gamma = \alpha - \gamma + j(\beta - \gamma)$$

$$= \frac{a_1 a_3 b_2 + a_3 b_1}{a_3 - j} - \frac{a_2 a_3 b_1 + a_2 b_3}{a_2 - j} + j\left(\frac{a_1 a_2 b_3 + a_1 b_2}{a_1 - j} - \frac{a_2 a_3 b_1 + a_2 b_3}{a_2 - j}\right)$$

$$= \frac{a_1 a_2 a_3 b_2 + a_2 a_3 b_1 - a_1 a_3 b_2 j - a_3 b_1 j - a_2 a_3^2 b_1 - a_2 a_3 b_3 + a_2 a_3 b_1 j + a_2 b_3 j}{(a_2 - j)(a_3 - j)}$$

$$+ j\frac{a_1 a_2^2 b_3 + a_1 a_2 b_2 - a_1 a_2 b_3 j - a_1 b_2 j - a_1 a_2 a_3 b_1 - a_1 a_2 b_3 + a_2 a_3 b_1 j + a_2 b_3 j}{(a_1 - j)(a_2 - j)}$$

$$= \frac{1}{a_2 - j}\left(\frac{b_2 j - a_2 a_3 b_1 j^2 - a_1 a_3 b_2 j - a_3 b_1 j - a_2 a_3^2 b_1 - a_2 a_3 b_3 + a_2 b_3 j}{a_3 - j}\right.$$

$$\left. + \frac{a_1 a_2^2 b_3 j + a_1 a_2 b_2 j + a_1 a_2 b_3 - a_1 b_2 j^2 - b_1 j^2 + a_2 a_3 b_1 j^2 + a_2 b_3 j^2}{a_1 - j}\right)$$

$$= \frac{1}{(a_1 - j)(a_2 - j)(a_3 - j)}(a_1 b_2 j - b_1 - a_1^2 a_3 b_2 j - a_1 a_3 b_1 j - a_1 a_2 a_3^2 b_1 - b_3 j$$

$$+ a_1 a_2 b_3 j - b_2 j^2 + a_2 a_3 b_1 + a_1 a_3 b_2 j^2 + a_3 b_1 j^2 + a_2 a_3^2 b_1 j + a_2 a_3 b_3 j - a_2 b_3 j^2$$

$$+ a_2 b_3 j^2 + b_2 j^2 + b_3 j - a_1 a_3 b_2 j^2 - a_3 b_1 j^2 + a_2 a_3 b_1 j^2 + a_2 a_3 b_3 j^2$$

$$- a_1 a_2^2 b_3 j^2 - a_1 a_2 b_2 j^2 - a_1 a_2 b_3 j + a_1 b_2 + b_1 - a_2 a_3 b_1 - a_2 b_3)$$

$$= \frac{1}{(a_1 - j)(a_2 - j)(a_3 - j)}(-a_1 b_2 j^2 - a_1^2 a_3 b_2 j - a_1 a_3 b_1 j - a_3 b_1 j$$

$$- a_2 a_3^2 b_1 - a_2 a_3 b_3 - a a_2^2 b_3 j^2 - a_1 a_2 b_2 j^2 - a_2 b_3)$$

$$= -\frac{1}{(a_1 - j)(a_2 - j)(a_3 - j)}(a_1^2 a_2^2 a_3^2 b_2 + a_1^3 a_2 a_3^2 b_2$$

$$+ a_1^2 a_2 a_3^2 b_1 + a_1 a_2 a_3^2 b_1 + a_2 a_3^2 b_1 + a_2 a_3 b_3 + a_1^3 a_2^4 a_3^3 b_3 + a_1^3 a_2^3 a_3^2 b_2 + a_2 b_3)$$

$$= -\frac{1}{(a_1 - j)(a_2 - j)(a_3 - j)}[a_2 a_3^2 b_1 (1 + a_1 + a_1^2) + a_1^3 a_2 a_3^2 b_2 (1 + a_2 + a_2^2)$$

$$+ a_2 b_3 (1 + a_3 + a_1^3 + a_1^3 a_2^3 a_3^2)]$$

$$= -\frac{a_2 a_3^2}{(a_1 - j)(a_2 - j)(a_3 - j)}[a_1^3 a_2^3 b_3 (1 + a_3 + a_3^2)$$

$$+ a_1^3 b_2 (1 + a_2 + a_2^2) + b_1 (1 + a_1 + a_1^2)]. \qquad \square$$

Theorem 2 (Morley). *The three points $A'(\alpha)$, $B'(\beta)$, $C'(\gamma)$ of the adjacent trisectors of the angles of any triangle ABC form an equilateral triangle.*

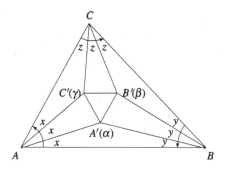

Figure 4.22.

Proof (Alain Connes). Let us consider the rotations $f_1 = r_{A,2x}$, $f_2 = r_{B,2y}$, $f_3 = r_{C,2z}$ with centers A, B, C and angles $x = \frac{1}{3}\hat{A}$, $y = \frac{1}{3}\hat{B}$, $z = \frac{1}{3}\hat{C}$ (Fig. 4.22).

Note that Fix $(f_1 \circ f_2) = \{A'\}$, Fix $(f_2 \circ f_3) = \{B'\}$, Fix $(f_3 \circ f_1) = \{C'\}$ (see Fig. 4.23).

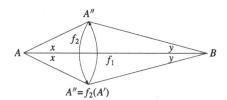

Figure 4.23.

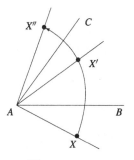

Figure 4.24.

To prove that triangle $A'B'C'$ is equilateral, it is sufficient to show, by Proposition 2, in Sect. 2.4 and above Theorem 1 in Sect. 4.13.6, that $f_1^3 \circ f_2^3 \circ f_3^3 = 1_{\mathbb{C}}$. The composition $s_{AC} \circ s_{AB}$ of reflections s_{AC} and s_{AB} across the lines AC and AB is a rotation about center A through angle $6x$ (Fig. 4.24).

Therefore, $f_1^3 = s_{AC} \circ s_{AB}$, and analogously, $f_2^3 = s_{BA} \circ s_{BC}$ and $f_3^3 = s_{CB} \circ s_{CA}$. It follows that

$$f_1^3 \circ f_2^3 \circ f_3^3 = s_{AC} \circ s_{AB} \circ s_{BA} \circ s_{BC} \circ s_{CB} \circ s_{CA} = 1_{\mathbb{C}}.$$

□

4.13.7 Homothecy

Given a fixed nonzero real number k, the mapping $h_k : \mathbb{C} \to \mathbb{C}$, $h_k(z) = kz$, is called the *homothety* of the complex plane with center O and magnitude k.

Figures 4.25 and 4.26 show the position of point $M'(h_k(z))$ in the cases $k > 0$ and $k < 0$.

Points $M(z)$ and $M'(h_k(z))$ are collinear with center O, which lies on the line segment MM' if and only if $k < 0$.

Moreover, the following relation holds:

$$|OM'| = |k||OM|.$$

Point M' is called the *homothetic* point of M with center O and magnitude k.

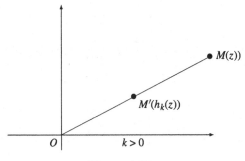

Figure 4.25.

It is clear that the composition of two homotheties h_{k_1} and h_{k_2} is also a homothety, that is,

$$h_{k_1} \circ h_{k_2} = h_{k_1 k_2}.$$

The set $\mathcal{H}$ of all homotheties of the complex plane is an abelian group with respect to the composition of mappings. The identity of the group $(\mathcal{H}, \circ)$ is $h_1 = 1_{\mathbb{C}}$, the homothety of magnitude 1.

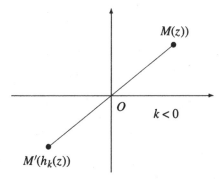

Figure 4.26.

Problem. *Let M be a point inside an equilateral triangle ABC and let M_1, M_2, M_3 be the feet of the perpendiculars from M to the sides BC, CA, AB, respectively. Find the locus of the centroid of the triangle $M_1 M_2 M_3$.*

Solution. Let 1, ε, ε^2 be the coordinates of points A, B, C, where $\varepsilon = \cos 120^\circ + i \sin 120^\circ$. Recall that

$$\varepsilon^2 + \varepsilon + 1 = 0 \text{ and } \varepsilon^3 = 1.$$

If m, m_1, m_2, m_3 are the coordinates of points M, M_1, M_2, M_3, we have

$$m_1 = \frac{1}{2}(1 + \varepsilon + m - \varepsilon \overline{m}),$$

$$m_2 = \frac{1}{2}(\varepsilon + \varepsilon^2 + m - \overline{m}),$$

$$m_3 = \frac{1}{2}(\varepsilon^2 + 1 + m - \varepsilon^2 \overline{m}).$$

Let g be the coordinate of the centroid of the triangle $M_1 M_2 M_3$. Then

$$g = \frac{1}{3}(m_1 + m_2 + m_3) = \frac{1}{6}(2(1 + \varepsilon + \varepsilon^2) + 3m - \overline{m}(1 + \varepsilon + \varepsilon^2)) = \frac{m}{2},$$

and hence $OG = \frac{1}{2}OM$.

The locus of G is the interior of the triangle obtained from ABC under a homothety of center O and magnitude $\frac{1}{2}$. In other words, the vertices of this triangle have coordinates $\frac{1}{2}$, $\frac{1}{2}\varepsilon$, $\frac{1}{2}\varepsilon^2$.

4.13.8 Problems

1. Prove that the composition of two isometries of the complex plane is an isometry.
2. An isometry of the complex plane has two fixed points A and B. Prove that every point M of line AB is a fixed point of the transformation.
3. Prove that every isometry of the complex plane is a composition of a rotation with a translation and possibly also with a reflection in the real axis.
4. Prove that the mapping $f : \mathbb{C} \to \mathbb{C}$, $f(z) = i \cdot \overline{z} + 4 - i$, is an isometry. Analyze f as in the previous problem.
5. Prove that the mapping $g : \mathbb{C} \to \mathbb{C}$, $g(z) = -iz + 1 + 2i$, is an isometry. Analyze g as in the previous problem.

Chapter 5
Olympiad-Caliber Problems

The use of complex numbers is helpful in solving Olympiad problems. In many instances, a rather complicated problem can be solved unexpectedly by employing complex numbers. Even though the methods of Euclidean geometry, coordinate geometry, vector algebra, and complex numbers look similar, in many situations the use of complex numbers has multiple advantages. This chapter will illustrate some classes of Olympiad-caliber problems for which the method of complex numbers works efficiently.

5.1 Problems Involving Moduli and Conjugates

Problem 1. *Let z_1, z_2, z_3 be complex numbers such that*

$$|z_1| = |z_2| = |z_3| = r > 0$$

and $z_1 + z_2 + z_3 \neq 0$. Prove that

$$\left| \frac{z_1 z_2 + z_2 z_3 + z_3 z_1}{z_1 + z_2 + z_3} \right| = r.$$

Solution. Observe that

$$z_1 \cdot \overline{z_1} = z_2 \cdot \overline{z_2} = z_3 \cdot \overline{z_3} = r^2.$$

Then

$$\left| \frac{z_1 z_2 + z_2 z_3 + z_3 z_1}{z_1 + z_2 + z_3} \right|^2 = \frac{z_1 z_2 + z_2 z_3 + z_3 z_1}{z_1 + z_2 + z_3} \cdot \frac{\overline{z_1 z_2} + \overline{z_2 z_3} + \overline{z_3 z_1}}{\overline{z_1} + \overline{z_2} + \overline{z_3}}$$

$$= \frac{z_1 z_2 + z_2 z_3 + z_3 z_1}{z_1 + z_2 + z_3} \cdot \frac{\dfrac{r^2}{z_1} \cdot \dfrac{r^2}{z_2} + \dfrac{r^2}{z_2} \cdot \dfrac{r^2}{z_3} + \dfrac{r^2}{z_3} \cdot \dfrac{r^2}{z_1}}{\dfrac{r^2}{z_1} + \dfrac{r^2}{z_2} + \dfrac{r^2}{z_3}} = r^2,$$

as desired.

Problem 2. *Let z_1, z_2 be complex numbers such that*

$$|z_1| = |z_2| = r > 0.$$

Prove that

$$\left(\frac{z_1 + z_2}{r^2 + z_1 z_2} \right)^2 + \left(\frac{z_1 - z_2}{r^2 - z_1 z_2} \right)^2 \geq \frac{1}{r^2}.$$

Solution. The desired inequality is equivalent to

$$\left(\frac{r(z_1 + z_2)}{r^2 + z_1 z_2} \right)^2 + \left(\frac{r(z_1 - z_2)}{r^2 - z_1 z_2} \right)^2 \geq 1.$$

Setting
$$z_1 = r(\cos 2x + i \sin 2x) \text{ and } z_2 = r(\cos 2y + i \sin 2y)$$

yields

$$\frac{r(z_1 + z_2)}{r^2 + z_1 z_2} = \frac{r^2(\cos 2x + i \sin 2x + \cos 2y + i \sin 2y)}{r^2(1 + \cos(2x + 2y) + i \sin(2x + 2y))} = \frac{\cos(x - y)}{\cos(x + y)}.$$

Similarly,

$$\frac{r(z_1 - z_2)}{r^2 - z_1 z_2} = \frac{\sin(y - x)}{\sin(y + x)}.$$

Thus

$$\left(\frac{r(z_1 + z_2)}{r^2 + z_1 z_2} \right)^2 + \left(\frac{r(z_1 - z_2)}{r^2 - z_1 z_2} \right)^2 = \frac{\cos^2(x - y)}{\cos^2(x + y)} + \frac{\sin^2(x - y)}{\sin^2(x + y)}$$
$$\geq \cos^2(x - y) + \sin^2(x - y) = 1,$$

as claimed.

Problem 3. *Let z_1, z_2, z_3 be complex numbers such that*

$$|z_1| = |z_2| = |z_3| = 1$$

and

$$\frac{z_1^2}{z_2 z_3} + \frac{z_2^2}{z_1 z_3} + \frac{z_3^2}{z_1 z_2} + 1 = 0.$$

Prove that

$$|z_1 + z_2 + z_3| \in \{1, 2\}.$$

Solution 1. The given equality can be written as

$$z_1^3 + z_2^3 + z_3^3 + z_1 z_2 z_3 = 0,$$

or

$$-4z_1 z_2 z_3 = z_1^3 + z_2^3 + z_3^3 - 3z_1 z_2 z_3$$
$$= (z_1 + z_2 + z_3)(z_1^2 + z_2^2 + z_3^2 - z_1 z_2 - z_2 z_3 - z_3 z_1).$$

Setting $z = z_1 + z_2 + z_3$ yields

$$z^3 - 3z(z_1 z_2 + z_2 z_3 + z_3 z_1) = -4z_1 z_2 z_3.$$

This is equivalent to

$$z^3 = z_1 z_2 z_3 \left[3z \left(\frac{1}{z_1} + \frac{1}{z_2} + \frac{1}{z_3} \right) - 4 \right].$$

The last relation can be written as

$$z^3 = z_1 z_2 z_3 [3z(\overline{z_1} + \overline{z_2} + \overline{z_3}) - 4], \text{ i.e., } z^3 = z_1 z_2 z_3 (3|z|^2 - 4).$$

Taking the absolute values of both sides yields $|z|^3 = |3|z|^2 - 4|$. If $|z| \geq \dfrac{2}{\sqrt{3}}$, then $|z|^3 - 3|z|^2 + 4 = 0$, implying $|z| = 2$. If $|z| < \dfrac{2}{\sqrt{3}}$, then $|z|^3 + 3|z|^2 - 4 = 0$, giving $|z| = 1$, as required.

Solution 2. It is not difficult to see that $|z_1^3 + z_2^3 + z_3^3| = 1$. From the algebraic identity

$$(u + v)(v + w)(w + u) = (u + v + w)(uv + vw + wu) - uvw$$

for $u = z_1^3$, $v = z_2^3$, $w = z_3^3$, it follows that

$$(z_1^3 + z_2^3)(z_2^3 + z_3^3)(z_3^3 + z_1^3) = (z_1^3 + z_2^3 + z_3^3)(z_1^3 z_2^3 + z_2^3 z_3^3 + z_3^3 z_1^3) - z_1^3 z_2^3 z_3^3$$
$$= z_1^3 z_2^3 z_3^3 (z_1^3 + z_2^3 + z_3^3) \left(\frac{1}{z_1^3} + \frac{1}{z_2^3} + \frac{1}{z_3^3} \right) - z_1^3 z_2^3 z_3^3$$
$$= z_1^3 z_2^3 z_3^3 (z_1^3 + z_2^3 + z_3^3)\overline{(z_1^3 + z_2^3 + z_3^3)} - z_1^3 z_2^3 z_3^3$$
$$= z_1^3 z_2^3 z_3^3 - z_1^3 z_2^3 z_3^3 = 0.$$

Suppose that $z_1^3 + z_2^3 = 0$. Then $z_1 + z_2 = 0$ or $z_1^2 - z_1 z_2 + z_2^2 = 0$, implying $z_1^2 + z_2^2 = -2 z_1 z_2$ or $z_1^2 + z_2^2 = z_1 z_2$.

On the other hand, from the given relation it follows that $z_3^3 = -z_1 z_2 z_3$, yielding $z_3^2 = -z_1 z_2$.

We have

$$|z_1 + z_2 + z_3|^2 = (z_1 + z_2 + z_3)\left(\frac{1}{z_1} + \frac{1}{z_2} + \frac{1}{z_3}\right)$$

$$= 3 + \left(\frac{z_1}{z_2} + \frac{z_2}{z_1}\right) + \left(\frac{z_1}{z_3} + \frac{z_3}{z_2}\right) + \left(\frac{z_2}{z_3} + \frac{z_3}{z_1}\right)$$

$$= 3 + \frac{z_1^2 + z_2^2}{z_1 z_2} + \frac{z_3^2 + z_1 z_2}{z_2 z_3} + \frac{z_3^2 + z_1 z_2}{z_3 z_1} = 3 + \frac{z_1^2 + z_2^2}{z_1 z_2}.$$

This leads to $|z_1 + z_2 + z_3|^2 = 1$ if $z_1^2 + z_2^2 = -2 z_1 z_2$ and $|z_1 + z_2 + z_3|^2 = 4$ if $z_1^2 + z_2^2 = z_1 z_2$. The conclusion follows.

Problem 4. *If $a, b \in \mathbb{C}$, then $|1 + a| + |1 + b| + |1 + ab| \geq 2$.*

Solution. If $|a| \geq 1$ we have

$$|1 + a| + |1 + b| + |1 + ab| \geq |(1 + a) - (1 + ab)| + |1 + b|$$
$$= |a| \cdot |1 - b| + |1 + b| \geq |1 - b| + |1 + b| \geq |(1 - b) + (1 + b)| = 2.$$

If $|a| \leq 1$, we have

$$|1 + a| + |1 + b| + |1 + ab| \geq |(1 + a) + (1 + ab)| + |1 + b|$$
$$= |2 + a(1 + b)| + |1 + b| \geq |2 + a(1 + b)| + |a| \cdot |1 + b|$$
$$= |2 + a(1 + b)| + |a(1 + b)| \geq |(2 + a(1 + b)) - a(1 + b)| = 2.$$

Problem 5. *Let $n > 0$ be an integer and let z be a complex number such that $|z| = 1$. Prove that*

$$n|1 + z| + |1 + z^2| + |1 + z^3| + \cdots + |1 + z^{2n}| + |1 + z^{2n+1}| \geq 2n.$$

Solution 1. We have

$$n|1 + z| + |1 + z^2| + |1 + z^3| + \cdots + |1 + z^{2n}| + |1 + z^{2n+1}|$$

$$= \sum_{k=1}^{n}(|1 + z| + |1 + z^{2k+1}|) + \sum_{k=1}^{n}|1 + z^{2k}|$$

$$\geq \sum_{k=1}^{n}|z - z^{2k+1}| + \sum_{k=1}^{n}|1 + z^{2k}| = \sum_{k=1}^{n}(|z||1 - z^{2k}| + |1 + z^{2k}|)$$

$$= \sum_{k=1}^{n}(|1 - z^{2k}| + |1 + z^{2k}|) \geq \sum_{k=1}^{n}|1 - z^{2k} + 1 + z^{2k}| = 2n,$$

as claimed.

Solution 2. We use induction on n.

For $n = 1$, we prove that $|1 + z| + |1 + z^2| + |1 + z^3| \geq 2$. Indeed,

$$2 = |1 + z + z^3 + 1 - z(1 + z^2)| \leq |1 + z| + |z^3 + 1| + |z||1 + z^2|$$
$$= |1 + z| + |1 + z^2| + |1 + z^3|.$$

Assume that the inequality is valid for some n, so

$$n|1 + z| + |1 + z^2| + \cdots + |1 + z^{2n+1}| \geq 2n.$$

We prove that

$$(n+1)|1 + z| + |1 + z^2| + \cdots + |1 + z^{2n+1}| + |1 + z^{2n+2}| + |1 + z^{2n+3}| \geq 2n + 2.$$

Using the inductive hypothesis yields

$$(n + 1)|1 + z| + |1 + z^2| + \cdots + |1 + z^{2n+2}| + |1 + z^{2n+3}|$$
$$\geq 2n + |1 + z| + |1 + z^{2n+2}| + |1 + z^{2n+3}|$$
$$= 2n + |1 + z| + |z||1 + z^{2n+2}| + |1 + z^{2n+3}|$$
$$\geq 2n + |1 + z - z(1 + z^{2n+2}) + 1 + z^{2n+3}| = 2n + 2,$$

as needed.

Problem 6. *Let z_1, z_2, z_3 be complex numbers such that*

(1) $|z_1| = |z_2| = |z_3| = 1$;
(2) $z_1 + z_2 + z_3 \neq 0$;
(3) $z_1^2 + z_2^2 + z_3^2 = 0$.

Prove that for all integers $n \geq 2$,

$$|z_1^n + z_2^n + z_3^n| \in \{0, 1, 2, 3\}.$$

Solution 1. Let

$$s_1 = z_1 + z_2 + z_3, \quad s_2 = z_1 z_2 + z_2 z_3 + z_3 z_1, \quad s_3 = z_1 z_2 z_3$$

and consider the cubic equation

$$z^3 - s_1 z^2 + s_2 z - s_3 = 0$$

with roots z_1, z_2, z_3.

Because $z_1^2 + z_2^2 + z_3^2 = 0$, we have

$$s_1^2 = 2s_2. \tag{1}$$

On the other hand,

$$s_2 = s_3 \left(\frac{1}{z_1} + \frac{1}{z_2} + \frac{1}{z_3} \right) = s_3(\overline{z_1} + \overline{z_2} + \overline{z_3}) = s_3 \cdot \overline{s_1}. \tag{2}$$

The relations (1) and (2) imply $s_1^2 = 2s_3 \cdot \overline{s_1}$ and, consequently, $|s_1|^2 = 2|s_3| \cdot |\overline{s_1}| = 2|s_1|$. Because $s_1 \neq 0$, we have $|s_1| = 2$, so $s_1 = 2\lambda$ with $|\lambda| = 1$.

From relations (1) and (2), it follows that $s_2 = \frac{1}{2}s_1^2 = 2\lambda^2$ and $s_3 = \frac{s_2}{\overline{s_1}} = \frac{2\lambda^2}{2\overline{\lambda}} = \lambda^3$.

The equation with roots z_1, z_2, z_3 becomes

$$z^3 - 2\lambda z^2 + 2\lambda^2 z - \lambda^3 = 0.$$

This is equivalent to

$$(z - \lambda)(z^2 - \lambda z + \lambda^2) = 0.$$

The roots are λ, $\lambda\varepsilon$, $-\lambda\varepsilon^2$, where $\varepsilon = \frac{1}{2} + i\frac{\sqrt{3}}{2}$.

Without loss of generality we may assume that $z_1 = \lambda$, $z_2 = \lambda\varepsilon$, $z_3 = -\lambda\varepsilon^2$. From the relations $\varepsilon^2 - \varepsilon + 1 = 0$ and $\varepsilon^3 = -1$, it follows that

$$E_n = |z_1^n + z_2^n + z_3^n| = |\lambda^n + \lambda^n \varepsilon^n + (-1)^n \lambda^n \varepsilon^{2n}|$$
$$= |1 + \varepsilon^n + (-1)^n \varepsilon^{2n}|.$$

It is not difficult to see that $E_{k+6} = E_k$ for all integers k and that the equalities

$$E_0 = 3, \ E_1 = 2, \ E_2 = 0, \ E_3 = 1, \ E_4 = 0, \ E_5 = 2,$$

settle the claim.

Solution 2. It is clear that z_1^2, z_2^2, z_3^2 are distinct. Otherwise, if, for example, $z_1^2 = z_2^2$, then $1 = |z_3^2| = |-(z_1^2 + z_2^2)| = 2|z_1^2| = 2$, a contradiction.

From $z_1^2 + z_2^2 + z_3^2 = 0$ it follows that z_1^2, z_2^2, z_3^2 are the coordinates of the vertices of an equilateral triangle. Hence we may assume that $z_2^2 = \varepsilon z_1^2$ and $z_3^2 = \varepsilon^2 z_1^2$, where $\varepsilon^2 + \varepsilon + 1 = 0$. Because $z_2^2 = \varepsilon^4 z_1^2$ and $z_3^2 = \varepsilon^2 z_1^2$, it follows that $z_2 = \pm\varepsilon^2 z_1$ and $z_3 = \pm\varepsilon z_1$. Then

$$|z_1^n + z_2^n + z_3^n| = |(1 + (\pm\varepsilon)^n + (\pm\varepsilon^2)^n)z_1^n| = |1 + (\pm\varepsilon)^n + (\pm\varepsilon^2)^n| \in \{0, 1, 2, 3\}$$

by the same argument used at the end of the previous solution.

Problem 7. *Find all complex numbers z such that*

$$|z - |z + 1|| = |z + |z - 1||.$$

Solution. We have

$$|z - |z + 1|| = |z + |z - 1||$$

if and only if

$$|z - |z + 1||^2 = |z + |z - 1||^2,$$

i.e.,

$$(z - |z + 1|) \cdot (\bar{z} - |z + 1|) = (z + |z - 1|) \cdot (\bar{z} + |z - 1|).$$

The last equation is equivalent to

$$z \cdot \bar{z} - (z + \bar{z})|z + 1| + |z + 1|^2 = z \cdot \bar{z} + (z + \bar{z}) \cdot |z - 1| + |z - 1|^2.$$

This can be written as

$$|z + 1|^2 - |z - 1|^2 = (z + \bar{z}) \cdot (|z + 1| + |z - 1|),$$

i.e.,

$$(z + 1)(\bar{z} + 1) - (z - 1)(\bar{z} - 1) = (z + \bar{z}) \cdot (|z + 1| + |z - 1|).$$

The last equation is equivalent to

$$2(z + \bar{z}) = (z + \bar{z}) \cdot (|z + 1| + |z - 1|), \text{ i.e., } z + \bar{z} = 0,$$

or $|z + 1| + |z - 1| = 2$.

The triangle inequality

$$2 = |(z + 1) - (z - 1)| \leq |z + 1| + |z - 1|$$

shows that the solutions to the equation $|z + 1| + |z - 1| = 2$ satisfy $z + 1 = t(1 - z)$, where t is a real number and $t \geq 0$.

It follows that $z = \dfrac{t - 1}{t + 1}$, so z is any real number such that $-1 \leq z \leq 1$.

The equation $z + \bar{z} = 0$ has the solutions $z = bi$, $b \in \mathbb{R}$. Hence, the solutions to the equation are

$$\{bi : b \in \mathbb{R}\} \cup \{a \in \mathbb{R} : a \in [-1, 1]\}.$$

Problem 8. *Let z_1, $z_2, \ldots, z_n$ be complex numbers such that $|z_1| = |z_2| = \cdots = |z_n| > 0$. Prove that*

$$\mathrm{Re} \left(\sum_{j=1}^{n} \sum_{k=1}^{n} \frac{z_j}{z_k} \right) = 0$$

if and only if

$$\sum_{k=1}^{n} z_k = 0.$$

(Romanian Mathematical Olympiad—Second Round, 1987)

Solution. Let

$$S = \sum_{j=1}^{n}\sum_{k=1}^{n} \frac{z_j}{z_k}.$$

Then

$$S = \left(\sum_{k=1}^{n} z_k\right) \cdot \left(\sum_{k=1}^{n} \frac{1}{z_k}\right),$$

and since $z_k \cdot \overline{z_k} = r^2$ for all k, we have

$$S = \left(\sum_{k=1}^{n} z_k\right) \cdot \left(\sum_{k=1}^{n} \frac{\overline{z_k}}{r^2}\right)$$

$$= \frac{1}{r^2}\left(\sum_{k=1}^{n} z_k\right)\left(\overline{\sum_{k=1}^{n} z_k}\right) = \frac{1}{r^2}\left|\sum_{k=1}^{n} z_k\right|^2.$$

Hence S is a real number, so $\mathrm{Re}S = S = 0$ if and only if $\sum_{k=1}^{n} z_k = 0$.

Problem 9. *Let λ be a real number and let $n \geq 2$ be an integer. Solve the equation*

$$\lambda(\overline{z} + z^n) = i(\overline{z} - z^n).$$

Solution. The equation is equivalent to

$$z^n(\lambda + i) = \overline{z}(-\lambda + i).$$

Taking the absolute values of both sides of the equation, we obtain $|z|^n = |\overline{z}| = |z|$; hence $|z| = 0$ or $|z| = 1$.

If $|z| = 0$, then $z = 0$, which satisfies the equation. If $|z| = 1$, then $\overline{z} = \dfrac{1}{z}$, and the equation may be rewritten as

$$z^{n+1} = \frac{-\lambda + i}{\lambda + i}.$$

Because $\left|\dfrac{-\lambda + i}{\lambda + i}\right| = 1$, there exists $t \in [0, 2\pi)$ such that

$$\frac{-\lambda + i}{\lambda + i} = \cos t + i\sin t.$$

Then
$$z_k = \cos \frac{t + 2k\pi}{n+1} + i \sin \frac{t + 2k\pi}{n+1}$$

for $k = 0, 1, \ldots, n$ are the other solutions to the equation (besides $z = 0$).

Problem 10. *Prove that*
$$\left| \frac{6z - i}{2 + 3iz} \right| \leq 1 \text{ if and only if } |z| \leq \frac{1}{3}.$$

Solution. We have
$$\left| \frac{6z - i}{2 + 3iz} \right| \leq 1 \text{ if and only if } |6z - i| \leq |2 + 3iz|.$$

The last inequality is equivalent to
$$|6z - i|^2 \leq |2 + 3iz|^2, \text{ i.e., } (6z - i)(6\overline{z} + i) \leq (2 + 3iz)(2 - 3i\overline{z}).$$

We obtain
$$36z \cdot \overline{z} + 6iz - 6i\overline{z} + 1 \leq 4 - 6i\overline{z} + 6iz + 9z\overline{z},$$

i.e., $27z \cdot \overline{z} \leq 3$. Finally, $z\overline{z} \leq \frac{1}{9}$, or equivalently, $|z| \leq \frac{1}{3}$, as desired.

Problem 11. *Let z be a complex number such that $z \in \mathbb{C} \backslash \mathbb{R}$ and*
$$\frac{1 + z + z^2}{1 - z + z^2} \in \mathbb{R}.$$

Prove that $|z| = 1$.

Solution. We have
$$\frac{1 + z + z^2}{1 - z + z^2} = 1 + 2\frac{z}{1 - z + z^2} \in \mathbb{R} \text{ if and only if } \frac{z}{1 - z + z^2} \in \mathbb{R}.$$

That is,
$$\frac{1 - z + z^2}{z} = \frac{1}{z} - 1 + z \in \mathbb{R}, \text{ i.e., } z + \frac{1}{z} \in \mathbb{R}.$$

The last relation is equivalent to
$$z + \frac{1}{z} = \overline{z} + \frac{1}{\overline{z}}, \text{ i.e., } (z - \overline{z})(1 - |z|^2) = 0.$$

We obtain $z = \overline{z}$ or $|z| = 1$.

Because z is not a real number, it follows that $|z| = 1$, as desired.

Problem 12. *Let z_1, z_2, ..., z_n be complex numbers such that $|z_1| = \cdots = |z_n| = 1$, and let*

$$z = \left(\sum_{k=1}^{n} z_k\right) \cdot \left(\sum_{k=1}^{n} \frac{1}{z_k}\right).$$

Prove that z is a real number and $0 \le z \le n^2$.

Solution. Note that $\overline{z_k} = \dfrac{1}{z_k}$ for all $k = 1, \ldots, n$. Because

$$\overline{z} = \left(\sum_{k=1}^{n} \overline{z_k}\right) \left(\sum_{k=1}^{n} \frac{1}{\overline{z_k}}\right) = \left(\sum_{k=1}^{n} \frac{1}{z_k}\right) \left(\sum_{k=1}^{n} z_k\right) = z,$$

it follows that z is a real number.

Let $z_k = \cos \alpha_k + i \sin \alpha_k$, where α_k are real numbers for $k = \overline{1, n}$. Then

$$z = \left(\sum_{k=1}^{n} \cos \alpha_k + i \sum_{k=1}^{n} \sin \alpha_k\right) \left(\sum_{k=1}^{n} \cos \alpha_k - i \sum_{k=1}^{n} \sin \alpha_k\right)$$

$$= \left(\sum_{k=1}^{n} \cos \alpha_k\right)^2 + \left(\sum_{k=1}^{n} \sin \alpha_k\right)^2 \ge 0.$$

On the other hand, we have

$$z = \sum_{k=1}^{n} (\cos^2 \alpha_k + \sin^2 \alpha_k) + 2 \sum_{1 \le i < j \le n} (\cos \alpha_i \cos \alpha_j + \sin \alpha_i \sin \alpha_j)$$

$$= n + 2 \sum_{1 \le i < j \le n} \cos(\alpha_i - \alpha_j) \le n + 2 \binom{n}{2} = n + 2 \frac{n(n-1)}{2} = n^2,$$

as desired.

Remark. An alternative solution to the inequalities $0 \le z \le n^2$ is as follows:

$$z = \left(\sum_{k=1}^{n} z_k\right) \left(\sum_{k=1}^{n} \frac{1}{z_k}\right) = \left(\sum_{k=1}^{n} z_k\right) \left(\sum_{k=1}^{n} \overline{z_k}\right)$$

$$= \left(\sum_{k=1}^{n} z_k\right) \overline{\left(\sum_{k=1}^{n} z_k\right)} = \left|\sum_{k=1}^{n} z_k\right|^2 \le \left(\sum_{k=1}^{n} |z_k|\right)^2 = n^2,$$

so $0 \le z \le n^2$.

Problem 13. *Let z_1, z_2, z_3 be complex numbers such that*

$$z_1 + z_2 + z_3 \ne 0 \ and \ |z_1| = |z_2| = |z_3|.$$

Prove that

$$\text{Re}\left(\frac{1}{z_1} + \frac{1}{z_2} + \frac{1}{z_3}\right) \cdot \text{Re}\left(\frac{1}{z_1 + z_2 + z_3}\right) \geq 0.$$

Solution. Let $r = |z_1| = |z_2| = |z_3| > 0$. Then

$$z_1\overline{z_1} = z_2\overline{z_2} = z_3\overline{z_3} = r^2$$

and

$$\frac{1}{z_1} + \frac{1}{z_2} + \frac{1}{z_3} = \frac{\overline{z_1} + \overline{z_2} + \overline{z_3}}{r^2} = \frac{\overline{z_1 + z_2 + z_3}}{r^2}.$$

On the other hand, we have

$$\frac{1}{z_1 + z_2 + z_3} = \frac{\overline{z_1 + z_2 + z_3}}{|z_1 + z_2 + z_3|^2},$$

and consequently,

$$\text{Re}\left(\frac{1}{z_1} + \frac{1}{z_2} + \frac{1}{z_3}\right) \cdot \text{Re}\left(\frac{1}{z_1 + z_2 + z_3}\right)$$
$$= \text{Re}\left(\frac{\overline{z_1 + z_2 + z_3}}{r^2}\right) \cdot \text{Re}\left(\frac{\overline{z_1 + z_2 + z_3}}{|z_1 + z_2 + z_3|^2}\right) = \frac{(\text{Re}(\overline{z_1 + z_2 + z_3}))^2}{r^2|z_1 + z_2 + z_3|^2} \geq 0,$$

as desired.

Problem 14. *Let x, y, z be complex numbers.*

(a) Prove that

$$|x| + |y| + |z| \leq |x + y - z| + |x - y + z| + |-x + y + z|.$$

(b) If x, y, z are distinct and the numbers $x + y - z$, $x - y + z$, $-x + y + z$
have equal absolute values, prove that

$$2(|x| + |y| + |z|) \leq |x + y - z| + |x - y + z| + |-x + y + z|.$$

Solution. Let

$$m = -x + y + z, \ n = x - y + z, \ p = x + y - z.$$

We have

$$x = \frac{n + p}{2}, \ y = \frac{m + p}{2}, \ z = \frac{m + n}{2}.$$

(a) Adding the inequalities

$$|x| \leq \frac{1}{2}(|n| + |p|), \ |y| \leq \frac{1}{2}(|m| + |p|), \ |z| \leq \frac{1}{2}(|m| + |n|)$$

yields

$$|x| + |y| + |z| \leq |m| + |n| + |p|,$$

as desired.

(b) Let A, B, C be the points with coordinates m, n, p and observe that the numbers m, n, p are distinct and that $|m| = |n| = |p| = R$, the circumradius of triangle ABC.

Let the origin of the complex plane be the circumcenter of triangle ABC. The orthocenter H of triangle ABC has the coordinate $h = m + n + p$. The desired inequality becomes

$$|h - m| + |h - n| + |h - p| \leq |m| + |n| + |p|,$$

or

$$AH + BH + CH \leq 3R.$$

This is equivalent to

$$\cos A + \cos B + \cos C \leq \frac{3}{2}. \tag{1}$$

Inequality (1) can be written as

$$2 \cos \frac{A + B}{2} \cos \frac{A - B}{2} + 1 - 2 \sin^2 \frac{C}{2} \leq \frac{3}{2},$$

or

$$0 \leq \left(2 \sin \frac{C}{2} - \cos \frac{A - B}{2} \right)^2 + \sin^2 \frac{A - B}{2},$$

which is clear. We have equality in (1) if and only if triangle ABC is equilateral, i.e., $m = a$, $n = a\varepsilon$, $p = a\varepsilon^2$, where a is a complex parameter and $\varepsilon = \cos \frac{2\pi}{3} + i \sin \frac{2\pi}{3}$. In this case, $x = -\frac{a}{2}$, $y = -\frac{a}{2}\varepsilon$, $z = -\frac{a}{2}\varepsilon^2$.

Problem 15. *Let z_0, z_1, z_2, $\ldots$, z_n be complex numbers such that*

$$(k + 1)z_{k+1} - i(n - k)z_k = 0$$

for all $k \in \{0, 1, 2, \ldots, n - 1\}$.

(1) Find z_0 such that

$$z_0 + z_1 + \cdots + z_n = 2^n.$$

(2) For the value of z_0 determined above, prove that

$$|z_0|^2 + |z_1|^2 + \cdots + |z_n|^2 < \frac{(3n + 1)^n}{n!}.$$

Solution.

(a) Use induction to prove that

$$z_k = i^k \binom{n}{k} z_0, \text{ for all } k \in \{0, 1, \ldots, n\}.$$

Then

$$z_0 + z_1 + \cdots + z_n = 2^n \text{ if and only if } z_0(1+i)^n = 2^n,$$

i.e., $z_0 = (1-i)^n$.

(b) Applying the AM-GM inequality, we have

$$|z_0|^2 + |z_1|^2 + \cdots + |z_n|^2 = |z_0|^2 \left(\binom{n}{0}^2 + \binom{n}{1}^2 + \cdots + \binom{n}{n}^2 \right)$$

$$= |z_0|^2 \cdot \binom{2n}{n} = 2^n \cdot \binom{2n}{n} = \frac{2^n}{n!} 2n(2n-1) \cdots (n+1)$$

$$< \frac{2^n}{n!} \left(\frac{2n + (2n-1) + \cdots + (n+1)}{n} \right)^n = \frac{(3n+1)^n}{n!},$$

as desired.

Problem 16. *Let z_1, z_2, z_3 be complex numbers such that*

$$z_1 + z_2 + z_3 = z_1 z_2 + z_2 z_3 + z_3 z_1 = 0.$$

Prove that $|z_1| = |z_2| = |z_3|$.

Solution 1. Substituting $z_1 + z_2 = -z_3$ in $z_1 z_2 + z_3(z_1 + z_2) = 0$ gives $z_1 z_2 = z_3^2$, so $|z_1| \cdot |z_2| = |z_3|^2$. Likewise, $|z_2| \cdot |z_3| = |z_1|^2$ and $|z_3||z_1| = |z_2|^2$. Then

$$|z_1|^2 + |z_2|^2 + |z_3|^2 = |z_1||z_2| + |z_2||z_3| + |z_3||z_1|,$$

i.e.,

$$(|z_1| - |z_2|)^2 + (|z_2| - |z_3|)^2 + (|z_3| - |z_1|)^2 = 0,$$

yielding $|z_1| = |z_2| = |z_3|$.

Solution 2. Using the relations between the roots and the coefficients, it follows that z_1, z_2, z_3 are the roots of the polynomial $z^3 - p$, where $p = z_1 z_2 z_3$. Hence $z_1^3 - p = z_2^3 - p = z_3^3 - p = 0$, implying $z_1^3 = z_2^3 = z_3^3$, and the conclusion follows.

Problem 17. *Prove that for all complex numbers z with $|z| = 1$, the following inequalities hold:*

$$\sqrt{2} \leq |1 - z| + |1 + z^2| \leq 4.$$

Solution. Setting $z = \cos t + i \sin t$ yields

$$|1 - z| = \sqrt{(1 - \cos t)^2 + \sin^2 t} = \sqrt{2 - 2\cos t} = 2\left|\sin \frac{t}{2}\right|$$

and

$$|1 + z^2| = \sqrt{(1 + \cos 2t)^2 + \sin^2 2t} = \sqrt{2 + 2\cos 2t}$$

$$= 2|\cos t| = 2\left|1 - 2\sin^2 \frac{t}{2}\right|.$$

It suffices to prove that $\dfrac{\sqrt{2}}{2} \leq |a| + |1 - 2a^2| \leq 2$ for $a = \sin \dfrac{t}{2} \in [-1, 1]$. We leave this to the reader.

Problem 18. *Let z_1, z_2, z_3, z_4 be distinct complex numbers such that*

$$\operatorname{Re}\frac{z_2 - z_1}{z_4 - z_1} = \operatorname{Re}\frac{z_2 - z_3}{z_4 - z_3} = 0.$$

(a) Find all real numbers x such that

$$|z_1 - z_2|^x + |z_1 - z_4|^x \leq |z_2 - z_4|^x \leq |z_2 - z_3|^x + |z_4 - z_3|^x.$$

(b) Prove that $|z_3 - z_1| \leq |z_4 - z_2|$.

Solution. Consider the points A, B, C, D with coordinates z_1, z_2, z_3, z_4, respectively. The conditions

$$\operatorname{Re}\frac{z_2 - z_1}{z_4 - z_1} = \operatorname{Re}\frac{z_2 - z_3}{z_4 - z_3} = 0$$

imply $\widehat{BAD} = \widehat{BCD} = 90°$. Then $|z_1 - z_2| = AB$ and $|z_1 - z_4| = AD$ are the lengths of the sides of the right triangle ABD with hypotenuse $BD = |z_2 - z_4|$.

The inequality $AB^x + AD^x \leq BD^x$ holds for $x \geq 2$.

Similarly, $|z_2 - z_3| = BC$ and $|z_4 - z_3| = CD$ are the sides of the right triangle BCD, so the inequality $BD^x \leq BC^x + CD^x$ holds for $x \leq 2$. Consequently, $x = 2$.

Finally, $AC = |z_3 - z_1| \leq BD = |z_4 - z_2|$, since AC is a chord in the circle of diameter BD.

Problem 19. *Let x and y be distinct complex numbers such that $|x| = |y|$. Prove that*

$$\frac{1}{2}|x + y| < |x|.$$

Solution 1. Let $x = a + ib$ and $y = c + id$, with a, b, c, $d \in \mathbb{R}$ and $a^2 + b^2 = c^2 + d^2$. The inequality is equivalent to

$$(a + c)^2 + (b + d)^2 < 4(a^2 + b^2),$$

or

$$(a - c)^2 + (b - d)^2 > 0,$$

which is clear, since $x \neq y$.

Solution 2. Consider points $X(x)$ and $Y(y)$. In triangle XOY, we have $OX = OY$. Hence $OM < OX$, where M is the midpoint of segment $[XY]$. The coordinate of point M is $\dfrac{x + y}{2}$, and the desired inequality follows.

Problem 20. *Consider the set*

$$A = \{z \in \mathbb{C} : z = a + bi, \ a > 0, \ |z| < 1\}.$$

Prove that for every $z \in A$, there is a number $x \in A$ such that

$$z = \frac{1 - x}{1 + x}.$$

Solution. Let $z \in A$. The equation $z = \dfrac{1 - x}{1 + x}$ has the root

$$x = \frac{1 - z}{1 + z} = \frac{1 - a - ib}{1 + a + ib},$$

where $a > 0$ and $a^2 + b^2 < 1$.

To prove that $x \in A$, it suffices to show that $|x| < 1$ and $\mathrm{Re}(x) > 0$. Indeed, we have

$$|x|^2 = \frac{(1 - a)^2 + b^2}{(1 + a)^2 + b^2} < 1 \text{ if and only if } (1 - a)^2 < (1 + a)^2,$$

i.e., $0 < 4a$, as needed.

Moreover, $\mathrm{Re}(x) = \dfrac{1 - |z|^2}{|1 + z|^2} > 0$, since $|z| < 1$.

Here are more problems involving moduli and conjugates of complex numbers.

Problem 21. *Consider the set*

$$A = \{z \in \mathbb{C} : |z| < 1\},$$

a real number a such that $|a| > 1$, and the function

$$f : A \to A, \ f(z) = \frac{1 + az}{z + a}.$$

Prove that f is bijective.

Problem 22. *Let z be a complex number such that $|z| = 1$ and both $\mathrm{Re}(z)$ and $\mathrm{Im}(z)$ are rational numbers. Prove that $|z^{2n} - 1|$ is rational for all integers $n \geq 1$.*

Problem 23. *Consider the function*

$$f : \mathbb{R} \to \mathbb{C}, \ f(t) = \frac{1 + ti}{1 - ti}.$$

Prove that f is injective and determine its range.

Problem 24. *Let z_1, $z_2 \in \mathbb{C}^*$ be such that $|z_1 + z_2| = |z_1| = |z_2|$. Compute $\dfrac{z_1}{z_2}$.*

Problem 25. *Prove that the following inequality holds for all complex numbers z_1, z_2, ..., z_n:*

$$(|z_1| + |z_2| + \cdots + |z_n| + |z_1 + z_2 + \cdots + z_n|)^2$$
$$\geq 2(|z_1|^2 + \cdots + |z_n|^2 + |z_1 + z_2 + \cdots + z_n|^2).$$

Problem 26. *Let z_1, z_2, ..., z_{2n} be complex numbers such that $|z_1| = |z_2| = \cdots = |z_{2n}|$ and $\arg z_1 \leq \arg z_2 \leq \cdots \leq \arg z_{2n} \leq \pi$. Prove that*

$$|z_1 + z_{2n}| \leq |z_2 + z_{2n-1}| \leq \cdots \leq |z_n + z_{n+1}|.$$

Problem 27. *Find all positive real numbers x and y satisfying the system of equations*

$$\sqrt{3x}\left(1 + \frac{1}{x + y}\right) = 2,$$
$$\sqrt{7y}\left(1 - \frac{1}{x + y}\right) = 4\sqrt{2}.$$

(1996 Vietnamese Mathematical Olympiad)

Problem 28. *Let z_1, z_2, z_3 be complex numbers. Prove that $z_1 + z_2 + z_3 = 0$ if and only if $|z_1| = |z_2 + z_3|$, $|z_2| = |z_3 + z_1|$ and $|z_3| = |z_1 + z_2|$.*

Problem 29. *Let z_1, z_2, ..., z_n be distinct complex numbers with the same modulus such that*

$$z_3 z_4 \ldots z_{n-1} z_n + z_1 z_4 \ldots z_{n-1} z_n + \cdots + z_1 z_2 \ldots z_{n-2} = 0.$$

Prove that

$$z_1 z_2 + z_2 z_3 + \cdots + z_{n-1} z_n = 0.$$

Problem 30. *Let a and z be complex numbers such that $|z + a| = 1$. Prove that*

$$|z^2 + a^2| \geq \frac{|1 - 2|a|^2|}{\sqrt{2}}.$$

Problem 31. *Find the geometric images of the complex numbers z for which*

$$z^n \cdot \mathrm{Re}(z) = \overline{z}^n \cdot \mathrm{Im}(z),$$

where n is an integer.

Problem 32. *Let a, b be real numbers with $a+b = 1$ and let z_1, z_2 be complex numbers with $|z_1| = |z_2| = 1$.*
   Prove that

$$|az_1 + bz_2| \geq \frac{|z_1 + z_2|}{2}.$$

Problem 33. *Let k, n be positive integers and let z_1, z_2, $\ldots$, z_n be nonzero complex numbers with the same modulus such that*

$$z_1^k + z_2^k + \cdots + z_n^k = 0.$$

Prove that

$$\frac{1}{z_1^k} + \frac{1}{z_2^k} + \cdots + \frac{1}{z_n^k} = 0.$$

Problem 34. *Find all pairs (a, b) of real numbers such that*

$$(a + bi)^5 = b + ai.$$

Problem 35. *For every value of $a \in \mathbb{R}$ find $\min |z^2 - az + a|$, where $z \in \mathbb{C}$ and $|z| \leq 1$.*

Problem 36. *Let a, b, c be three complex numbers such that*

$$a|bc| + b|ca| + c|ab| = 0.$$

Prove that

$$|(a - b)(b - c)(c - a)| \geq 3\sqrt{3}|abc|.$$

(Romanian Mathematical Olympiad—Final Round, 2008)

Problem 37. *Let a and b be two complex numbers. Prove the inequality*

$$|1 + ab| + |a + b| \geq \sqrt{|a^2 - 1||b^2 - 1|}.$$

(Romanian Mathematical Olympiad—District Round, 2008)

Problem 38. *Consider complex numbers a, b, and c such that $a + b + c = 0$ and $|a| = |b| = |c| = 1$. Prove that for every complex number z, $|z| \leq 1$, we have*

$$3 \leq |z - a| + |z - b| + |z - c| \leq 4.$$

(Romanian Mathematical Olympiad—Final Round, 2012)

5.2 Algebraic Equations and Polynomials

Problem 1. *Consider the quadratic equation*

$$a^2 z^2 + abz + c^2 = 0b$$

where a, b, $c \in \mathbb{C}^*$, and denote by z_1, z_2 its roots. Prove that if $\dfrac{b}{c}$ is a real number, then $|z_1| = |z_2|$ or $\dfrac{z_1}{z_2} \in \mathbb{R}$.

Solution. Let $t = \dfrac{b}{c} \in \mathbb{R}$. Then $b = tc$ and

$$\Delta = (ab)^2 - 4a^2 \cdot c^2 = a^2 c^2 (t^2 - 4).$$

If $|t| \geq 2$, the roots of the equation are

$$z_{1,2} = \frac{-tac \pm ac\sqrt{t^2 - 4}}{2a^2} = \frac{c}{2a}(-t \pm \sqrt{t^2 - 4}),$$

and it is obvious that $\dfrac{z_1}{z_2}$ is a real number.

If $|t| < 2$, the roots of the equation are

$$z_{1,2} = \frac{c}{2a}(-t \pm i\sqrt{4 - t^2});$$

hence $|z_1| = |z_2| = \dfrac{|c|}{|a|}$, as claimed.

Problem 2. *Let a, b, c, z be complex numbers such that $|a| = |b| = |c| > 0$ and $az^2 + bz + c = 0$. Prove that*

$$\frac{\sqrt{5} - 1}{2} \leq |z| \leq \frac{\sqrt{5} + 1}{2}.$$

Solution. Let $r = |a| = |b| = |c| > 0$. We have

$$|az^2| = |-bz - c| \leq |b||z| + |c|,$$

and hence $r|z^2| \leq r|z| + r$. It follows that $|z|^2 - |z| - 1 \leq 0$, so $|z| \leq \dfrac{1 + \sqrt{5}}{2}$.

On the other hand, $|c| = |-az^2 - bz| \leq |a||z|^2 + b|z|$, so that $|z|^2 + |z| - 1 \geq 0$. Thus $|z| \geq \dfrac{\sqrt{5} - 1}{2}$, and we are done.

Problem 3. *Let p, q be complex numbers such that $|p| + |q| < 1$. Prove that the moduli of the roots of the equation $z^2 + pz + q = 0$ are less than 1.*

Solution. Because $z_1 + z_2 = -p$ and $z_1 z_2 = q$, the inequality $|p| + |q| < 1$ implies $|z_1 + z_2| + |z_1 z_2| < 1$. But $||z_1| - |z_2|| \leq |z_1 + z_2|$; hence

$$|z_1| - |z_2| + |z_1||z_2| - 1 < 0 \text{ if and only if } (1 + |z_2|)(|z_1| - 1) < 0$$

and

$$|z_2| - |z_1| + |z_2||z_1| - 1 < 0 \text{ if and only if } (1 + |z_1|)(|z_2| - 1) < 0.$$

Consequently, $|z_1| < 1$ and $|z_2| < 1$, as desired.

Problem 4. *Let $f = x^2 + ax + b$ be a quadratic polynomial with complex coefficients with both roots having modulus 1. Prove that $g = x^2 + |a|x + |b|$ has the same property.*

Solution. Let x_1 and x_2 be the complex roots of the polynomial $f = x^2 + ax + b$ and let y_1 and y_2 be the complex roots of the polynomial $g = x^2 + |a|x + |b|$.

We have to prove that if $|x_1| = |x_2| = 1$, then $|y_1| = |y_2| = 1$.

Since $x_1 \cdot x_2 = b$ and $x_1 + x_2 = -a$, then $|b| = |x_1||x_2| = 1$ and $|a| \leq |x_1| + |x_2| = 2$.

The quadratic polynomial $g = x^2 + |a|x + 1$ has discriminant $\Delta = |a|^2 - 4 \leq 0$; hence

$$y_{1,2} = \frac{-|a| \pm i\sqrt{4 - |a|^2}}{2}.$$

It is easy to see that $|y_1| = |y_2| = 1$, as desired.

Problem 5. *Let a, b be nonzero complex numbers. Prove that the equation*

$$az^3 + bz^2 + \bar{b}z + \bar{a} = 0$$

has at least one root with absolute value equal to 1.

Solution. Observe that if z is a root of the equation, then $\dfrac{1}{\bar{z}}$ is also a root of the equation. Consequently, if z_1, z_2, z_3 are the roots of the equation, then $\dfrac{1}{\bar{z_1}}, \dfrac{1}{\bar{z_2}}, \dfrac{1}{\bar{z_3}}$ are the same roots, not necessarily in the same order.

If $z_k = \dfrac{1}{\bar{z_k}}$ for some $k = 1, 2, 3$, then $|z_k|^2 = z_k \bar{z_k} = 1$, and we are done. If $z_k \neq \dfrac{1}{\bar{z_k}}$ for all $k = 1, 2, 3$, we may consider without loss of generality that

$$z_1 = \frac{1}{\bar{z_2}}, z_2 = \frac{1}{\bar{z_3}}, z_3 = \frac{1}{\bar{z_1}}.$$

The first two equalities yield $z_1 \cdot \overline{z_2} \cdot z_2 \cdot \overline{z_3} = 1$; hence $|z_1| \cdot |z_2|^2 \cdot |z_3| = 1$. On the other hand, $z_1 z_2 z_3 = -\dfrac{a}{a}$, so $|z_1||z_2||z_3| = 1$. It follows that $|z_2| = 1$, as claimed.

Problem 6. Let $f = x^4 + ax^3 + bx^2 + cx + d$ be a polynomial with real coefficients and real roots. Prove that if $|f(i)| = 1$, then $a = b = c = d = 0$.

Solution. Let $x_1,\ x_2, x_3,\ x_4$ be the real roots of the polynomial f. Then

$$f = (x - x_1)(x - x_2)(x - x_3)(x - x_4),$$

and we have

$$f(i) = (-x_1 + i)(-x_2 + i)(-x_3 + i)(-x_4 + i);$$

hence

$$|f(i)| = |-x_1 + i| \cdot |-x_2 + i| \cdot |-x_3 + i| \cdot |-x_4 + i|$$
$$= \sqrt{1 + x_1^2} \cdot \sqrt{1 + x_2^2} \cdot \sqrt{1 + x_3^2} \cdot \sqrt{1 + x_4^2}.$$

Because $|f(i)| = 1$, we deduce that $x_1 = x_2 = x_3 = x_4 = 0$, and consequently $a = b = c = d = 0$, as desired.

Problem 7. Prove that if $11z^{10} + 10iz^9 + 10iz - 11 = 0$, then $|z| = 1$.

(1989 Putnam Mathematical Competition)

Solution. The equation can be rewritten as $z^9 = \dfrac{11 - 10iz}{11z + 10i}$. If $z = a + bi$, then

$$|z|^9 = \left| \frac{11 - 10iz}{11z + 10i} \right| = \frac{\sqrt{11^2 + 220b + 10^2(a^2 + b^2)}}{\sqrt{11^2(a^2 + b^2) + 220b + 10^2}}.$$

Let $f(a,\ b)$ and $g(a,\ b)$ denote the numerator and denominator of the right-hand side. If $|z| > 1$, then $a^2 + b^2 > 1$, so $g(a,\ b) > f(a,\ b)$, leading to $|z^9| < 1$, a contradiction. If $|z| < 1$, then $a^2 + b^2 < 1$, so $g(a,\ b) < f(a,\ b)$, yielding $|z^9| > 1$, again a contradiction. Hence $|z| = 1$.

Problem 8. Let $n \geq 3$ be an integer and let a be a nonzero real number. Show that every nonreal root z of the equation $x^n + ax + 1 = 0$ satisfies the inequality

$$|z| \geq \sqrt[n]{\frac{1}{n-1}}.$$

(Romanian Mathematical Olympiad—Final Round, 1995)

Solution. Let $z = r(\cos \alpha + i \sin \alpha)$ be a nonreal root of the equation, where $\alpha \in (0, 2\pi)$ and $\alpha \neq \pi$. Substituting back into the equation, we obtain $r^n \cos n\alpha + ra \cos \alpha + 1 + i(r^n \sin n\alpha + ra \sin \alpha) = 0$. Hence

$$r^n \cos n\alpha + ra \cos \alpha + 1 = 0 \text{ and } r^n \sin n\alpha + ra \sin \alpha = 0.$$

Multiplying the first relation by $\sin \alpha$, the second by $\cos \alpha$, and then subtracting them, we find that $r^n \sin(n-1)\alpha = \sin \alpha$. It follows that

$$r^n |\sin(n-1)\alpha| = |\sin \alpha|.$$

The inequality $|\sin k\alpha| \leq k|\sin \alpha|$ is valid for every positive integer k. The proof is based on a simple inductive argument on k.

Applying this inequality, from $r^n |\sin(n-1)\alpha| = |\sin \alpha|$, we obtain $|\sin \alpha| \leq r^n (n-1)|\sin \alpha|$. Because $\sin \alpha \neq 0$, it follows that $r^n \geq \dfrac{1}{n-1}$, i.e., $|z| \geq \sqrt[n]{\dfrac{1}{n-1}}$.

Problem 9. *Suppose P is a polynomial of even degree with complex coefficients. If all the roots of P are complex nonreal numbers with modulus 1, prove that*

$$P(1) \in \mathbb{R} \text{ if and only if } P(-1) \in \mathbb{R}.$$

Solution. It suffices to prove that $\dfrac{P(1)}{P(-1)} \in \mathbb{R}$.

Let $x_1, x_2, \ldots, x_{2n}$ be the roots of P. Then

$$P(x) = \lambda(x - x_1)(x - x_2) \cdots (x - x_{2n})$$

for some $\lambda \in \mathbb{C}^*$, and

$$\frac{P(1)}{P(-1)} = \frac{\lambda(1 - x_1)(1 - x_2) \cdots (1 - x_{2n})}{\lambda(-1 - x_1)(-1 - x_2) \cdots (-1 - x_{2n})} = \prod_{k=1}^{2n} \frac{1 - x_k}{1 + x_k}.$$

From the hypothesis, we have $|x_k| = 1$ for all $k = 1, 2, \ldots, 2n$. Then

$$\overline{\left(\frac{1 - x_k}{1 + x_k}\right)} = \frac{1 - \overline{x_k}}{1 + \overline{x_k}} = \frac{1 - \dfrac{1}{x_k}}{1 + \dfrac{1}{x_k}} = \frac{x_k - 1}{x_k + 1} = -\frac{1 - x_k}{1 + x_k},$$

whence

$$\overline{\left(\frac{P(1)}{P(-1)}\right)} = \prod_{k=1}^{2n} \overline{\left(\frac{1 - x_k}{1 + x_k}\right)} = \prod_{k=1}^{2n} \left(-\frac{1 - x_k}{1 + x_k}\right)$$

$$= (-1)^{2n} \prod_{k=1}^{2n} \frac{1 - x_k}{1 + x_k} = \frac{P(1)}{P(-1)}.$$

This proves that $\dfrac{P(1)}{P(-1)}$ is a real number, as desired.

Problem 10. *Consider the sequence of polynomials defined by $P_1(x) = x^2 - 2$ and $P_j(x) = P_1(P_{j-1}(x))$ for $j = 2, 3, \ldots$. Show that for every positive integer n, the roots of the equation $P_n(x) = x$ are all real and distinct.*

(18th IMO—Shortlist)

Solution. Put $x = z + z^{-1}$, where z is a nonzero complex number. Then $P_1(x) = x^2 - 2 = (z + z^{-1})^2 - 2 = z^2 + z^{-2}$. A simple inductive argument shows that for all positive integers n, we have $P_n(x) = z^{2^n} + z^{-2^n}$.

The equation $P_n(x) = x$ is equivalent to $z^{2^n} + z^{-2^n} = z + z^{-1}$. We obtain $z^{2^n} - z = z^{-1} - z^{-2^n}$, i.e., $z(z^{2^n-1} - 1) = z^{-2^n}(z^{2^n-1} - 1)$. It follows that $(z^{2^n-1} - 1)(z^{2^n+1} - 1) = 0$. Because $\gcd(2^n - 1, 2^n + 1) = 1$, the unique common root of the equations $z^{2^n-1} - 1 = 0$ and $z^{2^n+1} - 1 = 0$ is $z = 1$ (see Proposition 1 in Sect. 2.2.2). Moreover, for every root of the equation $(z^{2^n-1} - 1)(z^{2^n+1} - 1) = 0$, we have $|z| = 1$, i.e., $z^{-1} = \bar{z}$. Also, observe that for two roots z and w of $(z^{2^n-1} - 1)(z^{2^n+1} - 1) = 0$ that are different from 1, we have $z + z^{-1} = w + w^{-1}$ if and only if $(z - w)(1 - (zw)^{-1}) = 0$. This is equivalent to $zw = 1$, i.e., $w = z^{-1} = \bar{z}$, a contradiction to the fact that the unique common root of $z^{2^n-1} - 1 = 0$ and $z^{2^n+1} - 1 = 0$ is 1.

It is clear that the degree of the polynomial P_n is 2^n. As we have seen before, all the roots of $P_n(x) = x$ are given by $x = z + z^{-1}$, where $z = 1$, $z = \cos\dfrac{2k\pi}{2^n - 1} + i\sin\dfrac{2k\pi}{2^n - 1}$, $k = 1, \ldots, 2^n - 2$, and $z = \cos\dfrac{2s\pi}{2^n + 1} + i\sin\dfrac{2s\pi}{2^n + 1}$, $s = 1, \ldots, 2^n$.

Taking into account the symmetry of the expression $z + z^{-1}$, we see that the total number of these roots is $1 + \dfrac{1}{2}(2^n - 2) + \dfrac{1}{2}2^n = 2^n$, and all of them are real and distinct.

Here are other problems involving algebraic equations and polynomials.

Problem 11. *Let a, b, c be complex numbers with $a \neq 0$. Prove that if the roots of the equation $az^2 + bz + c = 0$ have equal moduli, then $\bar{a}b|c| = |a|\bar{b}c$.*

Problem 12. *Let z_1, z_2 be the roots of the equation $z^2 + z + 1 = 0$, and let z_3, z_4 be the roots of the equation $z^2 - z + 1 = 0$. Find all integers n such that $z_1^n + z_2^n = z_3^n + z_4^n$.*

Problem 13. *Consider the equation with real coefficients*

$$x^6 + ax^5 + bx^4 + cx^3 + bx^2 + ax + 1 = 0,$$

and denote by $x_1, x_2, \ldots, x_6$ the roots of the equation.
   Prove that

$$\prod_{k=1}^{6}(x_k^2 + 1) = (2a - c)^2.$$

Problem 14. *Let a and b be complex numbers and let $P(z) = az^2 + bz + i$. Prove that there exists $z_0 \in \mathbb{C}$ with $|z_0| = 1$ such that $|P(z_0)| \geq 1 + |a|$.*

Problem 15. *Find all polynomials f with real coefficients satisfying, for every real number x, the relation $f(x)f(2x^2) = f(2x^3 + x)$.*

(21st IMO—Shortlist)

Problem 16. *Find all complex numbers z such that*

$$(z - z^2)(1 + z + z^2)^2 = \frac{1}{7}.$$

(Mathematical Reflections, 2013)

Problem 17. *Determine all pairs (z, n) such that*

$$z + z^2 + \ldots + z^n = n|z|,$$

where $z \in \mathbb{C}$ and $|z| \in \mathbb{Z}_+$.

(Mathematical Reflections, 2008)

Problem 18. *Let a, b, c, d be nonzero complex numbers such that $ad - bc \neq 0$, and let n be a positive integer. Consider the equation*

$$(ax + b)^n + (cx + d)^n = 0.$$

(a) Prove that for $|a| = |c|$, the roots of the equation are situated on a line.
(b) Prove that for $|a| \neq |c|$, the roots of the equation are situated on a circle.
(c) Find the radius of the circle when $|a| \neq |c|$.

(Mathematical Reflections, 2010)

Problem 19. *Let n be a positive integer. Prove that a complex number of absolute value 1 is a solution to $z^n + z + 1 = 0$ if and only if $n = 3m + 2$ for some positive integer m.*

(Romanian Mathematical Olympiad—Final Round, 2007)

Problem 20. *Let a and b be two complex numbers. Prove that the following statements are equivalent:*

(1) The absolute values of the roots of the equation $x^2 - ax + b = 0$ are respectively equal to the absolute values of the roots of the equation

$$x^2 - bx + a = 0.$$

(2) $a^3 = b^3$ or $b = \overline{a}$.

(Romanian Mathematical Olympiad—District Round, 2011)

5.3 From Algebraic Identities to Geometric Properties

Problem 1. *Consider equilateral triangles ABC and $A'B'C'$, both in the same plane and having the same orientation. Show that the segments $[AA']$, $[BB']$, $[CC']$ can be the sides of a triangle.*

Solution. Let a, b, c be the coordinates of vertices A, B, C and let a', b', c' be the coordinates of vertices A', B', C'. Because triangles ABC and $A'B'C'$ are similar, we have the relation (see the remark in Sect. 3.3)

$$\begin{bmatrix} 1 & 1 & 1 \\ a & b & c \\ a' & b' & c' \end{bmatrix} = 0. \tag{1}$$

That is,

$$a'(b - c) + b'(c - a) + c'(a - b) = 0. \tag{2}$$

On the other hand, the following relation is clear:

$$a(b - c) + b(c - a) + c(a - b) = 0. \tag{3}$$

By subtracting relation (3) from relation (2), we obtain

$$(a' - a)(b - c) + (b' - b)(c - a) + (c' - c)(a - b) = 0. \tag{4}$$

Passing to moduli, it follows that

$$|a' - a||b - c| \le |b' - b||c - a| + |c' - c||a - b|. \tag{5}$$

Taking into account that $|b-c| = |c-a| = |a-b|$, we obtain $AA' \le BB' + CC'$. Similarly, we prove the inequalities $BB' \le CC' + AA'$ and $CC' \le AA' + BB'$, and the desired conclusion follows.

Remarks.

(1) If ABC and $A'B'C'$ are two similar triangles situated in the same plane and having the same orientation, then from (5), the inequality

$$AA' \cdot BC \le BB' \cdot CA + CC'. \, AB \tag{1}$$

follows. This is the *generalized Ptolemy inequality*. Ptolemy's inequality is obtained when the triangle $A'B'C'$ degenerates to a point.

(2) Taking into account the inequality (1), we have also $BB' \cdot CA \le CC' \cdot AB + AA' \cdot BC$ and $CC' \cdot AB \le AA' \cdot BC + BB' \cdot CA$. It follows that for any two similar triangles ABC and $A'B'C'$ with the same orientation and situated in the same plane, we can construct a triangle of side lengths $AA' \cdot BC$, $BB' \cdot CA$, $CC'. \, AB$.

(3) When the triangle $A'B'C'$ degenerates to the point M, it follows from the property in our problem that the segments MA, MB, MC are the sides of a triangle, which follows from Pompeiu's theorem (see also Sect. 4.9.1).

Problem 2. *Let P be an arbitrary point in the plane of a triangle ABC. Then*

$$\alpha \cdot PB \cdot PC + \beta \cdot PC \cdot PA + \gamma \cdot PA \cdot PB \geq \alpha\beta\gamma,$$

where α, β, γ are the sides of ABC.

Solution. Let us consider the origin of the complex plane at P and let a, b, c be the coordinates of the vertices of triangle ABC. From the algebraic identity

$$\frac{bc}{(a-b)(a-c)} + \frac{ca}{(b-c)(b-a)} + \frac{ab}{(c-a)(c-b)} = 1, \qquad (1)$$

it follows if we pass to absolute values that

$$\frac{|b||c|}{|a-b||a-c|} + \frac{|c||a|}{|b-c||b-a|} + \frac{|a||b|}{|c-a||c-b|} \geq 1. \qquad (2)$$

Taking into account that $|a| = PA$, $|b| = PB$, $|c| = PC$, and $|b-c| = \alpha$, $|c-a| = \beta$, $|a-b| = \gamma$, we see that the inequality (2) is equivalent to

$$\frac{PB \cdot PC}{\beta\gamma} + \frac{PC \cdot PA}{\gamma\alpha} + \frac{PA \cdot PB}{\alpha\beta} \geq 1,$$

which is the desired inequality.

Remarks.

(1) If P is the circumcenter O of triangle ABC, we can derive Euler's inequality $R \geq 2r$. Indeed, in this case, the inequality is equivalent to $R^2(\alpha + \beta + \gamma) \geq \alpha\beta\gamma$. Therefore,

$$R^2 \geq \frac{\alpha\beta\gamma}{\alpha+\beta+\gamma} = \frac{\alpha\beta\gamma}{2s} = \frac{4R}{2s} \cdot \frac{\alpha\beta\gamma}{4R} = 2R \cdot \frac{\text{area}[ABC]}{s} = 2Rr,$$

and hence $R \geq 2r$.

(2) If P is the centroid G of triangle ABC, we obtain the following inequality involving the medians m_α, m_β, m_γ:

$$\frac{m_\alpha m_\beta}{\alpha\beta} + \frac{m_\beta m_\gamma}{\beta\gamma} + \frac{m_\gamma m_\alpha}{\gamma\alpha} \geq \frac{9}{4},$$

with equality if and only if triangle ABC is equilateral. A good argument for the case of acute triangles is given in the next problem.

Problem 3. *Let ABC be an acute triangle and let P be a point in its interior. Prove that*

$$\alpha \cdot PB \cdot PC + \beta \cdot PC \cdot PA + \gamma \cdot PA \cdot PB = \alpha\beta\gamma$$

if and only if P is the orthocenter of triangle ABC.

(1998 Chinese Mathematical Olympiad)

Solution. Let P be the origin of the complex plane, and let a, b, c be the coordinates of A, B, C, respectively. The relation in the problem is equivalent to

$$|ab(a-b)| + |bc(b-c)| + |ca(c-a)| = |(a-b)(b-c)(c-a)|.$$

Let

$$z_1 = \frac{ab}{(a-c)(b-c)}, \quad z_2 = \frac{bc}{(b-a)(c-a)}, \quad z_3 = \frac{ca}{(c-b)(a-b)}.$$

It follows that

$$|z_1| + |z_2| + |z_3| = 1 \text{ and } z_1 + z_2 + z_3 = 1,$$

the latter from identity (1) in the previous problem.

We will prove that P is the orthocenter of triangle ABC if and only if z_1, z_2, z_3 are positive real numbers. Indeed, if P is the orthocenter, then since the triangle ABC is acute, it follows that P is in the interior of ABC. Hence there are positive real numbers r_1, r_2, r_3 such that

$$\frac{a}{b-c} = -r_1 i, \quad \frac{b}{c-a} = -r_2 i, \quad \frac{c}{a-b} = -r_3 i,$$

implying $z_1 = r_1 r_2 > 0$, $z_2 = r_2 r_3 > 0$, $z_3 = r_3 r_1 > 0$, and we are done. Conversely, suppose that z_1, z_2, z_3 are all positive real numbers. Because

$$-\frac{z_1 z_2}{z_3} = \left(\frac{b}{c-a}\right)^2, \quad -\frac{z_2 z_3}{z_1} = \left(\frac{c}{a-b}\right)^2, \quad -\frac{z_3 z_1}{z_2} = \left(\frac{a}{b-c}\right)^2,$$

it follows that $\dfrac{a}{b-c}$, $\dfrac{b}{c-a}$, $\dfrac{c}{a-b}$ are purely imaginary numbers, thus $AP \perp BC$ and $BP \perp CA$, showing that P is the orthocenter of triangle ABC.

Problem 4. *Let G be the centroid of triangle ABC and let R_1, R_2, R_3 be the circumradii of triangles GBC, GCA, GAB, respectively. Then*

$$R_1 + R_2 + R_3 \geq 3R,$$

where R is the circumradius of triangle ABC.

Solution. In Problem 2, consider P the centroid G of triangle ABC. Then

$$\alpha \cdot GB \cdot GC + \beta \cdot GC \cdot GA + \gamma \cdot GA \cdot GB \geq \alpha\beta\gamma, \tag{1}$$

where α, β, V are the lengths of the sides of triangle ABC.
 But

$$\alpha \cdot GB \cdot GC = 4R_1 \cdot \text{area}[GBC] = 4R_1 \cdot \frac{1}{3}\text{area}[ABC].$$

Likewise,

$$\beta \cdot GC \cdot GA = 4R_2 \cdot \frac{1}{3}\text{area}[ABC], \quad \gamma \cdot GA \cdot GB = 4R_3 \cdot \frac{1}{3}\text{area}[ABC].$$

Hence, the inequality (1) is equivalent to

$$\frac{4}{3}(R_1 + R_2 + R_3). \text{ area } [ABC] \geq 4R. \text{ area } [ABC],$$

i.e., $R_1 + R_2 + R_3 \geq 3R$.

Problem 5. *Let ABC be a triangle and let P be a point in its interior. Let R_1, R_2, R_3 be the radii of the circumcircles of triangles PBC, PCA, PAB, respectively. Lines PA, PB, PC intersect sides BC, CA, AB at A_1, B_1, C_1, respectively. Let*

$$k_1 = \frac{PA_1}{AA_1}, \quad k_2 = \frac{PB_1}{BB_1}, \quad k_3 = \frac{PC_1}{CC_1}.$$

Prove that $k_1 R_1 + k_2 R_2 + k_3 R_3 \geq R$, where R is the circumradius of triangle ABC.

(2004 Romanian IMO Team Selection Test)

Solution. Note that

$$k_1 = \frac{\text{area}[PBC]}{\text{area}[ABC]}, \quad k_2 = \frac{\text{area}[PCA]}{\text{area}[ABC]}, \quad k_3 = \frac{\text{area}[PAB]}{\text{area}[ABC]}.$$

But area $[ABC] = \dfrac{\alpha\beta\gamma}{4R}$ and area $[PBC] = \dfrac{\alpha \cdot PB \cdot PC}{4R_1}$. Two similar relations for area $[PCA]$ and area $[PAB]$ hold.
 The desired inequality is equivalent to

$$R\frac{\alpha \cdot PB \cdot PC}{\alpha\beta\gamma} + R\frac{\beta \cdot PC \cdot PA}{\alpha\beta\gamma} + R\frac{\gamma \cdot PA \cdot PB}{\alpha\beta\gamma} \geq R,$$

which reduces to the inequality in Problem 2.
 When triangle ABC is acute, it follows from Problem 3 that equality holds if and only if P is the orthocenter of ABC.

Problem 6. *The following inequality holds for every point M in the plane of triangle ABC:*

$$AM^3 \sin A + BM^3 \sin B + CM^3 \sin C \geq 6 \cdot MG. \text{ area } [ABC],$$

where G is the centroid of triangle ABC.

Solution. The identity

$$x^3(y-z) + y^3(z-x) + z^3(x-y) = (x-y)(y-z)(z-x)(x+y+z) \quad (1)$$

holds for all complex numbers x, y, z. Passing to the absolute value, we obtain the inequality

$$|x^3(y-z)| + |y^3(z-x)| + |z^3(x-y)| \geq |x-y||y-z||z-x||x+y+z|. \quad (2)$$

Let a, b, c, m be the coordinates of points A, B, C, M, respectively. In (2), consider $x = m - a$, $y = m - b$, $z = m - c$, and obtain

$$AM^3 \cdot \alpha + BM^3 \cdot \beta + CM^3 \cdot \gamma \geq 3\alpha\beta\gamma MG. \quad (3)$$

Using the formula area $[ABC] = \dfrac{\alpha\beta\gamma}{4R}$ and the law of sines, the desired inequality follows from (3).

Problem 7. *Let $ABCD$ be a cyclic quadrilateral inscribed in circle $\mathcal{C}(O; R)$ having sides of length α, β, γ, δ and diagonals of length d_1 and d_2. Then*

$$\text{area}[ABCD] \geq \frac{\alpha\beta\gamma\delta d_1 d_2}{8R^4}.$$

Solution. Take the center O to be the origin of the complex plane and consider a, b, c, d the coordinates of vertices A, B, C, D. From the well-known Euler identity

$$\sum_{\text{cyc}} \frac{a^3}{(a-b)(a-c)(a-d)} = 1, \quad (1)$$

by passing to the absolute value, it follows that

$$\sum_{\text{cyc}} \frac{|a|^3}{|a-b||a-c||a-d|} \geq 1. \quad (2)$$

The inequality (2) is equivalent to

$$\sum_{\text{cyc}} \frac{R^3}{AB \cdot AC \cdot AD} \geq 1, \quad (3)$$

or

$$\sum_{\text{cyc}} R^3 \cdot BD \cdot CD \cdot BC \geq \alpha\beta\gamma\delta d_1 d_2. \tag{4}$$

But we have the known relation $BD \cdot CD \cdot BC = 4R$. area $[BCD]$ and three other such relations. The inequality (4) can be written in the form

$$4R^4(\text{area}[ABC] + \text{area}[BCD] + \text{area}[CDA] + \text{area}[DAB]) \geq \alpha\beta\gamma\delta d_1 d_2,$$

or equivalently, $8R^4$ area$[ABCD] \geq \alpha\beta\gamma\delta d_1 d_2$.

Problem 8. *Let a, b, c be distinct complex numbers such that*

$$(a - b)^7 + (b - c)^7 + (c - a)^7 = 0.$$

Prove that a, b, c are the coordinates of the vertices of an equilateral triangle.

Solution 1. Setting $x = a - b$, $y = b - c$, $z = c - a$ implies $x + y + z = 0$ and $x^7 + y^7 + z^7 = 0$. Since $z \neq 0$, we may set $\alpha = \dfrac{x}{z}$ and $\beta = \dfrac{y}{z}$. Hence $\alpha + \beta = -1$ and $\alpha^7 + \beta^7 = -1$. Then the given relation becomes

$$\alpha^6 - \alpha^5\beta + \alpha^4\beta^2 - \alpha^3\beta^3 + \alpha^2\beta^4 - \alpha\beta^5 + \beta^6 = 1. \tag{1}$$

Let $s = \alpha + \beta = -1$ and $p = ab$. The relation (1) becomes

$$(\alpha^6 + \beta^6) - p(\alpha^4 + \beta^4) + p^2(\alpha^2 + \beta^2) - p^3 = 1. \tag{2}$$

Because $\alpha^2 + \beta^2 = s^2 - 2p = 1 - 2p$,

$$\alpha^4 + \beta^4 = (\alpha^2 + \beta^2)^2 - 2\alpha^2\beta^2 = (1 - 2p)^2 - 2p^2 = 1 - 4p + 2p^2,$$
$$\alpha^6 + \beta^6 = (\alpha^2 + \beta^2)((\alpha^4 + \beta^4) - \alpha^2\beta^2) = (1 - 2p)(1 - 4p + p^2),$$

the equality (2) is equivalent to

$$(1 - 2p)(1 - 4p + p^2) - p(1 - 4p + 2p^2) + p^2(1 - 2p) - p^3 = 1.$$

That is, $1 - 4p + p^2 - 2p + 8p^2 - 2p^3 - p + 4p^2 - 2p^3 + p^2 - 2p^3 - p^3 = 1$; i.e., $-7p^3 + 14p^2 - 7p + 1 = 1$. We obtain $7p(p - 1)^2 = 0$, and hence $p = 0$ or $p = 1$.

If $p = 0$, then $\alpha = 0$ or $\beta = 0$, and consequently, $x = 0$ or $y = 0$. It follows that $a = b$ or $b = c$, which is false; hence $p = 1$.

From $\alpha\beta = 1$ and $\alpha + \beta = -1$, we deduce that α and β are the roots of the quadratic equation $x^2 + x + 1 = 0$. Thus $\alpha^3 = \beta^3 = 1$ and $|\alpha| = |\beta| = 1$. Therefore, $|x| = |y| = |z|$, or $|a - b| = |b - c| = |c - a|$, as claimed.

Solution 2. Let $x = a - b$, $y = b - c$, $z = c - a$. Because $x + y + z = 0$ and $x^7 + y^7 + z^7 = 0$, we find that $(x + y)^7 - x^7 - y^7 = 0$. This is equivalent to $7xy(x + y)(x^2 + xy + y^2)^2 = 0$.

But $xyz \neq 0$, so $x^2 + xy + y^2 = 0$, i.e., $x^3 = y^3$. From symmetry, $x^3 = y^3 = z^3$, whence $|x| = |y| = |z|$.

Problem 9. *Let M be a point in the plane of the square $ABCD$ and let $MA = x$, $MB = y$, $MC = z$, $MD = t$. Prove that the numbers xy, yz, zt, tx are the sides of a quadrilateral.*

Solution. Consider the complex plane such that $1, i, -1, -i$ are the coordinates of vertices A, B, C, D of the square. If z is the coordinate of point M, then we have the identity

$$1(z-i)(z+1) + i(z+1)(z+i) - 1(z+i)(z-1) - i(z-1)(z-i) = 0. \quad (1)$$

Subtracting the first term of the sum from both sides yields

$$i(z+1)(z+i) - 1(z+i)(z-1) - i(z-1)(z-i) = -1(z-i)(z+1),$$

and using the triangle inequality, we obtain

$$|z-i||z+1| + |z+1||z+i| + |z+i||z-1| \geq |z-1||z-i|,$$

or $yz + zt + tx \geq xy$.

In the same manner, we prove that

$$xy + zt + tx \geq yz, \quad xy + yz + tx \geq zt$$

and $xy + yz + zt \geq tx$, as needed.

Problem 10. *Let z_1, z_2, z_3 be distinct complex numbers such that $|z_1| = |z_2| = |z_3| = R$. Prove that*

$$\frac{1}{|z_1 - z_2||z_1 - z_3|} + \frac{1}{|z_2 - z_1||z_2 - z_3|} + \frac{1}{|z_3 - z_1||z_3 - z_2|} \geq \frac{1}{R^2}.$$

Solution 1. The following identity is easy to verify:

$$\frac{z_1^2}{(z_1 - z_2)(z_1 - z_3)} + \frac{z_2^2}{(z_2 - z_1)(z_2 - z_3)} + \frac{z_3^2}{(z_3 - z_1)(z_3 - z_2)} = 1.$$

Passing to the absolute value, we find that

$$1 = \left| \sum_{\text{cyc}} \frac{z_1^2}{(z_1 - z_2)(z_1 - z_3)} \right| \leq \sum_{\text{cyc}} \frac{|z_1|^2}{|z_1 - z_2||z_1 - z_3|}$$

$$= R^2 \sum_{\text{cyc}} \frac{1}{|z_1 - z_2||z_1 - z_3|},$$

which is the desired inequality.

Solution 2. Let

$$\alpha = |z_2 - z_3|, \ \beta = |z_3 - z_1|, \ \gamma = |z_1 - z_2|.$$

From Problem 29 in Sect. 1.1.9, we have

$$\alpha\beta + \beta\gamma + \gamma\alpha \le 9R^2.$$

Using the inequality

$$(\alpha\beta + \beta\gamma + \gamma\alpha) \left(\frac{1}{\alpha\beta} + \frac{1}{\beta\gamma} + \frac{1}{\gamma\alpha} \right) \ge 9$$

yields

$$\frac{1}{\alpha\beta} + \frac{1}{\beta\gamma} + \frac{1}{\gamma\alpha} \ge \frac{9}{\alpha\beta + \beta\gamma + \gamma\alpha} \ge \frac{1}{R^2},$$

as desired.

Remark. Consider the triangle with vertices at z_1, z_2, z_3 and whose circumcenter is the origin of the complex plane. Then the circumradius R equals $|z_1| = |z_2| = |z_3|$, and the side lengths are

$$\alpha = |z_2 - z_3|, \ \beta = |z_1 - z_3|, \ \gamma = |z_1 - z_2|.$$

The above inequality is equivalent to

$$\frac{1}{\alpha\beta} + \frac{1}{\beta\gamma} + \frac{1}{\gamma\alpha} \ge \frac{1}{R^2},$$

i.e.,

$$\alpha + \beta + \gamma \ge \frac{\alpha\beta\gamma}{R^2} = \frac{4K}{R} = \frac{4sr}{R}.$$

We obtain $R \ge 2r$, i.e., Euler's inequality for a triangle.

Problem 11. *Let ABC be a triangle and let P be a point in its plane.*

(1) Prove that

$$\alpha \cdot PA^3 + \beta \cdot PB^3 + \gamma \cdot PC^3 \ge 3\alpha\beta\gamma \cdot PG,$$

where G is the centroid of ABC.
(2) Prove that

$$R^2(R^2 - 4r^2) \ge 4r^2[8R^2 - (\alpha^2 + \beta^2 + \gamma^2)].$$

Solution.

(1) The identity

$$x^3(y - z) + y^3(z - x) + z^3(x - y) = (x - y)(y - z)(z - x)(x + y + z) \quad (1)$$

holds for all complex numbers x, y, z. Passing to absolute values, we obtain

$$|x|^3|y-z| + |y|^3|z-x| + |z|^3|x-y| \geq |x-y||y-z||z-x||x+y+z|.$$

Let a, b, c, z_P be the coordinates of A, B, C, P, respectively. In the equation above, take $x = zp - a$, $y = zp - b$, $z = zp - c$ and obtain the desired inequality.

(2) If P is the circumcenter O of triangle ABC, after some elementary transformations, the previous inequality becomes $R^2 \geq 6r \cdot OG$. Squaring both sides yields $R^4 \geq 36r^2 \cdot OG^2$. Using the well-known relation $OG^2 = R^2 - \frac{1}{9}(\alpha^2 + \beta^2 + \gamma^2)$, we obtain $R^4 \geq 36R^2r^2 - 4r^2(\alpha^2 + \beta^2 + \gamma^2)$, and the conclusion follows.

Remark. The inequality (2) improves Euler's inequality for the class of obtuse triangles. This is equivalent to proving that $\alpha^2 + \beta^2 + \gamma^2 < 8R^2$ in every such triangle. The last relation can be written as $\sin^2 A + \sin^2 B + \sin^2 C < 2$, or $\cos^2 A + \cos^2 B - \sin^2 C > 0$. That is,

$$\frac{1 + \cos 2A}{2} + \frac{1 + \cos 2B}{2} - 1 + \cos^2 C > 0,$$

which reduces to $\cos(A + B)\cos(A - B) + \cos^2 C > 0$. This is equivalent to $\cos C[\cos(A - B) + \cos(A + B)] < 0$, i.e., $\cos A \cos B \cos C < 0$.

Here are some other problems involving this topic.

Problem 12. *Let a, b, c, d be distinct complex numbers with $|a| = |b| = |c| = |d|$ and $a + b + c + d = 0$.*

Then the geometric images of a, b, c, d are the vertices of a rectangle.

Problem 13. *The complex numbers z_i, $i = 1, 2, 3, 4, 5$, have the same nonzero modulus, and*

$$\sum_{i=1}^{5} z_i = \sum_{i=1}^{5} z_i^2 = 0.$$

Prove that z_1, z_2, $\ldots$, z_5 are the coordinates of the vertices of a regular pentagon.

(Romanian Mathematical Olympiad—Final Round, 2003)

Problem 14. *Let ABC be a triangle.*

(a) Prove that if M is any point in its plane, then

$$AM \ \sin A \leq BM \sin B + CM \sin C.$$

(b) Let A_1, B_1, C_1 be points on the sides BC, AC and AB, respectively, such that the angles of the triangle $A_1B_1C_1$ are in this order α, β, γ. Prove that

$$\sum_{cyc} AA_1 \sin \alpha \le \sum_{cyc} BC \sin \alpha.$$

(Romanian Mathematical Olympiad—Second Round, 2003)

Problem 15. *Let M and N be points inside triangle ABC such that*

$$\widehat{MAB} = \widehat{NAC} \text{ and } \widehat{MBA} = \widehat{NBC}.$$

Prove that

$$\frac{AM \cdot AN}{AB \cdot AC} + \frac{BM \cdot BN}{BA \cdot BC} + \frac{CM \cdot CN}{CA \cdot CB} = 1.$$

(39th IMO—Shortlist)

5.4 Solving Geometric Problems

Problem 1. *On each side of a parallelogram, a square is drawn external to the figure. Prove that the centers of the squares are the vertices of another square.*

Solution. Consider the complex plane with origin at the intersection point of the diagonals and let $a, b, -a, -b$ be the coordinates of the vertices A, B, C, D, respectively.

Using the rotation formulas, we obtain

$$b = z_{O_1} + (a - z_{O_1})(-i) \text{ or } z_{O_1} = \frac{b + ai}{1 + i}.$$

Likewise,

$$z_{O_2} = \frac{a - bi}{1 + i}, \ z_{O_3} = \frac{-b - ai}{1 + i}, \ z_{O_4} = \frac{-a + bi}{1 + i}.$$

It follows that

$$\widehat{O_4O_1O_2} = \arg \frac{z_{O_2} - z_{O_1}}{z_{O_4} - z_{O_1}} = \arg \frac{a - bi - b - ai}{-a + bi - b - ai} = \arg i = \frac{\pi}{2},$$

so $O_1O_2 = O_1O_4$, and

$$\widehat{O_2O_3O_4} = \arg \frac{z_{O_4} - z_{O_4}}{z_{O_2} - z_{O_3}} = \arg \frac{-a + bi + b + ai}{a - bi + b + ai} = \arg i = \frac{\pi}{2},$$

so $O_3O_4 = O_3O_2$. Therefore, $O_1O_2O_3O_4$ is a square.

Problem 2. *Given a point on the circumcircle of a cyclic quadrilateral, prove that the products of the distances from the point to any pair of opposite sides or to the diagonals are equal.*

(Pappus's theorem)

Solution. Let a, b, c, d be the coordinates of the vertices A, B, C, D of the quadrilateral and consider the complex plane with origin at the circumcenter of $ABCD$. Without loss of generality assume that the circumradius equals 1.

The equation of line AB is

$$\begin{vmatrix} a & \bar{a} & 1 \\ b & \bar{b} & 1 \\ z & \bar{z} & 1 \end{vmatrix} = 0.$$

This is equivalent to

$$z(\bar{a} - \bar{b}) - \bar{z}(a - b) = \bar{a}b - a\bar{b}, \text{ i.e., } z + ab\bar{z} = a + b.$$

Let point M_1 be the foot of the perpendicular from a point M on the circumcircle to the line AB. If m is the coordinate of point M, then (see the proposition in Sect. 4.5)

$$Z_{M_1} = \frac{m - ab\bar{m} + a + b}{2}$$

and

$$d(M, \ AB) = |m - m_1| = \left| m - \frac{m - ab\bar{m} + a + b}{2} \right| = \left| \frac{(m - a)(m - b)}{2m} \right|,$$

since $m\bar{m} = 1$.

Likewise,

$$d(M, \ BC) = \left| \frac{(m - b)(m - c)}{2m} \right|, \ d(M, \ CD) = \left| \frac{(m - c)(m - d)}{2m} \right|,$$

$$d(M, \ DA) = \left| \frac{(m - d)(m - a)}{2m} \right|, \ d(M, \ AC) = \left| \frac{(m - a)(m - c)}{2m} \right|,$$

and

$$d(M, \ BD) = \left| \frac{(m - b)(m - d)}{2m} \right|.$$

Thus,

$$d(M, AB) \cdot d(M, \ CD) = d(M, \ BC) \cdot d(M, \ DA) = d(M, \ AC) \cdot d(M, \ BD),$$

as claimed.

Problem 3. *Three equal circles $C_1(O_1;\ r)$, $C_2(O_2;r)$, and $C_3(O_3;r)$ have a common point O. Circles C_1 and C_2, C_2 and C_3, C_3 and C_1, meet again at points A, B, C respectively. Prove that the circumradius of triangle ABC is equal to r.*

(Tzitzeica's[1] "five-lei-coin problem")

Solution. Consider the complex plane with origin at point O and let z_1, z_2, z_3 be the coordinates of the centers O_1, O_2, O_3, respectively. It follows that points A, B, C have the coordinates $z_1 + z_2$, $z_2 + z_3$, $z_3 + z_1$, and hence

$$AB = |(z_1 + z_2) - (z_2 + z_3)| = |z_1 - z_3| = O_1 O_3.$$

Likewise, $BC = O_1 O_2$ and $AC = O_2 O_3$; hence triangle ABC and $O_1 O_2 O_3$ are congruent. Consequently, their circumradii are equal. Since $OO_1 = OO_2 = OO_3 = r$, the circumradii of triangles $O_1 O_2 O_3$ and ABC are both equal to r, as desired.

Problem 4. *On the sides AB and BC of triangle ABC, draw squares with centers D and E such that points C and D lie on the same side of line AB and points A and E lie on opposite sides of line BC. Prove that the angle between lines AC and DE is equal to $45°$.*

Solution. The rotation about E through angle $90°$ maps point C to point B; hence

$$z_B = z_E + (z_C - z_E)i \text{ and } z_E = \frac{z_B - z_C i}{1 - i}.$$

Similarly, $z_D = \dfrac{z_B - z_A i}{1 - i}$.

The angle between the lines AC and DE is equal to

$$\arg \frac{z_C - z_A}{z_E - z_D} = \arg \frac{(z_C - z_A)(1 - i)}{z_B - z_C i - z_B + z_A^i} = \arg \frac{1 - i}{-i} = \arg(1 + i) = \frac{\pi}{4},$$

as desired (Fig. 5.1).

Remark. If on the sides AB and BC of the triangle ABC, we draw rectangles with centers D and E, satisfying the same conditions, then the angle between lines AC and DE is equal to $90° - \widehat{BAD}$.

Problem 5. *On the sides AB and BC of triangle ABC, equilateral triangles ABN and ACM are drawn external to the figure. If P, Q, R are the midpoints of segments BC, AM, AN, respectively, prove that triangle PQR is equilateral.*

Solution. Consider the complex plane with origin at A and denote by the corresponding lowercase letter the coordinate of a point denoted by an uppercase letter (Fig. 5.2).

---

[1] Gheorghe Tzitzeica (1873–1939), Romanian mathematician, made important contributions in geometry.

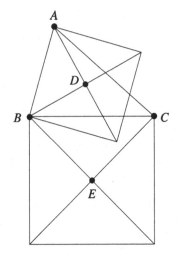

Figure 5.1.

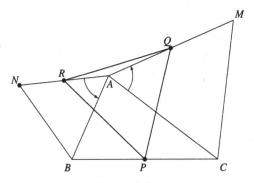

Figure 5.2.

The rotation about center A through angle $60°$ maps points N and C to B and M, respectively. Setting $\varepsilon = \cos 60° + i \sin 60°$, we have $b = n \cdot \varepsilon$ and $m = c \cdot \varepsilon$. Thus

$$p = \frac{b+c}{2}, \quad q = \frac{m}{2} = \frac{c \cdot \varepsilon}{2}, \quad r = \frac{n}{2} = \frac{b}{2\varepsilon} = \frac{b\varepsilon^5}{2} = -\frac{b\varepsilon^2}{2}.$$

To prove that triangle PQR is equilateral, using Proposition 1 in Sect. 3.4, it suffices to observe that

$$p^2 + q^2 + r^2 = pq + qr + rp.$$

Problem 6. *Let $AA'BB'CC'$ be a hexagon inscribed in the circle $\mathcal{C}(O; R)$ such that*

$$AA' = BB' = CC' = R.$$

If M, N, P are midpoints of sides $A'B$, $B'C$, $C'A$ respectively, prove that triangle MNP is equilateral.

Solution. Consider the complex plane with origin at the circumcenter O and let a, b, c, a', b', c' be the coordinates of the vertices A, B, C, A', B', C', respectively. If $\varepsilon = \cos 60° + i \sin 60°$, then

$$a' = a \cdot \varepsilon, \ b' = b \cdot \varepsilon, \ c' = c \cdot \varepsilon.$$

The points M, N, P have the coordinates

$$m = \frac{a\varepsilon + b}{2}, \ n = \frac{b\varepsilon + c}{2}, \ p = \frac{c\varepsilon + a}{2}.$$

It is easy to observe that

$$m^2 + n^2 + p^2 = mn + np + pm;$$

therefore, MNP is an equilateral triangle (see Proposition 1 in Sect. 3.4).

Problem 7. *On the sides AB and AC of triangle ABC, squares $ABDE$ and $ACFG$ are drawn external to the figure. If M is the midpoint of side BC, prove that $AM \perp EG$ and $EG = 2AM$.*

Solution. Consider the complex plane with origin at A and let b, c, g, e, m be the coordinates of points B, C, G, E, M (Fig. 5.3).

Observe that $g = ci$, $e = -bi$, $m = \dfrac{b+c}{2}$; hence

$$\frac{m - a}{g - e} = \frac{-(b+c)}{2i(b+c)} = \frac{i}{2} \in i\mathbb{R}^*$$

and

$$|m - a| = \frac{1}{2}|e - g|.$$

Thus, $AM \perp EG$ and $2AM = EG$.

Problem 8. *The sides AB, BC, and CA of the triangle ABC are divided into three equals parts by points M, N; P, Q; and R, S, respectively. Equilateral triangles MND, PQE, RSF are constructed exterior to triangle ABC. Prove that triangle DEF is equilateral.*

Solution. Denote by the corresponding lowercase letters the coordinates of the points denoted by uppercase letters. Then

$$m = \frac{2a + b}{3}, \ n = \frac{a + 2b}{3}, \ p = \frac{2b + c}{3},$$
$$q = \frac{b + 2c}{3}, \ r = \frac{2c + a}{3}, \ s = \frac{c + 2a}{3}.$$

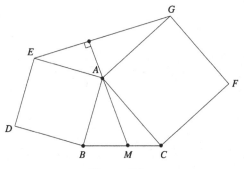

Figure 5.3.

The point D is obtained from point M by a rotation with center N and angle 60° (Fig. 5.4).

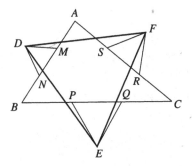

Figure 5.4.

Hence

$$d = n + (m - n)\varepsilon = \frac{a + 2b + (a - b)\varepsilon}{3},$$

where $\varepsilon = \cos 60° + i \sin 60°$. Likewise,

$$e = q + (p - q)\varepsilon = \frac{b + 2c + (b - c)\varepsilon}{3}$$

and

$$f = s + (r - s)\varepsilon = \frac{c + 2a + (c - a)\varepsilon}{3}.$$

Since

$$\frac{f - d}{e - d} = \frac{c + a - 2b + (b + c - 2a)\varepsilon}{2c - a - b + (2b - a - c)\varepsilon}$$

$$= \frac{\varepsilon(b + c - 2a + (c + a - 2b)(-\varepsilon^2))}{2c - a - b + (2b - a - c)\varepsilon}$$

$$= \frac{\varepsilon(b + c - 2a) + (c + a - 2b)(\varepsilon - 1))}{2c - a - b + (2b - a - c)\varepsilon} = \varepsilon,$$

we have $\widehat{FDE} = 60°$ and $FD = FE$, so triangle DEF is equilateral.

Problem 9. *Let $ABCD$ be a square of side length a and consider a point P on the incircle of the square. Find the value of*

$$PA^2 + PB^2 + PC^2 + PD^2.$$

Solution. Consider the complex plane such that points A, B, C, D have coordinates

$$z_A = \frac{a\sqrt{2}}{2}, \; z_B = \frac{a\sqrt{2}}{2}i, \; z_C = -\frac{a\sqrt{2}}{2}, \; z_D = -\frac{a\sqrt{2}}{2}i.$$

Let $z_P = \dfrac{a}{2}(\cos x + i \sin x)$ be the coordinate of point P (Fig. 5.5).

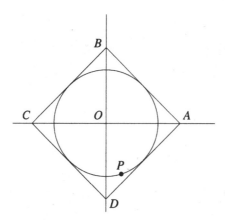

Figure 5.5.

Then

$$PA^2+PB^2+PC^2+PD^2=|z_A-z_P|^2+|z_B-z_P|^2+|z_c - z_P|^2 + |z_D - z_P|^2$$

$$= \sum_{\text{cyc}}(z_A - z_P)(\overline{z_A} - \overline{z_P}) = 4\frac{a^2}{2} + 2\frac{a\sqrt{2}}{2} \cdot \frac{a}{2}\left(2\cos x + 2\cos\left(x + \frac{\pi}{2}\right) + \right.$$

$$+2\cos(x + \pi) + 2\cos\left(x + \frac{3\pi}{2}\right)\right) + 4\frac{a^2}{4} = 2a^2 + 0 + a^2 = 3a^2.$$

Problem 10. *On the sides AB and AD of the triangle ABD draw externally squares $ABEF$ and $ADGH$ with centers O and Q, respectively. If M is the midpoint of the side BD, prove that OMQ is an isosceles triangle with a right angle at M.*

Solution. Let a, b, d be the coordinates of the points A, B, D, respectively (Fig. 5.6).

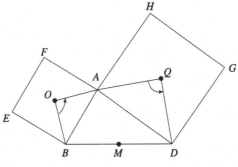

Figure 5.6.

The rotation formula gives

$$\frac{a - z_O}{b - z_O} = \frac{d - z_Q}{a - z_Q} = i,$$

so

$$z_O = \frac{b + a + (a - b)i}{2} \text{ and } z_Q = \frac{a + d + (d - a)i}{2}.$$

The coordinate of the midpoint M of segment $[BD]$ is $Z_M = \dfrac{b + d}{2}$; hence

$$\frac{z_O - z_M}{z_Q - z_M} = \frac{a - d + (a - b)i}{a - b + (d - a)i} = i.$$

Therefore, $QM \perp OM$ and $OM = QM$, as desired.

Problem 11. *On the sides of a convex quadrilateral $ABCD$, equilateral triangles ABM and CDP are drawn external to the figure, and equilateral triangles BCN and ADQ are drawn internal to the figure. Describe the shape of the quadrilateral $MNPQ$.*

(23rd IMO—Shortlist)

Solution. Denote by the corresponding lowercase letter the coordinate of a point denoted by an uppercase letter (Fig. 5.7).

Using the rotation formula, we obtain

$$m = a + (b - a)\varepsilon, \ n = c + (b - c)\varepsilon,$$
$$p = c + (d - c)\varepsilon, \ q = a + (d - a)\varepsilon,$$

where

$$\varepsilon = \cos 60° + i \sin 60°.$$

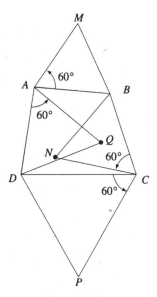

Figure 5.7.

It is easy to see that

$$m + p = a + c + (b + d - a - c)\varepsilon = n + q,$$

whence $MNPQ$ is a parallelogram or points M, N, P, Q are collinear.

Problem 12. *On the sides of a triangle ABC draw externally the squares $ABMM'$, $ACNN'$, and $BCPP'$. Let A', B', C' be the midpoints of the segments $M'N'$, $P'M$, PN, respectively.*
Prove that triangles ABC and $A'B'C'$ have the same centroid.

Solution. Denote by the corresponding lowercase letter the coordinate of a point denoted by an uppercase letter (Fig. 5.8).
Using the rotation formula, we obtain

$$n' = a + (c - a)i \text{ and } m' = a + (b - a)(-i);$$

hence

$$a' = \frac{m' + n'}{2} = \frac{2a + (c - b)i}{2}.$$

Likewise,

$$b' = \frac{2b + (a - c)i}{2} \text{ and } c' = \frac{2c + (b - a)i}{2}.$$

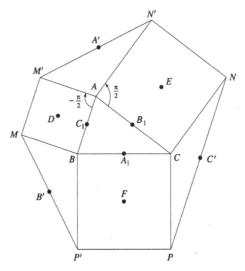

Figure 5.8.

Triangles $A'B'C'$ and ABC have the same centroid if and only if

$$\frac{a'+b'+c'}{3} = \frac{a+b+c}{3}.$$

Since

$$a'+b'+c' = \frac{2a+2b+2c+(c-b+a-c+b-a)i}{2} = a+b+c,$$

the conclusion follows.

Problem 13. *Let ABC be an acute triangle. On the same side of line AC as point B, draw isosceles triangles DAB, BCE, AFC with right angles at A, C, F, respectively.*

Prove that the points D, E, F are collinear.

Solution. Denote by the corresponding lowercase letters the coordinates of the points denoted by uppercase letters. The rotation formula gives

$$d = a+(b-a)(-i), \ e = c+(b-c)i, \ a = f+(c-f)i.$$

Then

$$f = \frac{a-ci}{1-i} = \frac{a+c+(a-c)i}{2} = \frac{d+e}{2},$$

so points F, D, E are collinear.

Problem 14. *On sides AB and CD of the parallelogram $ABCD$, draw external equilateral triangles ABE and CDF. On the sides AD and BC, draw external squares of centers G and H.*

Prove that $EHFG$ is a parallelogram.

Solution. Denote by the corresponding lowercase letter the coordinate of a point denoted by an uppercase letter.

Since $ABCD$ is a parallelogram, we have $a + c = b + d$ (Fig. 5.9).

The rotations with $90°$ and centers G and H map the points A and C into D and B, respectively. Then $d - g = (a - g)i$ and $b - h = (c - h)i$, whence

$$g = \frac{d - ai}{1 - i} \quad \text{and} \quad h = \frac{b - ci}{1 - i}.$$

The rotations with angle $60°$ and centers E and F map the points B and D into A and C, respectively. Then $a - e = (b - e)\varepsilon$ and $c - f = (d - f)\varepsilon$, where $\varepsilon = \cos 60° + i \sin 60°$. Hence $e = \dfrac{a - b\varepsilon}{1 - \varepsilon}$ and $f = \dfrac{c - d\varepsilon}{1 - \varepsilon}$.

Observe that

$$g + h = \frac{d + b - (a + c)i}{1 - i} = \frac{(a + c) - (a + c)i}{1 - i} = a + c$$

and

$$e + f = \frac{a + c - (b + d)\varepsilon}{1 - \varepsilon} = \frac{a + c - (a_c)\varepsilon}{1 - \varepsilon} = a + c;$$

hence $EHFG$ is a parallelogram.

Problem 15. *Let ABC be a right triangle with $\hat{C} = 90°$ and let D be the foot of the altitude from C. If M and N are the midpoints of the segments $[DC]$ and $[BD]$, prove that lines AM and CN are perpendicular.*

Solution 1. Consider the complex plane with origin at point C, and let a, b, d, m, n be the coordinates of points A, B, D, M, N, respectively (Fig. 5.10).

Triangles ABC and CDB are similar with the same orientation; hence

$$\frac{a - d}{d - 0} = \frac{0 - d}{d - b} \quad \text{or} \quad d = \frac{ab}{a + b}.$$

Then

$$m = \frac{d}{2} = \frac{ab}{2(a + b)} \quad \text{and} \quad n = \frac{b + d}{2} = \frac{2ab + b^2}{2(a + b)}.$$

Thus

$$\arg \frac{m - a}{n - 0} = \arg \frac{\dfrac{ab}{2(a + b)} - a}{\dfrac{2ab + b^2}{2(a + b)}} = \arg\left(-\frac{a}{b}\right) = \frac{\pi}{2},$$

so $AM \perp CN$.

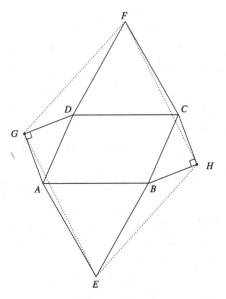

Figure 5.9.

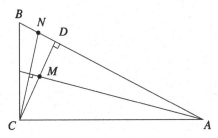

Figure 5.10.

Solution 2. Using the properties of the real product in Proposition 1, Sect. 4.1, and taking into account that $CA \perp CB$, we have

$$(m - a) \cdot (n - c) = \left(\frac{ab}{2(a+b)} - a \right) \cdot \left(\frac{2ab + b^2}{2(a+b)} \right)$$

$$= \left(a \frac{2a + b}{2(a+b)} \right) \cdot \left(b \frac{2a+b}{2(a+b)} \right) = \left| \frac{2a+b}{2(a+b)} \right|^2 (a \cdot b) = 0.$$

The conclusion follows from Proposition 2 in Sect. 4.1.

Problem 16. *Let ABC be an equilateral triangle with circumradius equal to 1. Prove that for every point P on the circumcircle, we have*

$$PA^2 + PB^2 + PC^2 = 6.$$

Solution. Consider the complex plane such that the coordinates of points A, B, C are the cube roots of unity $1, \varepsilon, \varepsilon^2$, respectively, and let z be the coordinate of point P. Then $|z| = 1$, and we have

$$
\begin{aligned}
PA^2 + PB^2 + PC^2 &= |z - 1|^2 + |z - \varepsilon|^2 + |z - \varepsilon^2|^2 \\
&= (z - 1)(\bar{z} - 1) + (z - \varepsilon)(\bar{z} - \bar{\varepsilon}) + (z - \varepsilon^2)(\bar{z} - \bar{\varepsilon}^2) \\
&= 3|z|^2 - (1 + \varepsilon + \varepsilon^2)\bar{z} - (1 + \bar{\varepsilon} + \bar{\varepsilon}^2)z + 1 + |\varepsilon|^2 + |\varepsilon^2|^2 \\
&= 3 - 0 \cdot \bar{z} - 0 \cdot z + 1 + 1 + 1 = 6,
\end{aligned}
$$

as desired.

Problem 17. *Point B lies inside the segment $[AC]$. Equilateral triangles ABE and BCF are constructed on the same side of line AC. If M and N are the midpoints of segments AF and CE, prove that triangle BMN is equilateral.*

Solution. Denote by the corresponding lowercase letter the coordinate of a point denoted by an uppercase letter. The point E is obtained from point B by a rotation with center A and angle $60°$; hence

$$e = a + (b - a)\varepsilon, \text{ where } \varepsilon = \cos 60° + i \sin 60°.$$

Likewise, $f = b + (c - b)\varepsilon$.

The coordinates of points M and N are

$$m = \frac{a + b + (c - b)\varepsilon}{2} \text{ and } n = \frac{c + a + (b - a)\varepsilon}{2}.$$

It suffices to prove that $\dfrac{m - b}{n - b} = \varepsilon$. Indeed, we have

$$m - b = (n - b)\varepsilon$$

if and only if

$$a - b + (c - b)\varepsilon = (c + a - 2b)\varepsilon + (b - a)\varepsilon^2.$$

That is,

$$a - b = (a - b)\varepsilon + (b - a)(\varepsilon - 1),$$

as needed.

Problem 18. *Let $ABCD$ be a square with center O and let M, N be the midpoints of segments BO, CD respectively.*

Prove that triangle AMN is an isosceles right triangle.

Solution. Consider the complex plane with center at O such that $1i, -1, -i$ are the coordinates of points A, B, C, D, respectively (Fig. 5.11).

The points M and N have coordinates $m = \dfrac{i}{2}$ and $n = \dfrac{-1 - i}{2}$, so

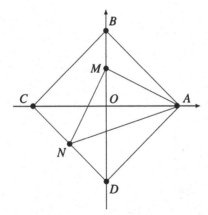

Figure 5.11.

$$\frac{a-m}{n-m} = \frac{1-\dfrac{i}{2}}{\dfrac{-1-i}{2} - \dfrac{i}{2}} = \frac{2-i}{-1-2i} = i.$$

Then $AM \perp MN$ and $AM = NM$, as needed.

Problem 19. *In the plane of the nonequilateral triangle $A_1A_2A_3$ consider points B_1, B_2, B_3 such that triangles $A_1A_2B_3$, $A_2A_3B_1$ and $A_3A_1B_2$ are similar with the same orientation.*

Prove that triangle $B_1B_2B_3$ is equilateral if and only if triangles $A_1A_2B_3$, $A_2A_3B_1$, $A_3A_1B_2$ are isosceles with bases A_1A_2, A_2A_3, A_3A_1 and base angles equal to $30°$.

Solution. Triangles $A_1A_2B_3$, $A_2A_3B_1$, $A_3A_1B_2$ are similar with the same orientation; hence $\dfrac{b_3 - a_2}{a_1 - a_2} = \dfrac{b_1 - a_3}{a_2 - a_3} = \dfrac{b_2 - a_1}{a_3 - a_1} = z$. Then

$$b_3 = a_2 + z(a_1 - a_2), \; b_1 = a_3 + z(a_2 - a_3), \; b_2 = a_1 + z(a_3 - a_1).$$

Triangle $B_1B_2B_3$ is equilateral if and only if

$$b_1 + \varepsilon b_2 + \varepsilon^2 b_3 = 0 \text{ or } b_1 + \varepsilon b_3 + \varepsilon^2 b_2 = 0.$$

Assume that the first is valid.

Then we have

$$b_1 + \varepsilon b_2 + \varepsilon^2 b_3 = 0 \text{ if and only if}$$

$$a_3 + z(a_2 - a_3) + \varepsilon a_1 + \varepsilon z(a_3 - a_1) + \varepsilon^2 a_2 + \varepsilon^2 z(a_1 - a_2) = 0, \text{ i.e.,}$$

$$a_3 + \varepsilon a_1 + \varepsilon^2 a_2 + z(a_2 - a_3 + \varepsilon a_3 - \varepsilon a_1 + \varepsilon^2 a_1 - \varepsilon^2 a_2) = 0.$$

The last relation is equivalent to

$$z[a_2(1 - \varepsilon)(1 + \varepsilon) - a_1\varepsilon(1 - \varepsilon) - a_3(1 - \varepsilon)] = -(a_3 + \varepsilon a_1 + \varepsilon^2 a_2),$$

i.e.,

$$z = +\frac{a_3 + \varepsilon a_1 + \varepsilon^2 a_2}{(1 - \varepsilon)(a_3 + \varepsilon a_1 + \varepsilon^2 a_2)} = \frac{1}{1 - \varepsilon} = \frac{1}{\sqrt{3}}(\cos 30° + i\sin 30°),$$

which shows that triangles $A_1A_2B_3$, $A_2A_3B_1$, and $A_3A_1B_2$ are isosceles with angles of 30°.

Notice that $a_3 + \varepsilon a_1 + \varepsilon^2 a_2 \neq 0$, since triangle $A_1A_2A_3$ is not equilateral.

Problem 20. *The diagonals AC and CE of a regular hexagon $ABCDEF$ are divided by interior points M and N, respectively, such that*

$$\frac{AM}{AC} = \frac{CN}{CE} = r.$$

Determine r knowing that points B, M, and N are collinear.

(23rd IMO)

Solution. Consider the complex plane with origin at the center of the regular hexagon such that 1, ε, ε^2, ε^3, ε^4, ε^5 are the coordinates of the vertices B, C, D, E, F, A, where

$$\varepsilon = \cos\frac{\pi}{3} + i\sin\frac{\pi}{3} = \frac{1 + i\sqrt{3}}{2}.$$

Since

$$\frac{MC}{MA} = \frac{NE}{NC} = \frac{1 - r}{r},$$

the coordinates of points M and N are

$$m = \varepsilon r + \varepsilon^5(1 - r)$$

and

$$n = \varepsilon^2 r + \varepsilon(1 - r),$$

respectively (Fig. 5.12).

The points B, M, N are collinear if and only if $\dfrac{m - 1}{n - 1} \in \mathbb{R}^*$. We have

$$m - 1 = \varepsilon r + \varepsilon^5(1 - r) - 1 = \varepsilon r - \varepsilon^2(1 - r) - 1$$
$$= \frac{1 + i\sqrt{3}}{2}r - \frac{-1 + i\sqrt{3}}{2}(1 - r) = -\frac{1}{2} + \frac{i\sqrt{3}}{2}(2r - 1)$$

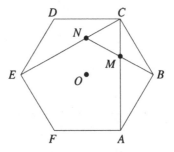

Figure 5.12.

and

$$n - 1 = \varepsilon^3 r + \varepsilon(1 - r) - 1 = -r + \frac{1 + i\sqrt{3}}{2}(1 - r) - 1$$

$$= -\frac{1}{2} - \frac{3r}{2} + \frac{i\sqrt{3}}{2}(1 - r);$$

hence

$$\frac{m - 1}{n - 1} = \frac{-1 + i\sqrt{3}(2r - 1)}{-(1 + 3r) + i\sqrt{3}(1 - r)} \in \mathbb{R}^*$$

if and only if

$$\sqrt{3}(1 - r) - (1 + 3r) \cdot \sqrt{3}(2r - 1) = 0.$$

This is equivalent to $1 - r = 6r^2 - r - 1$, i.e., $r^2 = \frac{1}{3}$. It follows that $r = \frac{1}{\sqrt{3}}$.

Problem 21. *Let G be the centroid of quadrilateral $ABCD$. Prove that if lines GA and GD are perpendicular, then AD is congruent to the line segment joining the midpoints of sides AD and BC.*

Solution. Consider a, b, c, d, g the coordinates of points A, B, C, D, G, respectively. Using properties of the real product of complex numbers, we have

$$GA \perp GD \text{ if and only if } (a - g) \cdot (d - g) = 0, \text{ i.e.,}$$

$$\left(a - \frac{a + b + c + d}{4} \right) \cdot \left(d - \frac{a + b + c + d}{4} \right) = 0.$$

That is,

$$(3a - b - c - d) \cdot (3d - a - b - c) = 0,$$

and we obtain

$$[a - b - c + d + 2(a - d)] \cdot [a - b - c + d - 2(a - d)] = 0.$$

The last relation is equivalent to

$$(a + d - b - c) \cdot (a + d - b - c) = 4(a - d) \cdot (a - d), \text{ i.e.,}$$

$$\left| \frac{a+d}{2} - \frac{b+c}{2} \right|^2 = |a - d|^2. \tag{1}$$

Let M and N be the midpoints of the sides AD and BC. The coordinates of points M and N are $\dfrac{a+d}{2}$ and $\dfrac{b+c}{2}$; hence relation (1) shows that $MN = AD$, and we are done.

Problem 22. *Consider a convex quadrilateral $ABCD$ with nonparallel opposite sides AD and BC. Let G_1, G_2, G_3, G_4 be the centroids of the triangles BCD, ACD, ABD, ABC, respectively. Prove that if $AG_1 = BG_2$ and $CG_3 = DG_4$, then $ABCD$ is an isosceles trapezoid.*

Solution. Denote by the corresponding lowercase letter the coordinate of a point denoted by an uppercase letter. Setting $s = a + b + c + d$ yields

$$g_1 = \frac{b+c+d}{3} = \frac{s-a}{3}, \quad g_2 = \frac{s-b}{3}, \quad g_3 = \frac{s-c}{3}, \quad g_4 = \frac{s-d}{3}.$$

The relation $AG_1 = BG_2$ can be written as

$$|a - g_1| = |b - g_2|, \text{ that is, } |4a - s| = |4b - s|.$$

Using the real product of complex numbers, we see that the last relation is equivalent to

$$(4a - s) \cdot (4a - s) = (4b - s) \cdot (4b - s), \text{ i.e.,}$$

$$16|a|^2 - 8a \cdot s = 16|b|^2 - 8b \cdot s.$$

We obtain

$$2(|a|^2 - |b|^2) = (a - b) \cdot s. \tag{1}$$

Likewise, we have

$$CG_3 = DG_4 \text{ if and only if } 2(|c|^2 - |d|^2) = (c - d) \cdot s. \tag{2}$$

Subtracting the relations (1) and (2) gives

$$2(|a|^2 - |b|^2 - |c|^2 + |d|^2) = (a - b - c + d) \cdot (a + b + c + d).$$

That is,

$$2(|a|^2 - |b|^2 - |c|^2 + |d|^2) = |a + d|^2 - |b + c|^2, \text{ i.e.,}$$

$$2(a\bar{a} - b\bar{b} - c\bar{c} + d\bar{d}) = a\bar{c} + a\bar{d} + \bar{a}d + d\bar{d} - b\bar{b} - b\bar{c} - \bar{b}c - c\bar{c}.$$

We obtain

$$a\bar{a} - a\bar{d} - \bar{a}d + d\bar{d} = b\bar{b} - b\bar{c} - \bar{b}c + c\bar{c}, \text{ i.e.,}$$

$$|a - d|^2 = |b - c|^2.$$

Hence

$$AD = BC. \tag{3}$$

Adding relations (1) and (2) gives

$$2(|a|^2 - |b|^2 - |d|^2 + |c|^2) = (a - b - d + c) \cdot (a + b + c + d),$$

and similarly, we obtain

$$AC = BD. \tag{4}$$

From relations (3) and (4), we deduce that $AB \,\|\, CD$, and consequently, $ABCD$ is an isosceles trapezoid.

Problem 23. *Prove that in every quadrilateral $ABCD$,*

$$AC^2 \cdot BD^2 = AB^2 \cdot CD^2 + AD^2 \cdot BC^2 - 2AB \cdot BC \cdot CD \cdot DA \cdot \cos(A + C).$$

(Bretschneider relation, or a first generalization of Ptolemy's theorem)

Solution. Let z_A, z_B, z_C, z_D be the coordinates of the points A, B, C, D in the complex plane with origin at A and point B on the positive real axis (see Fig. 5.13).
Using the identities

$$(z_A - z_C)(z_B - z_D) = -(z_A - z_B)(z_D - z_C) - (z_A - z_D)(z_C - z_B)$$

and

$$(\overline{z_A - z_C})(\overline{z_B - z_D}) = -(\overline{z_A - z_B})(\overline{z_D - z_C}) - (\overline{z_A - z_D})(\overline{z_C - z_B}),$$

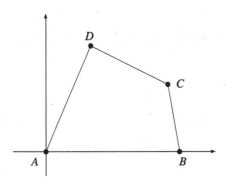

Figure 5.13.

by multiplication we obtain

$$AC^2 \cdot BD^2 = AB^2 \cdot DC^2 + AD \cdot BC^2 + z + \bar{z},$$

where

$$z = (z_A - z_B)(z_D - z_C)(\overline{z_A - z_D})(\overline{z_C - z_B}).$$

It suffices to prove that

$$z + \bar{z} = -2AB \cdot BC \cdot CD \cdot DA \cdot \cos(A + C).$$

We have

$$z_A - z_B = AB(\cos \pi + i \sin \pi),$$
$$z_D - z_C = DC[\cos(2\pi - B - C) + i \sin(2\pi - B - C)],$$
$$\overline{z_A - z_D} = DA[\cos(\pi - A) + i \sin(\pi - A)],$$

and

$$\overline{z_C - z_D} = BC[\cos(\pi + B) + i \sin(\pi + B)].$$

Then

$$z + \bar{z} = 2\mathrm{Re}z = 2AB \cdot BC \cdot CD \cdot DA \cos(5\pi - A - C)$$
$$= -2AB \cdot BC \cdot CD \cdot DA \cdot \cos(A + C),$$

and we are done.

Remark. Since $\cos(A + C) \geq -1$, this relation gives Ptolemy's inequality

$$AC \cdot BD \leq AB \cdot DC + AD \cdot BC,$$

with equality only for cyclic quadrilaterals.

Problem 24. *Let $ABCD$ be a quadrilateral and $AB = a$, $BC = b$, $CD = c$, $DA = d$, $AC = d_1$, and $BD = d_2$.*
Prove that

$$d_2^2[a^2 d^2 + b^2 c^2 - 2abcd \cos(B - D)] = d_1^2[a^2 b^2 + c^2 d^2 - 2abcd \cos(A - C)].$$

(A second generalization of Ptolemy's theorem)

Solution. Let z_A, z_B, z_C, z_D be the coordinates of the points A, B, C, D in the complex plane with origin at D and point C on the positive real axis (see Figure 5.13 but with different notation).
 Multiplying the identities

$$(z_B - z_D)[(z_A - z_B)(z_A - z_D) - (z_C - z_B)(z_C - z_D)]$$
$$= (z_C - z_A) \cdot [(z_B - z_A)(z_B - z_C) - (z_D - z_A)(z_D - z_C)]$$

and

$$(\overline{z_B - z_D})[(\overline{z_A - z_B})(\overline{z_A - z_D}) - (\overline{z_C - z_B})(\overline{z_C - z_D})]$$
$$= (\overline{z_C - z_A}) \cdot [(\overline{z_B - z_A})(\overline{z_B - z_C}) - (\overline{z_D - z_A})(\overline{z_D - z_C})]$$

yields

$$d_2^2[a^2 \cdot d^2 + b^2 \cdot c^2 - (z_A - z_B)(z_A - z_D)(\overline{z_C - z_B})(\overline{z_C - z_D})$$
$$- (z_C - z_B)(z_C - z_D)(\overline{z_A - z_B})(\overline{z_A - z_D})]$$
$$= d_1^2[a^2 \cdot b^2 + c^2 \cdot d^2 - (z_B - z_A)(z_B - z_C)(\overline{z_D - z_A})(\overline{z_D - z_C})$$
$$- (z_D - z_A)(z_D - z_C)(\overline{z_B - z_A})(\overline{z_B - z_C})].$$

It suffices to prove that

$$2\mathrm{Re}(z_A - z_B)(z_A - z_D)(\overline{z_C - z_B})(\overline{z_C - z_D}) = 2abcd\cos(B - D)$$

and

$$2\mathrm{Re}(z_B - z_A)(z_B - z_C)(\overline{z_B - z_A})(\overline{z_D - z_c}) = 2abcd\cos(A - C).$$

We have

$$z_B - z_A = a[\cos(\pi + A + D) + i\sin(\pi + A + D)],$$
$$z_B - z_C = b[\cos(\pi - C) + i\sin(\pi - C)],$$
$$\overline{z_D - z_A} = d[\cos(\pi - D) + i\sin(\pi - D)],$$
$$\overline{z_D - z_C} = c[\cos\pi + i\sin\pi],$$
$$z_A - z_B = a[\cos(A + D) + i\sin(A + D)],$$
$$z_A - z_D = d(\cos D + i\sin D),$$
$$\overline{z_C - z_B} = b(\cos B + i\sin B),$$
$$\overline{z_C - z_D} = c(\cos 0 + i\sin 0);$$

hence

$$2\mathrm{Re}(z_A - z_B)(z_A - z_D)(\overline{z_C - z_B})(\overline{z_C - z_D})$$
$$= 2abcd\cos(A + D + D + C) = 2abcd\cos(2\pi - B + D) = 2abcd\cos(B - D)$$

and

$$2\mathrm{Re}(z_B - z_A)(z_B - z_C)(\overline{z_D - z_A})(\overline{z_D - z_C})$$
$$= 2abcd\cos(\pi + A + D + \pi - C + \pi - D + \pi)$$
$$= 2abcd\cos(4\pi + A - C) = 2abcd\cos(A - C),$$

as desired.

Remark. If $ABCD$ is a cyclic quadrilateral, then $B + D = A + C = \pi$. It follows that

$$\cos(B - A) = \cos(2B - \pi) = -\cos 2B$$

and

$$\cos(A - C) = \cos(2A - \pi) = -\cos 2A.$$

The relation becomes

$$d_2^2[(ad + bc)^2 - 2abcd(1 - \cos 2B)] = d_1^2[(ab + cd)^2 - 2abcd(1 - \cos 2A)].$$

This is equivalent to

$$d_2^2(ad + bc)^2 - 4abcdd_2^2 \sin^2 B = d_1^2(ab + cd)^2 - 2abcdd_1^2 \sin^2 A. \qquad (1)$$

The law of sines applied to the triangles ABC and ABD with circumradii R gives $d_1 = 2R \sin B$ and $d_2 = 2R \sin A$, hence $d_1 \sin A = d_2 \sin B$. The relation (1) is equivalent to

$$d_2^2(ad + bc)^2 = d_1^2(ab + cd)^2,$$

and consequently,

$$\frac{d_2}{d_1} = \frac{ab + cd}{ad + bc}. \qquad (2)$$

Relation (2) is known as *Ptolemy's second theorem*.

Problem 25. *In a plane, three equilateral triangles OAB, OCD, and OEF are given. Prove that the midpoints of the segments BC, DE, and FA are the vertices of an equilateral triangle.*

Solution. Consider the complex plane with origin at O and assume that triangles OAB, OCD, OEF are positively oriented. Denote by the corresponding lowercase letter the coordinate of a point denoted by an uppercase letter.

Let $\varepsilon = \cos 60° + i \sin 60°$. Then

$$b = a\varepsilon, \ d = c\varepsilon, \ f = e\varepsilon$$

and

$$m = \frac{b + c}{2} = \frac{a\varepsilon + c}{2}, \ n = \frac{d + e}{2} = \frac{c\varepsilon + e}{2}, \ p = \frac{f + a}{2} = \frac{e\varepsilon + a}{2}.$$

Triangle MNP is equilateral if and only if

$$m + \omega n + \omega^2 p = 0,$$

where

$$\omega = \cos 120° + i \sin 120° = \varepsilon^2.$$

Because

$$m + \varepsilon^2 n + \varepsilon^4 p = m + \varepsilon^2 n - \varepsilon p = \frac{1}{2}(a\varepsilon + c - c + e\varepsilon^2 - e\varepsilon^2 - \varepsilon a) = 0,$$

we are done.

We invite the reader to solve the following problems using complex numbers.

Problem 26. *Let ABC be a triangle such that $AC^2 + AB^2 = 5BC^2$. Prove that the medians from the vertices B and C are perpendicular.*

Problem 27. *On the sides BC, CA, AB of triangle ABC, the points A', B', C' are chosen such that*

$$\frac{A'B}{A'C} = \frac{B'C}{B'A} = \frac{C'A}{C'B} = k.$$

Consider the points A'', B'', C'' on the segments $B'C'$, $C'A'$, $A'B'$ such that

$$\frac{A''C'}{A''B'} = \frac{C''B'}{C''A'} = \frac{B''A'}{B''C'} = k.$$

Prove that triangles ABC and $A''B''C''$ are similar.

Problem 28. *Prove that the following inequality is true in every triangle:*

$$\frac{R}{2r} \geq \frac{m_\alpha}{h_\alpha}.$$

Equality holds only for equilateral triangles.

Problem 29. *Let $ABCD$ be a quadrilateral inscribed in the circle $\mathcal{C}(O;\ R)$. Prove that*

$$AB^2 + BC^2 + CD^2 + DA^2 = 8R^2$$

if and only if $AC \perp BD$ or one of the diagonals is a diameter of $\mathcal{C}(O; R)$.

Problem 30. *On the sides of the convex quadrilateral $ABCD$, equilateral triangles ABM, BCN, CDP and DAQ are drawn external to the figure. Prove that quadrilaterals $ABCD$ and $MNPQ$ have the same centroid.*

Problem 31. *Let $ABCD$ be a quadrilateral and consider the rotations $\mathcal{R}_1$, $\mathcal{R}_2$, $\mathcal{R}_3$, $\mathcal{R}_4$ with centers A, B, C, D through angle α in the same direction.*

Points M, N, P, Q are the images of points A, B, C, D under the rotations $\mathcal{R}_2$, $\mathcal{R}_3$, $\mathcal{R}_4$, $\mathcal{R}_1$, respectively.

Prove that the midpoints of the diagonals of the quadrilaterals $ABCD$ and $MNPQ$ are the vertices of a parallelogram.

Problem 32. *Prove that in every cyclic quadrilateral $ABCD$, the following hold:*

(a) $AD + BC \cos(A + B) = AB \cos A + CD \cos D$.
(b) $BC \sin(A + B) = AB \sin\ A - CD \sin D$.

Problem 33. *Let O_9, I, G be the nine-point center, the incenter, and the centroid, respectively, of a triangle ABC. Prove that lines O_9G and AI are perpendicular if and only if $\hat{A} = \dfrac{\pi}{3}$.*

Problem 34. *Two circles ω_1 and ω_2 are given in the plane, with centers O_1 and O_2, respectively. Let M_1' and M_2' be two points on ω_1 and ω_2, respectively, such that the lines O_1M_1' and O_2M_2' intersect. Let M_1 and M_2 be points on ω_1 and ω_2, respectively, such that when measured clockwise, the angles $\widehat{M_1'O_1M_1}$ and $\widehat{M_2'O_2M_2}$ are equal.*

(a) Determine the locus of the midpoint of $[M_1M_2]$.
(b) Let P be the point of intersection of lines O_1M_1 and O_2M_2 The circum-circle of triangle M_1PM_2 intersects the circumcircle of triangle O_1PO_2 at P and another point Q. Prove that Q is fixed, independent of the locations of M_1 and M_2.

(2000 Vietnamese Mathematical Olympiad)

Problem 35. *Isosceles triangles $A_3A_1O_2$ and $A_1A_2O_3$ are constructed externally along the sides of a triangle $A_1A_2A_3$ with $O_2A_3 = O_2A_1$ and $O_3A_1 = O_3A_2$. Let O_1 be a point on the opposite side of line A_2A_3 from A_1, with $\widehat{O_1A_3A_2} = \frac{1}{2}\widehat{A_1O_3A_2}$ and $\widehat{O_1A_2A_3} = \frac{1}{2}\widehat{A_1O_2A_3}$, and let T be the foot of the perpendicular from O_1 to A_2A_3. Prove that $A_1O_1 \perp O_2O_3$ and that*

$$\frac{A_1O_1}{O_2O_3} = 2\frac{O_1T}{A_2A_3}.$$

(2000 Iranian Mathematical Olympiad)

Problem 36. *A triangle $A_1A_2A_3$ and a point P_0 are given in the plane. We define $A_s = A_{s-3}$ for all $s \geq 4$. We construct a sequence of points P_0, P_1, P_2, ... such that P_{k+1} is the image of P_k under the rotation with center A_{k+1} through the angle $120°$ clockwise ($k = 0$, 1, 2, ...). Prove that if $P_{1986} = P_0$, then the triangle $A_1A_2A_3$ is equilateral.*

(27th IMO)

Problem 37. *Two circles in a plane intersect. Let A be one of the points of intersection. Starting simultaneously from A, two points move with constant speed, each point traveling along its own circle in the same direction. After one revolution, the two points return simultaneously to A. Prove that there exists a fixed point P in the plane such that at every time, the distances from P to the moving points are equal.*

(21st IMO)

Problem 38. *Inside the square ABCD, the equilateral triangles ABK, BCL, CDM, DAN are inscribed. Prove that the midpoints of the segments KL, LM, MN, NK and the midpoints of the segments AK, BK, BL, CL, CM, DM, DN, AN are the vertices of a regular dodecagon.*

(19th IMO)

Problem 39. *Let ABC be an equilateral triangle and let M be a point in the interior of angle $\widehat{BAC}$. Points D and E are the images of points B and C under the rotations with center M and angle $120°$, counterclockwise and clockwise, respectively. Prove that the fourth vertex of the parallelogram with sides MD and ME is the reflection of point A across point M.*

Problem 40. *Prove that the following inequality holds for every point M inside parallelogram ABCD:*

$$MA \cdot MC + MB \cdot MD \geq AB \cdot BC.$$

Problem 41. *Let ABC be a triangle, H its orthocenter, O its circumcenter, and R its circumradius. Let D be the reflection of A across BC, let E be that of B across CA, and F that of C across AB. Prove that D, E, and F are collinear if and only if $OH = 2R$.*

(39th IMO—Shortlist)

Problem 42. *Let ABC be a triangle such that $\widehat{ACB} = 2\widehat{ABC}$. Let D be the point on the side BC such that $CD = 2BD$. The segment AD is extended to E so that $AD = DE$. Prove that*

$$\widehat{ECB} + 180° = 2\widehat{EBC}.$$

(39th IMO—Shortlist)

Problem 43. *Let P be point situated in the interior of a circle. Two variable perpendicular lines through P intersect the circle at A and B. Find the locus of the midpoint of the segment AB.*

(Mathematical Reflections, 2010)

Problem 44. *Let ABC be a triangle and consider the points $M \in (BC)$, $N \in (CA)$, $P \in (AB)$ such that*

$$\frac{AP}{PB} = \frac{BM}{MC} = \frac{CN}{NA}.$$

Prove that if MNP is an equilateral triangle, then ABC is an equilateral triangle as well.

(Romanian Mathematical Olympiad—District Round, 2006)

Problem 45. *Consider the triangle ABC and the points $D \in (BC)$, $E \in (CA)$, $F \in (AB)$, such that*

$$\frac{BD}{DC} = \frac{CE}{EA} = \frac{AF}{FB}.$$

Prove that if the circumcenter of triangles DEF and ABC coincide, then the triangle ABC is equilateral.

(Romanian Mathematical Olympiad—Final Round, 2008)

Problem 46. *Exterior to a nonequilateral triangle ABC, consider the similar triangles (in this order) ABM, BCN, and CAP such that the triangle MNP is equilateral. Find the angles of the triangles ABM, BCN, and CAP.*

(Romanian Mathematical Olympiad—Final Round, 2010)

5.5 Solving Trigonometric Problems

Problem 1. *Prove that*

$$\cos \frac{\pi}{11} + \cos \frac{3\pi}{11} + \cos \frac{5\pi}{11} + \cos \frac{7\pi}{11} + \cos \frac{9\pi}{11} = \frac{1}{2}.$$

Solution. Setting $z = \cos \dfrac{\pi}{11} + i \sin \dfrac{\pi}{11}$ implies that

$$z + z^3 + z^5 + z^7 + z^9 = \frac{z^{11} - z}{z^2 - 1} = \frac{-1 - z}{z^2 - 1} = \frac{1}{1 - z}.$$

Taking the real parts of both sides of the equality gives the desired result.

Problem 2. *Compute the product $P = \cos 20° \cdot \cos 40° \cdot \cos 80°$.*

Solution 1. Setting $z = \cos 20° + i \sin 20°$ implies $z^9 = -1$, $\overline{z} = \cos 20° - i \sin 20°$, and $\cos 20° = \dfrac{z^2 + 1}{2z}$, $\cos 40° = \dfrac{z^4 + 1}{2z^2}$, $\cos 80° = \dfrac{z^8 + 1}{2z^4}$. Then

$$P = \frac{(z^2+1)(z^4+1)(z^8+1)}{8z^7} = \frac{(z^2-1)(z^2+1)(z^4+1)(z^8+1)}{8z^7(z^2-1)}$$

$$= \frac{z^{16}-1}{8(z^9-z^7)} = \frac{-z^7-1}{8(-1-z^7)} = \frac{1}{8}.$$

Solution 2. This is a classical problem with a classical solution. Let $S = \cos 20° \cos 40° \cos 80°$. Then

$$S \sin 20° = \sin 20° \cos 20° \cos 40° \cos 80°$$

$$= \frac{1}{2} \sin 40° \cos 40° \cos 80°$$

$$= \frac{1}{4} \sin 80° \cos 80°$$

$$= \frac{1}{8} \cos 160° = \frac{1}{8} \sin 20°.$$

So $S = \dfrac{1}{8}$.

Note that this classical solution is contrived, with no motivation. The solution using complex numbers, however, is a straightforward computation.

Problem 3. *Let x, y, z be real numbers such that*

$$\sin x + \sin y + \sin z = 0 \ \ and \ \ \cos x + \cos y + \cos z = 0.$$

Prove that

$$\sin 2x + \sin 2y + \sin 2z = 0 \ \ and \ \ \cos 2x + \cos 2y + \cos 2z = 0.$$

Solution. Setting $z_1 = \cos x + i \sin x$, $z_2 = \cos y + i \sin y$, $z_3 = \cos z + i \sin z$, we have $z_1 + z_2 + z_3 = 0$ and $|z_1| = |z_2| = |z_3| = 1$.

We have

$$z_1^2 + z_2^2 + z_3^2 = (z_1 + z_2 + z_3)^2 - 2(z_1 z_2 + z_2 z_3 + z_3 z_1)$$

$$= -2z_1 z_2 z_3 \left(\frac{1}{z_1} + \frac{1}{z_2} + \frac{1}{z_3} \right) = -2z_1 z_2 z_3 (\overline{z}_1 + \overline{z}_2 + \overline{z}_3)$$

$$= -2z_1 z_2 z_3 \overline{(z_1 + z_2 + z_3)} = 0.$$

Thus $(\cos 2x + \cos 2y + \cos 2z) + i(\sin 2x + \sin 2y + \sin 2z) = 0$, and the conclusion is obvious.

Problem 4. *Prove that*

$$\cos^2 10° + \cos^2 50° + \cos^2 70° = \frac{3}{2}.$$

Solution. Setting $z = \cos 10° + i \sin 10°$, we have $z^9 = i$ and

$$\cos 10° = \frac{z^2 + 1}{2z}, \quad \cos 50° = \frac{z^{10} + 1}{2z^5}, \quad \cos 70° = \frac{z^{14} + 1}{2z^7}.$$

The identity is equivalent to

$$\left(\frac{z^2 + 1}{2z}\right)^2 + \left(\frac{z^{10} + 1}{2z^5}\right)^2 + \left(\frac{z^{14} + 1}{2z^7}\right)^2 = \frac{3}{2}.$$

That is,

$$z^{16} + 2z^{14} + z^{12} + z^{24} + 2z^{14} + z^4 + z^{28} + 2z^{14} + 1 = 6z^{14}, \text{ i.e.,}$$

$$z^{28} + z^{24} + z^{16} + z^{12} + z^4 + 1 = 0.$$

Using relation $z^{18} = -1$, we obtain

$$z^{16} + z^{12} - z^{10} - z^6 + z^4 + 1 = 0,$$

or equivalently,

$$(z^4 + 1)(z^{12} - z^6 + 1) = 0.$$

That is,

$$\frac{(z^4 + 1)(z^{18} + 1)}{z^6 + 1} = 0,$$

which is obvious.

Problem 5. *Solve the equation*

$$\cos x + \cos 2x - \cos 3x = 1.$$

Solution. Setting $z = \cos x + i \sin x$ yields

$$\cos x = \frac{z^2 + 1}{2z}, \quad \cos 2x = \frac{z^4 + 1}{2z^2}, \quad \cos 3x = \frac{z^6 + 1}{2z^3}.$$

The equation may be rewritten as

$$\frac{z^2 + 1}{2z} + \frac{z^4 + 1}{2z^2} - \frac{z^6 + 1}{2z^3} = 1, \text{ i.e., } z^4 + z^2 + z^5 + z - z^6 - 1 - 2z^3 = 0.$$

This is equivalent to

$$(z^6 - z^5 - z^4 + z^3) + (z^3 - z^2 - z + 1) = 0,$$

or

$$(z^3 + 1)(z^3 - z^2 - z + 1) = 0.$$

Finally, we obtain

$$(z^3 + 1)(z - 1)^2(z + 1) = 0.$$

Thus, $z = 1$ or $z = -1$ or $z^3 = -1$, and consequently, $x \in \{2k\pi | k \in \mathbb{Z}\}$ or $x \in \{\pi + 2k\pi | k \in \mathbb{Z}\}$ or $x \in \left\{\dfrac{\pi + 2k\pi}{3} | k \in \mathbb{Z}\right\}$. Therefore,

$$x \in \{k\pi | k \in \mathbb{Z}\} \cup \left\{\dfrac{2k+1}{3}\pi | k \in \mathbb{Z}\right\}.$$

Problem 6. *Compute the sums*

$$S = \sum_{k=1}^{n} q^k \cdot \cos kx \ \text{and} \ T = \sum_{k=1}^{n} q^k \cdot \sin kx.$$

Solution. We have

$$1 + S + iT = \sum_{k=0}^{n} q^k (\cos kx + i \sin kx) = \sum_{k=0}^{n} q^k (\cos x + i \sin x)^k$$

$$= \frac{1 - q^{n+1}(\cos x + i \sin x)^{n+1}}{1 - q \cos x - iq \sin x}$$

$$= \frac{1 - q^{n+1}[\cos(n+1)x + i \sin(n+1)x]}{1 - q \cos x - iq \sin x}$$

$$= \frac{[1 - q^{n+1} \cos(n+1)x - iq^{n+1} \sin(n+1)x][1 - q \cos x + iq \sin x]}{q^2 - 2q \cos x + 1};$$

hence

$$1 + S = \frac{q^{n+2} \cos nx - q^{n+1} \cos(n+1)x - q \cos x + 1}{q^2 - 2q \cos x + 1}$$

and

$$T = \frac{q^{n+2} \sin nx - q^{n+1} \sin(n+1)x + q \sin x}{q^2 - 2q \cos x + 1}.$$

Remark. If $q = 1$, then we obtain the well-known formulas

$$\sum_{k=1}^{n} \cos kx = \frac{\sin \dfrac{nx}{2} \cos \dfrac{(n+1)x}{2}}{\sin \dfrac{x}{2}} \ \text{and} \ \sum_{k=1}^{n} \sin kx = \frac{\sin \dfrac{nx}{2} \sin \dfrac{(n+1)x}{2}}{\sin \dfrac{x}{2}}.$$

Indeed, we have

$$\sum_{k=1}^{n} \cos kx = \frac{\cos nx - \cos(n+1)x - (1 - \cos x)}{2(1 - \cos x)}$$

$$= \frac{2 \sin \dfrac{x}{2} \sin \dfrac{(2n+1)x}{2} - 2 \sin^2 \dfrac{x}{2}}{4 \sin^2 \dfrac{x}{2}}$$

$$= \frac{\sin \dfrac{(2n+1)x}{2} - \sin \dfrac{x}{2}}{2 \sin \dfrac{x}{2}} = \frac{\sin \dfrac{nx}{2} \cos \dfrac{(n+1)x}{2}}{\sin \dfrac{x}{2}}$$

and

$$\sum_{k=1}^{n} \sin kx = \frac{\sin nx - \sin(n+1)x + \sin x}{2(1 - \cos x)}$$

$$= \frac{2 \sin \dfrac{x}{2} \cos \dfrac{x}{2} - 2 \sin \dfrac{x}{2} \cos \dfrac{(2n+1)x}{2}}{4 \sin^2 \dfrac{x}{2}}$$

$$= \frac{\cos \dfrac{x}{2} - \cos \dfrac{(2n+1)x}{2}}{2 \sin \dfrac{x}{2}} = \frac{\sin \dfrac{nx}{2} \sin \dfrac{(n+1)x}{2}}{\sin \dfrac{x}{2}}.$$

Problem 7. *The points A_1, A_2, $\ldots$, A_{10} are equally distributed on a circle of radius R (in that order). Prove that $A_1 A_4 - A_1 A_2 = R$.*

Solution. Let $z = \cos \dfrac{\pi}{10} + i \sin \dfrac{\pi}{10}$. Without loss of generality, we may assume that $R = 1$. We need to show that $2 \sin \dfrac{3\pi}{10} - 2 \sin \dfrac{\pi}{10} = 1$.

In general, if $z = \cos a + i \sin a$, then $\sin a = \frac{z^2 - 1}{2iz}$, and we have to prove that $\dfrac{z^6 - 1}{iz^3} - \dfrac{z^2 - 1}{iz} = 1$. This reduces to $z^6 - z^4 + z^2 - 1 = iz^3$. Because $z^5 = i$, the previous relation is equivalent to $z^8 - z^6 + z^4 - z^2 + 1 = 0$. But this is true, because $(z^8 - z^6 + z^4 - z^2 + 1)(z^2 + 1) = z^{10} + 1 = 0$ and $z^2 + 1 \neq 0$.

Problem 8. *Show that*

$$\cos \frac{\pi}{7} - \cos \frac{2\pi}{7} + \cos \frac{3\pi}{7} = \frac{1}{2}.$$

(5th IMO)

Solution. Let $z = \cos \dfrac{\pi}{7} + i \sin \dfrac{\pi}{7}$. Then $z^7 + 1 = 0$. Because $z \neq -1$ and $z^7 + 1 = (z+1)(z^6 - z^5 + z^4 - z^3 + z^2 - z + 1) = 0$, it follows that the second factor in the above product is zero. The condition is equivalent to

$$z(z^2 - z + 1) = \frac{1}{1 - z^3}.$$

The given sum is

$$\cos \frac{\pi}{7} - \cos \frac{2\pi}{7} + \cos \frac{3\pi}{7} = \operatorname{Re}(z^3 - z^2 + z).$$

Therefore, we have to prove that $\operatorname{Re}\left(\dfrac{1}{1 - z^3}\right) = \dfrac{1}{2}$. This follows from the following well-known lemma.

Lemma. If $z = \cos t + i \sin t$ and $z \neq 1$, then $\operatorname{Re}\dfrac{1}{1 - z} = \dfrac{1}{2}$.

Proof.

$$\frac{1}{1 - z} = \frac{1}{1 - (\cos t + i \sin t)} = \frac{1}{(1 - \cos t) - i \sin t}$$

$$= \frac{1}{2 \sin^2 \dfrac{t}{2} - 2i \sin \dfrac{t}{2} \cos \dfrac{t}{2}} = \frac{1}{2 \sin \dfrac{t}{2}\left(\sin \dfrac{t}{2} - i \cos \dfrac{t}{2}\right)}$$

$$= \frac{\sin \dfrac{t}{2} + i \cos \dfrac{t}{2}}{2 \sin \dfrac{t}{2}} = \frac{1}{2} + i \frac{\cos \dfrac{t}{2}}{2 \sin \dfrac{t}{2}}.$$

Problem 9. *Prove that the average of the numbers $k \sin k°$ ($k = 2, 4, 6, \ldots, 180$) is $\cot 1°$.*

(1996 USA Mathematical Olympiad)

Solution. Set $z = \cos t + i \sin t$. From the identity

$$z + 2z^2 + \cdots + nz^n = (z + \cdots + z^n) + (z^2 + \cdots + z^n) + \cdots + z^n$$

$$= \frac{1}{z - 1}[(z^{n+1} - z) + (z^{n+1} - z^2) + \cdots + (z^{n+1} - z^n)]$$

$$= \frac{nz^{n+1}}{z - 1} - \frac{z^{n+1} - z}{(z - 1)^2},$$

we derive the formulas

$$\sum_{k=1}^{n} k \cos kt = \frac{(n + 1) \sin \dfrac{(2n + 1)t}{2}}{2 \sin \dfrac{t}{2}} - \frac{1 - \cos(n + 1)t}{4 \sin^2 \dfrac{t}{2}}, \tag{1}$$

$$\sum_{k=1}^{n} k \sin kt = \frac{\sin(n + 1)t}{4 \sin^2 \dfrac{t}{2}} - \frac{n \cos \dfrac{(2n + 1)t}{2}}{2 \sin \dfrac{t}{2}}. \tag{2}$$

Using relation (2), one obtains

$$2\sin 2° + 4\sin 4° + \cdots + 178\sin 178° = 2(\sin 2° + 2\sin 2 \cdot 2° + \cdots + 89\sin 89 \cdot 2°)$$
$$= 2\left(\frac{\sin 90 \cdot 2°}{4\sin^2 1°} - \frac{90\cos 179°}{2\sin 1°}\right) = -\frac{90\cos 179°}{\sin 1°} = 90\cot 1°.$$

Finally,

$$\frac{1}{90}(2\sin 2° + 4\sin 4° + \cdots + 178\sin 178° + 180\sin 180°) = \cot 1°.$$

Problem 10. *Let n be a positive integer. Find real numbers a_0 and a_{kl}, $k, l = \overline{1, n}$, $k > l$, such that*

$$\frac{\sin^2 nx}{\sin^2 x} = a_0 + \sum_{1 \le l < k \le n} a_{kl} \cos 2(k - l)x$$

for all real numbers $x \ne m\pi$, $m \in \mathbb{Z}$.

(Romanian Mathematical Regional Contest "Grigore Moisil," 1995)

Solution. Using the identities

$$S_1 = \sum_{j=1}^{n} \cos 2jx = \frac{\sin nx \cos(n+1)x}{\sin x}$$

and

$$S_2 = \sum_{j=1}^{n} \sin 2jx = \frac{\sin nx \sin(n+1)x}{\sin x},$$

we obtain

$$S_1^2 + S_2^2 = \left(\frac{\sin nx}{\sin x}\right)^2.$$

On the other hand,

$$S_1^2 + S_2^2 = (\cos 2x + \cos 4x + \cdots + \cos 2nx)^2$$
$$+ (\sin 2x + \sin 4x + \cdots + \sin 2nx)^2$$
$$= n + 2 \sum_{1 \le l < k \le n} (\cos 2kx \cos 2lx + \sin 2kx \sin 2lx)$$
$$= n + 2 \sum_{1 \le l < k \le n} \cos 2(k - l)x;$$

hence

$$\left(\frac{\sin nx}{\sin x}\right)^2 = n + 2 \sum_{1 \le l < k \le n} \cos 2(k - l)x.$$

Set $a_0 = n$ and $a_{kl} = 2$, $1 \le l < k \le n$, and the problem is solved.

Here are some more problems.

Problem 11. *Sum the following two n-term series for $\theta = 30°$:*

(i) $1 + \dfrac{\cos\theta}{\cos\theta} + \dfrac{\cos(2\theta)}{\cos^2\theta} + \dfrac{\cos(3\theta)}{\cos^3\theta} + \cdots + \dfrac{\cos((n-1)\theta)}{\cos^{n-1}\theta}$, *and*

(ii) $\cos\theta\cos\theta + \cos^2\theta\cos(2\theta) + \cos^3\theta\cos(3\theta) + \cdots + \cos^n\theta\cos(n\theta)$.

(Crux Mathematicorum, 2003)

Problem 12. *Prove that*

$$1 + \cos^{2n}\left(\frac{\pi}{n}\right) + \cos^{2n}\left(\frac{2\pi}{n}\right) + \cdots + \cos^{2n}\left(\frac{(n-1)\pi}{n}\right)$$
$$= n \cdot 4^{-n}\left(2 + \binom{2n}{n}\right),$$

for all integers $n \ge 2$.

Problem 13. *For every integer $p \ge 0$, there are real numbers $a_0, a_1, \ldots, a_p$ with $a_p \ne 0$ such that*

$$\cos 2p\alpha = a_0 + a_1 \sin^2\alpha + \cdots + a_p \cdot (\sin^2\alpha)^p, \text{ for all } \alpha \in \mathbb{R}.$$

Problem 14. *Let*

$$x = \frac{\displaystyle\sum_{n=1}^{44} \cos n^6}{\displaystyle\sum_{n=1}^{44} \sin n^6}.$$

What is the greatest integer that does not exceed $100x$?

(1997, AIME Problem 11)

Problem 15. *Prove that*

$$\sum_{k=0}^{n}\binom{n}{k}\cos[(n-k)x + ky] = \left(2\cos\frac{x-y}{2}\right)^n \cos n\frac{x+y}{2}$$

for all positive integers n and all real numbers x and y.

(Mathematical Reflections, 2009)

Problem 16. *Let k be a fixed positive integer and let*

$$S_n^{(j)} = \binom{n}{j} + \binom{n}{j+k} + \binom{n}{j+2k} + \ldots, \quad j = 0, 1, \ldots, k-1.$$

Prove that

$$\left(S_n^{(0)} + S_n^{(1)} \cos \frac{2\pi}{k} + \ldots + S_n^{(k-1)} \cos \frac{2(k-1)\pi}{k}\right)^2$$

$$+ \left(S_n^{(1)} \sin \frac{2\pi}{k} + S_n^{(2)} \sin \frac{4\pi}{k} + \ldots + S_n^{(k-1)} \sin \frac{2(k-1)\pi}{k}\right)^2 = \left(2 \cos \frac{\pi}{k}\right)^{2n}.$$

(Mathematical Reflections, 2010)

Problem 17.

(a) Let z_1, z_2, z_3, z_4 be distinct complex numbers of zero sum, having equal absolute values. Prove that the points of complex coordinates z_1, z_2, z_3, z_4 are the vertices of a rectangle.

(b) Let x, y, z, t be real numbers such that $\sin x + \sin y + \sin z + \sin t = 0$ and $\cos x + \cos y + \cos z + \cos t = 0$. Prove that for every integer n,

$$\sin(2n+1)x + \sin(2n+1)y + \sin(2n+1)z + \sin(2n+1)t = 0.$$

(Romanian Mathematical Olympiad—District Round, 2011)

5.6 More on the nth Roots of Unity

Problem 1. *Let $n \geq 3$ and $k \geq 2$ be positive integers and consider the complex numbers*

$$z = \cos \frac{2\pi}{n} + i \sin \frac{2\pi}{n}$$

and

$$\theta = 1 - z + z^2 - z^3 + \cdots + (-1)^{k-1} z^{k-1}.$$

(a) If k is even, prove that $\theta^n = 1$ if and only if n is even and $\dfrac{n}{2}$ divides $k-1$ or $k+1$.

(b) If k is odd, prove that $\theta^n = 1$ if and only if n divides $k-1$ or $k+1$.

Solution. Since $z \neq -1$, we have

$$\theta = \frac{1 + (-1)^{k+1} z^k}{1 + z}.$$

(a) If k is even, then

$$\theta = \frac{1 - z^k}{1 + z} = \frac{1 - \cos\dfrac{2k\pi}{n} - i\sin\dfrac{2k\pi}{n}}{1 + \cos\dfrac{2\pi}{n} + i\sin\dfrac{2\pi}{n}} = \frac{\sin\dfrac{k\pi}{n}\left(\sin\dfrac{k\pi}{n} - i\cos\dfrac{k\pi}{n}\right)}{\cos\dfrac{\pi}{n}\left(\cos\dfrac{\pi}{n} + i\sin\dfrac{\pi}{n}\right)}$$

$$= -i\frac{\sin\dfrac{k\pi}{n}}{\cos\dfrac{\pi}{n}}\left(\cos\frac{(k-1)\pi}{n} + i\sin\frac{(k-1)\pi}{n}\right),$$

and

$$|\theta| = \left|\frac{\sin\dfrac{k\pi}{n}}{\cos\dfrac{\pi}{n}}\right|.$$

We have

$$|\theta| = 1 \text{ if and only if } \left|\sin\frac{k\pi}{n}\right| = \left|\cos\frac{\pi}{n}\right|.$$

That is,

$$\sin^2\frac{k\pi}{n} = \cos^2\frac{\pi}{n} \text{ or } \cos\frac{2k\pi}{n} + \cos\frac{2\pi}{n} = 0.$$

The last relation is equivalent to

$$\cos\frac{(k+1)\pi}{n}\cos\frac{(k-1)\pi}{n} = 0, \text{ i.e., } \frac{2(k+1)}{n} \in 2\mathbb{Z}+1,$$

or $\dfrac{2(k-1)}{n} \in 2\mathbb{Z}+1$. This is equivalent to the statement that n is even and $\dfrac{n}{2}$ divides $k+1$ or $k-1$. Hence, it suffices to prove that $\theta^n = 1$ is equivalent to $|\theta| = 1$.

The direct implication is obvious. Conversely, if $|\theta| = 1$, then $n = 2t$, $t \in \mathbb{Z}_+$, and t divides $k + 1$ or $k - 1$. Since k is even, the numbers $k + 1$, $k - 1$ are odd; hence $t = 2l + 1$ and $n = 4l + 2$, $l \in \mathbb{Z}$. Then

$$\theta = \pm i\left(\cos\frac{(k-1)\pi}{n} + i\sin\frac{(k-1)\pi}{n}\right)$$

and

$$\theta^n = -\cos(k-1)\pi = 1,$$

as desired.

(b) If k is odd, then

$$\theta = \frac{1 + z^k}{1 + z} = \frac{1 + \cos \dfrac{2k\pi}{n} + i \sin \dfrac{2k\pi}{n}}{1 + \cos \dfrac{2\pi}{n} + i \sin \dfrac{2\pi}{n}} = \frac{\cos \dfrac{k\pi}{n} \left(\cos \dfrac{k\pi}{n} + i \sin \dfrac{k\pi}{n} \right)}{\cos \dfrac{\pi}{n} \left(\cos \dfrac{\pi}{n} + i \sin \dfrac{\pi}{n} \right)}$$

$$= \frac{\cos \dfrac{k\pi}{n}}{\cos \dfrac{\pi}{n}} \left(\cos \frac{k-1}{n}\pi + i \sin \frac{k-1}{n}\pi \right).$$

We have

$$|\theta| = 1 \text{ if and only if } \left| \cos \frac{k\pi}{n} \right| = \left| \cos \frac{\pi}{n} \right|.$$

That is,

$$\cos^2 \frac{k\pi}{n} = \cos^2 \frac{\pi}{n} \text{ so } \cos \frac{2k\pi}{n} = \cos \frac{2\pi}{n}.$$

It follows that

$$\sin \frac{(k+1)\pi}{n} \sin \frac{(k-1)\pi}{n} = 0,$$

i.e., n divides $k + 1$ or $k - 1$.

It suffices to prove that $\theta^n = 1$ is equivalent to $|\theta| = 1$. Since the direct implication is obvious, let us prove the converse. If $|\theta| = 1$, then $k \pm 1 = nt$, $t \in \mathbb{Z}$. Then $k = nt \pm 1$ and

$$\theta = (-1)^t \left(\cos \frac{(k-1)\pi}{n} + i \sin \frac{(k-1)\pi}{n} \right).$$

It follows that

$$\theta^n = (-1)^{k \pm 1}(\cos(k-1)\pi + i \sin(k-1)\pi) = (-1)^{k \pm 1}(-1)^{k-1} = 1,$$

as desired.

Problem 2. *Consider the cube root of unity*

$$\varepsilon = \cos \frac{2\pi}{3} + i \sin \frac{2\pi}{3}.$$

Compute

$$(1 + \varepsilon)(1 + \varepsilon^2) \cdots (1 + \varepsilon^{1987}).$$

Solution. Notice that $\varepsilon^3 = 1$, $\varepsilon^2 + \varepsilon + 1 = 0$ and $1987 = 662 \cdot 3 + 1$. Then

$$(1 + \varepsilon)(1 + \varepsilon^2) \cdots (1 + \varepsilon^{1987})$$

$$= \prod_{k=0}^{661} [(1 + \varepsilon^{3k+1})(1 + \varepsilon^{3k+2})(1 + \varepsilon^{3k+3})](1 + \varepsilon^{1987})$$

$$= \prod_{k=0}^{661} [(1 + \varepsilon)(1 + \varepsilon^2)(1 + 1)](1 + \varepsilon) = (1 + \varepsilon)[2(1 + \varepsilon + \varepsilon^2 + \varepsilon^3)]^{662}$$

$$= (1 + \varepsilon)[2(0 + 1)]^{662} = 2^{662}(1 + \varepsilon)$$

$$= 2^{662}(-\varepsilon^2) = 2^{662} \frac{1 + i\sqrt{3}}{2} = 2^{661}(1 + i\sqrt{3}).$$

Problem 3. *Let $\varepsilon \neq 1$ be a cube root of unity. Compute*

$$(1 - \varepsilon + \varepsilon^2)(1 - \varepsilon^2 + \varepsilon^4) \cdots (1 - \varepsilon^n + \varepsilon^{2n}).$$

Solution. Notice that $1 + \varepsilon + \varepsilon^2 = 0$ and $\varepsilon^3 = 1$. Hence $1 - \varepsilon + \varepsilon^2 = -2\varepsilon$ and $1 + \varepsilon - \varepsilon^2 = -2\varepsilon^2$.

Then

$$1 - \varepsilon^n + \varepsilon^{2n} = \begin{cases} 1, & \text{if } n \equiv 0 \pmod 3, \\ -2\varepsilon, & \text{if } n \equiv 1 \pmod 3, \\ -2\varepsilon^2, & \text{if } n \equiv 2 \pmod 3, \end{cases}$$

and the product of any three consecutive factors of the given product equals

$$1 \cdot (-2\varepsilon) \cdot (-2\varepsilon^2) = 2^2.$$

Therefore,

$$(1 - \varepsilon + \varepsilon^2)(1 - \varepsilon^2 + \varepsilon^4) \cdots (1 - \varepsilon^n + \varepsilon^{2n})$$

$$= \begin{cases} 2^{\frac{2n}{3}}, & \text{if } n \equiv 0 \pmod 3, \\ -2^{2[\frac{n}{3}]+1}\varepsilon, & \text{if } n \equiv 1 \pmod 3, \\ 2^{2[\frac{n}{3}]+2}, & \text{if } n \equiv 2 \pmod 3. \end{cases}$$

Problem 4. *Prove that the complex number*

$$z = \frac{2 + i}{2 - i}$$

has modulus equal to 1, but z is not an nth root of unity for any positive integer n.

Solution. Obviously, $|z| = 1$. Assume for the sake of a contradiction that there is an integer $n \geq 1$ such that $z^n = 1$.

Then $(2 + i)^n = (2 - i)^n$, and writing $2 + i = (2 - i) + 2i$, it follows that

$$(2 - i)^n = (2 + i)^n$$

$$= (2 - i)^n + \binom{n}{1}(2 - i)^{n-1}2i + \cdots + \binom{n}{n-1}(2 - i)(2i)^{n-1} + (2i)^n.$$

This is equivalent to

$$(2i)^n = (-2 + i)\left[\binom{n}{1}(2i - 1)^{n-2}2i + \cdots + \binom{n}{n-1}(2i)^{n-1}\right]$$

$$= (-2 + i)(a + bi),$$

with a, $b \in \mathbb{Z}$.

Taking the modulus of both sides of the equality gives $2^n = 5(a^2 + b^2)$, a contradiction.

Problem 5. *Let U_n be the set of nth roots of unity. Prove that the following statements are equivalent:*

(a) there is $\alpha \in U_n$ such that $1 + \alpha \in U_n$;
(b) there is $\beta \in U_n$ such that $1 - \beta \in U_n$.

(Romanian Mathematical Olympiad—Second Round, 1990)

Solution. Assume that there exists $\alpha \in U_n$ such that $1 + \alpha \in U_n$. Setting $\beta = \dfrac{1}{1+\alpha}$, we have $\beta^n = \left(\dfrac{1}{1+\alpha}\right)^n = \dfrac{1}{(1+\alpha)^n} = 1$, and hence $\beta \in U_n$. On the other hand, $1 - \beta = \dfrac{\alpha}{\alpha + 1}$ and $(1 - \beta)^n = \dfrac{\alpha^n}{(1+\alpha)^n} = 1$; hence $1 - \beta \in U_n$, as desired.

Conversely, if β, $1 - \beta \in U_n$, set $\alpha = \dfrac{1-\beta}{\beta}$. Since $\alpha^n = \dfrac{(1-\beta)^n}{\beta^n} = 1$ and $(1 + \alpha)^n = \dfrac{1}{\beta^n} = 1$, we have $\alpha \in U_n$ and $1 + \alpha \in U_n$, as desired.

Remark. The statements (a) and (b) are equivalent to $6|n$. Indeed, if α, $1 + \alpha \in U_n$, then $|\alpha| = |1+\alpha| = 1$. It follows that $1 = |1+\alpha|^2 = (1+\alpha)(1+\overline{\alpha}) = 1 + \alpha + \overline{\alpha} + |\alpha|^2 = 1 + \alpha + \overline{\alpha} + 1 = 2 + \alpha + \dfrac{1}{\alpha}$, i.e., $\alpha = -\frac{1}{2} \pm i\frac{\sqrt{3}}{2}$; hence

$$1 + \alpha = \frac{1}{2} \pm i\frac{\sqrt{3}}{2} = \cos\frac{2\pi}{6} \pm i\sin\frac{2\pi}{6}.$$

Since $(1 + \alpha)^n = 1$, it follows that 6 divides n.

Conversely, if n is a multiple of 6, then both $\alpha = -\dfrac{1}{2} + i\dfrac{\sqrt{3}}{2}$ and $1 + \alpha = \dfrac{1}{2} + i\dfrac{\sqrt{3}}{2}$ belong to U_n.

Problem 6. *Let $n \geq 3$ be a positive integer and let $\varepsilon \neq 1$ be an nth root of unity.*

(1) Show that $|1 - \varepsilon| > \dfrac{2}{n-1}$.

(2) If k is a positive integer such that n does not divide k, then

$$\left| \sin \frac{k\pi}{n} \right| > \frac{1}{n-1}.$$

(Romanian Mathematical Olympiad—Final Round, 1988)

Solution.

(1) We have $\varepsilon^n - 1 = (\varepsilon - 1)(\varepsilon^{n-1} + \cdots + \varepsilon + 1)$; hence taking into account that $\varepsilon \neq 1$, we obtain $\varepsilon^{n-1} + \cdots + \varepsilon + 1 = 0$. The last relation is equivalent to $(\varepsilon^{n-1} - 1) + \cdots + (\varepsilon - 1) = -n$, i.e., $(\varepsilon - 1)[\varepsilon^{n-2} + 2\varepsilon^{n-3} + \cdots + (n - 2)\varepsilon + (n - 1)] = -n$. Passing to the absolute value, we find that

$$n = |\varepsilon - 1||\varepsilon^{n-2} + 2\varepsilon^{n-3} + \cdots + (n-1)| \leq |\varepsilon - 1|(|\varepsilon^{n-2}| + 2|\varepsilon|^{n-3} + \cdots + (n-1)).$$

Therefore,

$$n \leq |1 - \varepsilon|(1 + 2 + \cdots + (n - 1)) = |1 - \varepsilon|\frac{n(n-1)}{2},$$

i.e., we obtain the inequality $|1 - \varepsilon| \geq \dfrac{2}{n-1}$. Moreover, equality is not possible, since the geometric images of $1, \varepsilon, \ldots, \varepsilon^{n-1}$ are not collinear.

(2) Consider $\varepsilon = \cos \dfrac{2k\pi}{n} + i \sin \dfrac{2k\pi}{n}$ and obtain

$$1 - \varepsilon = 1 - \cos \frac{2k\pi}{n} - i \sin \frac{2k\pi}{n}.$$

Hence

$$|1 - \varepsilon|^2 = \left(1 - \cos \frac{2k\pi}{n}\right)^2 + \sin^2 \frac{2k\pi}{n} = 2 - 2\cos \frac{2k\pi}{n} = 4 \sin^2 \frac{k\pi}{n}.$$

Applying the inequality in (1), we see that the desired inequality follows.

Problem 7. *Let U_n be the set of nth roots of unity. Prove that*

$$\prod_{\varepsilon \in U_n} \left(\varepsilon + \frac{1}{\varepsilon}\right) = \begin{cases} 0 & \text{if } n \equiv 0 \ (\mathrm{mod}\, 4), \\ 2, & \text{if } n \equiv 1 \ (\mathrm{mod}\, 2), \\ -4, & \text{if } n \equiv 2 \ (\mathrm{mod}\, 4), \\ 2, & \text{if } n \equiv 3 \ (\mathrm{mod}\, 4). \end{cases}$$

Solution. Consider the polynomial

$$f(x) = X^n - 1 = \prod_{\varepsilon \in U_n} (X - \varepsilon).$$

Denoting by P_n the product in our problem, we have

$$P_n = \prod_{\varepsilon \in U_n} \left(\varepsilon + \frac{1}{\varepsilon} \right) = \prod_{\varepsilon \in U_n} \frac{\varepsilon^2 + 1}{\varepsilon} = \frac{\prod\limits_{\varepsilon \in U_n} (\varepsilon + i)(\varepsilon - i)}{\prod\limits_{\varepsilon \in U_n} \varepsilon}$$

$$= \frac{\prod\limits_{\varepsilon \in U_n} (i + \varepsilon) \prod\limits_{\varepsilon \in U_n} (-i + \varepsilon)}{(-1)^n f(0)} = \frac{f(-1) \cdot f(i)}{(-1)^{n-1}} = \frac{[(-i)^n - 1](i^n - 1)}{(-1)^{n-1}}.$$

If $n \equiv 0 \pmod 4$, then $i^n = 1$ and $P_n = 0$.
    If $n \equiv 1 \pmod 2$, then $(-1)^{n-1} = 1$ and

$$P_n = (-i^n - 1)(i^n - 1) = -(i^{2n} - 1) = -((-1)^n - 1) = -(-1 - 1) = 2.$$

If $n \equiv 2 \pmod 4$, then $(-1)^{n-1} = -1$, $(-i)^n = i^n = i^2 = -1$, $i^n = -1$; hence

$$P_n = \frac{(-1 - 1)(-1 - 1)}{-1} = -4.$$

If $n \equiv 3 \pmod 4$, then $(-1)^{n-1} = 1$ and

$$P_n = (-i^n - 1)(i^n - 1) = (i^3 - 1)(-i^3 - 1) = -(i^6 - 1) = -((-1)^3 - 1) = 2,$$

and we are done.

Problem 8. *Let*

$$\omega = \cos \frac{2\pi}{2n+1} + i \sin \frac{2\pi}{2n+1}, \quad n \geq 0,$$

and let

$$z = \frac{1}{2} + \omega + \omega^2 + \cdots + \omega^n.$$

Prove the following:

(a) $\operatorname{Im}(z^{2k}) = \operatorname{Re}(z^{2k+1}) = 0$ *for all* $k \in \mathbb{N}$.
(b) $(2z + 1)^{2n+1} + (2z - 1)^{2n+1} = 0$.

Solution. We have $\omega^{2n+1} = 1$ and

$$1 + \omega + \omega^2 + \cdots + \omega^{2n} = 0.$$

Then

$$\frac{1}{2} + \omega + \omega^2 + \cdots + \omega^n + \omega^n(\omega + \omega^2 + \cdots + \omega^n) + \frac{1}{2} = 0,$$

or

$$z + \omega^n \left(z - \frac{1}{2} \right) + \frac{1}{2} = 0,$$

whence

$$z = \frac{1}{2} \cdot \frac{\omega^n - 1}{\omega^n + 1}.$$

(a) We have $\overline{z} = \frac{1}{2} \frac{\frac{1}{\omega^n} - 1}{\frac{1}{\omega^n} + 1} = -z$. Thus $z^{2k} = \overline{z^{2k}}$ and $z^{2k+1} = -\overline{z^{2k+1}}$. The

conclusion follows from these two equalities.

(b) From the relation

$$z + \omega^n \left(z - \frac{1}{2} \right) + \frac{1}{2} = 0,$$

we obtain $2z + 1 = -\omega^n(2z - 1)$. Taking into account that $\omega^{2n+1} = 1$, we obtain $(2z + 1)^{2n+1} = -(2z - 1)^{2n+1}$, and we are done.

Problem 9. *Let n be an odd positive integer and $\varepsilon_0, \varepsilon_1, \ldots, \varepsilon_{n-1}$ the complex roots of unity of order n. Prove that*

$$\prod_{k=0}^{n-1} (a + b\varepsilon_k^2) = a^n + b^n$$

for all complex numbers a and b.

(Romanian Mathematical Olympiad—Second Round, 2000)

Solution. If $ab = 0$, then the claim is obvious, so consider the case that $a \neq 0$ and $b \neq 0$.

We start with a useful lemma.

Lemma. *If $\varepsilon_0, \varepsilon_1, \ldots, \varepsilon_{n-1}$ are the complex roots of unity of order n, where n is an odd integer, then*

$$\prod_{k=0}^{n-1} (A + B\varepsilon_k) = A^n + B^n$$

for all complex numbers A and B.

Proof. Using the identity

$$x^n - 1 = \prod_{k=0}^{n-1} (x - \varepsilon_k)$$

for $x = -\frac{A}{B}$ yields

$$-\left(\frac{A^n}{B^n} + 1 \right) = -\prod_{k=0}^{n-1} \left(\frac{A}{B} + \varepsilon_k \right),$$

and the conclusion follows.   □

Because n is odd, the function $f : U_n \to U_n$ is bijective. To prove this, it suffices to show that it is injective. Indeed, assume that $f(x) = f(y)$. It follows that $(x - y)(x + y) = 0$. If $x + y = 0$, then $x^n = (-y)^n$, i.e., $1 = -1$, a contradiction. Hence $x = y$.

From the lemma we have

$$\prod_{k=0}^{n-1}(a + b\varepsilon_k^2) = \prod_{j=0}^{n-1}(a + b\varepsilon_j) = a^n + b^n.$$

Problem 10. *Let n be an even positive integer such that $\frac{n}{2}$ is odd and let $\varepsilon_0, \varepsilon_1, \ldots, \varepsilon_{n-1}$ be the complex roots of unity of order n. Prove that*

$$\prod_{k=0}^{n-1}(a + b\varepsilon_k^2) = (a^{\frac{n}{2}} + b^{\frac{n}{2}})^2$$

for arbitrary complex numbers a and b.

(Romanian Mathematical Olympiad—Second Round, 2000)

Solution. If $b = 0$, the claim is obvious. If not, let $n = 2(2s + 1)$. Consider a complex number α such that $\alpha^2 = \dfrac{a}{b}$ and the polynomial

$$f = X^n - 1 = (X - \varepsilon_0)(X - \varepsilon_1)\cdots(X - \varepsilon_{n-1}).$$

We have

$$f\left(\frac{\alpha}{i}\right) = \left(\frac{1}{i}\right)^a (\alpha - i\varepsilon_0)\cdots(\alpha - i\varepsilon_{n-1})$$

and

$$f\left(-\frac{\alpha}{i}\right) = \left(\frac{-1}{i}\right)^a (\alpha + i\varepsilon_0)\cdots(\alpha + i\varepsilon_{n-1});$$

hence

$$f\left(\frac{\alpha}{i}\right) f\left(-\frac{\alpha}{i}\right) = (\alpha^2 + \varepsilon_0^2)\cdots(\alpha^2 + \varepsilon_{n-1}^2).$$

Therefore,

$$\prod_{k=0}^{n-1}(a + b\varepsilon_k^2) = b^n \prod_{k=0}^{n-1}\left(\frac{a}{b} + \varepsilon_k^2\right) = b^n \prod_{k=0}^{n-1}(\alpha^2 + \varepsilon_k^2)$$

$$= b^n f\left(\frac{\alpha}{i}\right) f\left(-\frac{\alpha}{i}\right) = b^n[(\alpha^2)^{2s+1} + 1]^2 = b^n\left[\left(\frac{a}{b}\right)^{2s+1} + 1\right]^2$$

$$= b^{2(2s+1)}\left(\frac{a^{2s+1} + b^{2s+1}}{b^{2s+1}}\right)^2 = (a^{\frac{n}{2}} + b^{\frac{n}{2}})^2.$$

The following problems also involve nth roots of unity.

Problem 11. *For all positive integers k, define*

$$U_k = \{z \in \mathbb{C} \mid z^k = 1\}.$$

Prove that for integers m and n with $0 < m < n$, we have

$$U_1 \cup U_2 \cup \cdots \cup U_m \subset U_{n-m+1} \cup U_{n-m+2} \cup \cdots \cup U_n.$$

(Romanian Mathematical Regional Contest "Grigore Moisil," 1997)

Problem 12. *Let a, b, c, d, α be complex numbers such that $|a| = |b| \neq 0$ and $|c| = |d| \neq 0$. Prove that all roots of the equation*

$$c(bx + a\alpha)^n - d(ax + b\overline{\alpha})^n = 0, \ n \geq 1,$$

are real numbers.

Problem 13. *Suppose that $z \neq 1$ is a complex number such that $z^n = 1$, $n \geq 1$. Prove that*

$$|nz - (n+2)| \leq \frac{(n+1)(2n+1)}{6} |z - 1|^2.$$

(Crux Mathematicorum, 2003)

Problem 14. *Let M be a set of complex numbers such that if x, $y \in M$, then $\dfrac{x}{y} \in M$. Prove that if the set M has n elements, then M is the set of the nth roots of 1.*

Problem 15. *A finite set A of complex numbers has the property that $z \in A$ implies $z^n \in A$ for every positive integer n.*

(a) Prove that $\sum\limits_{z \in A} z$ is an integer.

(b) Prove that for every integer k, one can choose a set A that satisfies the above condition and $\sum\limits_{z \in A} z = k$.

(Romanian Mathematical Olympiad—Final Round, 2003)

Problem 16. *Let $n \geq 3$ be an odd integer. Evaluate $\sum\limits_{k=1}^{\frac{n-1}{2}} \sec \dfrac{2k\pi}{n}$.*

(Mathematical Reflections)

Problem 17. *Let n be an odd positive integer and let z be a complex number such that $z^{2n-1} - 1 = 0$. Evaluate*

$$\prod_{k=0}^{n-1} \left(z^{2^k} + \frac{1}{z^{2^k}} - 1 \right).$$

(Mathematical Reflections)

Problem 18. *The expression* $\sin 2° \sin 4° \sin 6° \ldots \sin 90°$ *is equal to* $p\sqrt{5}/2^{50}$, *where p is an integer. Find p.*

Problem 19. *The polynomial* $P(x) = (1 + x + x^2 + \ldots + x^{17})^2 - x^{17}$ *has 34 complex roots of the form*

$$z_k = r_k[\cos(2\pi a_k) + i\sin(2\pi a_k)], \quad k = 1, 2, 3, \ldots, 34,$$

with $0 < a_1 \le a_2 \le a_3 \le \ldots \le a_{34} < 1$ *and* $r_k > 0$. *Given that* $a_1 + a_2 + a_3 + a_4 + a_5 = m/n$, *where m and n are relatively prime positive integers, find $m + n$.*

(2004 AIME I, Problem 13)

Problem 20. *The sets* $A = \{z : z^{18} = 1\}$ *and* $B = \{w : w^{48} = 1\}$ *are both sets of complex roots of unity. The set* $C = \{zw : a \in A \text{ and } w \in B\}$ *is also a set of complex roots of unity. How many distinct elements are in C?*

(1990 AIME, Problem 10)

Problem 21. *Let $n \ge 3$ be an integer and* $z = \cos\dfrac{2\pi}{n} + i\sin\dfrac{2\pi}{n}$. *Consider the sets*

$$A = \{1, z, z^2, \ldots, z^{n-1}\}$$

and

$$B = \{1, 1 + z, 1 + z + z^2, \ldots, 1 + z + \ldots + z^{n-1}\}.$$

Determine $A \cap B$.

(Romanian Mathematical Olympiad—District Round, 2008)

5.7 Problems Involving Polygons

Problem 1. *Let z_1, z_2, $\ldots$, z_n be distinct complex numbers such that* $|z_1| = |z_2| = \cdots = |z_n|$. *Prove that*

$$\sum_{1 \le i < j \le n} \left| \frac{z_i + z_j}{z_i - z_j} \right|^2 \ge \frac{(n-1)(n-2)}{2}.$$

Solution. Consider the points $A_1, A_2, \ldots, A_n$ with coordinates $z_1, z_2, \ldots, z_n$. The polygon $A_1 A_2 \cdots A_n$ is inscribed in the circle with center at the origin and radius $R = |z_1|$.

The coordinate of the midpoint A_{ij} of the segment $[A_i A_j]$ is equal to $\dfrac{z_i + z_j}{2}$, for $1 \le i < j \le n$. Hence

$$|z_i + z_j|^2 = 4OA_{ij}^2 \text{ and } |z_i - z_j|^2 = A_i A_j^2.$$

Moreover, $4OA_{ij}^2 = 4R^2 - A_iA_j^2$.

The sum

$$\sum_{1\leq i<j\leq n} \left| \frac{z_i + z_j}{z_i - z_j} \right|^2$$

equals

$$\sum_{1\leq i<j\leq n} \frac{4OA_{ij}^2}{A_iA_j^2} = \sum_{1\leq i<j\leq n} \frac{4R^2 - A_iA_j^2}{A_iA_j^2} = 4R^2 \sum_{1\leq i<j\leq n} \frac{1}{A_iA_j^2} - \binom{n}{2}.$$

The *AM–HM* (arithmetic mean–harmonic mean) inequality gives

$$\sum_{1\leq i<j\leq n} \frac{1}{A_iA_j^2} \geq \frac{\left(\binom{n}{2}\right)^2}{\sum_{1\leq i<j\leq n} A_iA_j^2}.$$

Since $\sum_{1\leq i<j\leq n} A_iA_j^2 \leq n^2 \cdot R^2$, it follows that

$$\sum_{1\leq i<j\leq n} \left| \frac{z_i + z_j}{z_i - z_j} \right|^2 \geq 4R^2 \frac{\left(\binom{n}{2}\right)^2}{\sum_{1\leq i<j\leq n} A_iA_j} - \binom{n}{2}$$

$$\geq \frac{4\left(\binom{n}{2}\right)^2}{n^2} - \binom{n}{2} = \frac{(4\binom{n}{2} - n^2) \cdot \binom{n}{2}}{n^2} = \frac{(n-1)(n-2)}{2},$$

as claimed.

Problem 2. *Let $A_1A_2\cdots A_n$ be a polygon and let a_1, a_2, $\ldots$, a_n be the coordinates of the vertices A_1, A_2, $\ldots$, A_n. If $|a_1| = |a_2| = \cdots = |a_n| = R$, prove that*

$$\sum_{1\leq i<j\leq n} |a_i + a_j|^2 \geq n(n-2)R^2.$$

Solution. We have

$$\sum_{1 \leq i < j \leq n} |a_i + a_j|^2 = \sum_{1 \leq i < j \leq n} (a_i + a_j)(\overline{a_i} + \overline{a_j})$$

$$= \sum_{1 \leq i < j \leq n} (|a_i|^2 + |a_j|^2 + a_i \overline{a_j} + \overline{a_i} a_j)$$

$$= 2R^2 \binom{n}{2} + \sum_{i \neq j} a_i \overline{a_j} = n(n-1)R^2 + \sum_{i=1}^{n} \sum_{j=1}^{n} a_i \overline{a_j} - \sum_{i=1}^{n} a_i \overline{a_i}$$

$$= n(n-1)R^2 + \left(\sum_{i=1}^{n} a_i\right) \left(\sum_{i=1}^{n} \overline{a_i}\right) - nR^2$$

$$= n(n-2)R^2 + \left|\sum_{i=1}^{n} a_i\right|^2 \geq n(n-2)R^2,$$

as desired.

Problem 3. *Let z_1, z_2, $\ldots$, z_n be the coordinates of the vertices of a regular polygon with circumcenter at the origin of the complex plane. Prove that there are i, j, $k \in \{1, 2, \ldots, n\}$ such that $z_i + z_j = z_k$ if and only if 6 divides n.*

Solution. Let $\varepsilon = \cos \frac{2\pi}{n} + i \sin \frac{2\pi}{n}$. Then $z_p = z_1 \cdot \varepsilon^{p-1}$, for all $p = \overline{1, n}$.
We have $z_i + z_j = z_k$ if and only if $1 + \varepsilon^{j-i} = \varepsilon^{k-i}$, i.e.,

$$2 \cos \frac{(j-i)\pi}{n} \left[\cos \frac{(j-i)\pi}{n} + i \sin \frac{(j-i)\pi}{n} \right] = \cos \frac{2(k-i)\pi}{n} + i \sin \frac{2(k-i)\pi}{n}.$$

The last relation is equivalent to

$$\frac{(j-i)\pi}{n} = \frac{\pi}{3} = \frac{2(k-i)\pi}{n}, \text{ i.e., } n = 6(k-i) = 3(j-i);$$

hence 6 divides n.
Conversely, if 6 divides n, let

$$i = 1, \ j = \frac{n}{3} + 1, \ k = \frac{n}{6} + 1,$$

and we have $z_i + z_l = z_k$, as desired.

Problem 4. *Let z_1, z_2, $\ldots$, z_n be the coordinates of the vertices of a regular polygon. Prove that*

$$z_1^2 + z_2^2 + \cdots + z_n^2 = z_1 z_2 + z_2 z_3 + \cdots + z_n z_1.$$

Solution. Without loss of generality, we may assume that the center of the polygon is the origin of the complex plane.
Let $z_k = z_1 \varepsilon^{k-1}$, where

$$\varepsilon = \cos \frac{2\pi}{n} + i \sin \frac{2\pi}{n}, \ k = 1, \ldots, n.$$

The right-hand side is equal to

$$z_1 z_2 + z_2 z_3 + \cdots + z_n z_1 = \sum_{k=1}^{n} z_i z_{k+1}$$

$$= \sum_{k=1}^{n} z_1^2 \varepsilon^{2k-1} = z_1^2 \cdot \varepsilon \cdot \frac{1 - \varepsilon^{2n}}{1 - \varepsilon^2} = 0.$$

On the other hand,

$$z_1^2 + z_2^2 + \cdots + z_n^2 = \sum_{k=1}^{n} z_i^2 = \sum_{k=1}^{n} z_1^2 \varepsilon^{2k-2} = z_1^2 \frac{1 - \varepsilon^{2n}}{1 - \varepsilon^2} = 0,$$

and we are done.

Problem 5. Let $n \geq 4$ and let a_1, a_2, ..., a_n be the coordinates of the vertices of a regular polygon. Prove that

$$a_1 a_2 + a_2 a_3 + \cdots + a_n a_1 = a_1 a_3 + a_2 a_4 + \cdots + a_n a_2.$$

Solution. Assume that the center of the polygon is the origin of the complex plane and $a_k = a_1 \varepsilon^{k-1}$, $k = 1$, ..., n, where

$$\varepsilon = \cos \frac{2\pi}{n} + i \sin \frac{2\pi}{n}.$$

The left-hand side of the equality is

$$a_1 a_2 + a_2 a_3 + \cdots + a_n a_1 = a_1^2 \sum_{k=1}^{n} \varepsilon^{2k-1} = a_1^2 \varepsilon \frac{1 - \varepsilon^{2n}}{1 - \varepsilon^2} = 0.$$

The right-hand side of the equality is

$$a_1^2 \sum_{k=1}^{n} \varepsilon^{2k} = a_1^2 \varepsilon^2 \frac{1 - \varepsilon^{2n}}{1 - \varepsilon^2} = 0,$$

and we are done.

Problem 6. Let z_1, z_2, ..., z_n be distinct complex numbers such that

$$|z_1| = |z_2| = \cdots = |z_n| = 1.$$

Consider the following statements:

(a) z_1, z_2, ..., z_n are the coordinates of the vertices of a regular polygon.
(b) $z_1^n + z_2^n + \cdots + z_n^n = n(-1)^{n+1} z_1 z_2 \ldots z_n$.
  Decide with proof whether the implications (a) $\Rightarrow$ (b) and (b) $\Rightarrow$ (a) are true.

Solution. We condider first the implication (a) $\Rightarrow$ (b).

Let $\varepsilon = \cos\dfrac{2\pi}{n} + i\sin\dfrac{2\pi}{n}$. Since $z_1, z_2, \ldots, z_n$ are coordinates of the vertices of a regular polygon, without loss of generality we may assume that

$$z_k = z_1\varepsilon^{k-1} \text{ for } k = \overline{1, n}.$$

Then relation (b) becomes

$$z_1^n(1 + \varepsilon^n + \varepsilon^{2n} + \cdots + \varepsilon^{n(n-1)}) = n(-1)^{n+1}z_1^n\varepsilon^{1+2+\cdots+(n-1)}.$$

This is equivalent to

$$n = n(-1)^{n+1}\varepsilon^{\frac{n(n-1)}{2}}, \text{ i.e.,}$$

$$1 = (-1)^{n+1}\left(\cos\frac{n(n-1)}{2}\cdot\frac{2\pi}{n} + i\sin\frac{n(n-1)}{2}\cdot\frac{2\pi}{n}\right).$$

We obtain

$$1 = (-1)^{n+1}(\cos(n-1)\pi + i\sin(n-1)\pi), \text{ i.e., } 1 = (-1)^{n+1}(-1)^{n-1},$$

which is valid. Therefore, the implication (a) $\Rightarrow$ (b) holds.

We prove now that the implication (b) $\Rightarrow$ (a) is also valid.

Observe that

$$|n\cdot(-1)^{n+1}z_1z_2\ldots z_n| = n|z_1|\cdot|z_2|\cdots|z_n| = n;$$

hence

$$|z_1^n + z_2^n + \cdots + z_n^n| = n.$$

Using the triangle inequality, we obtain

$$n = |z_1^n + z_2^n + \cdots + z_n^n| \leq |z_1^n| + |z_2^n| + \cdots + |z_n^n| = \underbrace{1 + 1 + \cdots + 1}_{n \text{ times}} = n;$$

hence the numbers $z_1^n, z_2^n, \ldots, z_n^n$ have the same argument. Since $|z_1^n| = |z_2^n| = \cdots = |z_n^n| = 1$, it follows that $z_1^n = z_2^n = \cdots = z_n^n = a$, where a is a complex number with $|a| = 1$. The numbers $z_1, z_2, \ldots, z_n$ are distinct. Therefore, they are the nth roots of a, and consequently the coordinates of the vertices of a regular polygon.

Problem 7. *Let A, B, C be three consecutive vertices of a regular n-gon and consider the point M on the circumcircle such that points B and M lie on opposite sides of the line AC.*

Prove that $MA + MC = 2MB\cos\dfrac{\pi}{n}.$

(A generalization of the Van Schouten theorem; see the first remark below)

Solution. Consider the complex plane with origin at the center of the polygon and let 1 be the coordinate of A_1.

If $\varepsilon = \cos \dfrac{2\pi}{n} + i \sin \dfrac{2\pi}{n}$, then ε^{k-1} is the coordinate of A_k, then $k = \overline{1, n}$.

Without loss of generality, assume that $A = A_1$, $B = A_2$, and $C = A_3$. Let $z_M = \cos t + i \sin t, t \in [0, 2\pi)$ be the coordinate of the point M. Since points B and M are separated by the line AC, it follows that $\dfrac{4\pi}{n} < t$.

Then

$$MA = |z_M - 1| = \sqrt{(\cos t - 1)^2 + \sin^2 t} = \sqrt{2 - 2\cos t} = 2\sin \frac{t}{2},$$

$$MB = |z_M - \varepsilon| = 2\sin\left(\frac{t}{2} - \frac{\pi}{n}\right),$$

and

$$MC = |z_M - \varepsilon^2| = 2\sin\left(\frac{t}{2} - \frac{2\pi}{n}\right).$$

The equality

$$MA + MC = 2MB \cos\frac{\pi}{n}$$

is equivalent to

$$2\sin\frac{t}{2} + 2\sin\left(\frac{t}{2} - \frac{2\pi}{n}\right) = 4\sin\left(\frac{t}{2} - \frac{\pi}{n}\right)\cos\frac{\pi}{n},$$

which follows from the sum-to-product formula on the left-hand side.

Remarks.

(1) If $n = 3$, then we obtain Van Schouten's theorem: *For every point M on the circumcircle of the equilateral triangle ABC such that M belongs to the arc $\overset{\frown}{AC}$, we have*

$$MA + MC = MB.$$

Note that this result also follows from Ptolemy's theorem.

(2) If $n = 4$, then for every point M on the circumcircle of the square $ABCD$ such that B and M lie on opposite sides of the line AC, we have the relation

$$MA + MC = \sqrt{2}MB.$$

Problem 8. *Let P be a point on the circumcircle of square $ABCD$. Find all integers $n > 0$ such that the sum*

$$S_n(P) = PA^n + PB^n + PC^n + PD^n$$

is constant with respect to the point P.

Solution. Consider the complex plane with origin at the center of the square such that A, B, C, D have coordinates $1, i, -1, -i$, respectively.

Let $z = a + bi$ be the coordinate of point P, where a, $b \in \mathbb{R}$ with $a^2 + b^2 = 1$. The sum $S_n(P)$ is equal to

$$S_n(P) = [(a-1)^2 + b^2]^{\frac{n}{2}} + [a^2 + (b-1)^2]^{\frac{n}{2}} + [(a+1)^2 + b^2]^{\frac{n}{2}} + [a^2 + (b+1)^2]^{\frac{n}{2}}$$
$$= 2^{\frac{n}{2}} \left[(1+a)^{\frac{n}{2}} + (1-a)^{\frac{n}{2}} + (1+b)^{\frac{n}{2}} + (1-b)^{\frac{n}{2}} \right].$$

Set $P = A(1,0)$. Then $S_n(A) = 2^{\frac{n+2}{2}} + 2^n$. For $P = E\left(\frac{\sqrt{2}}{2}, \frac{\sqrt{2}}{2}\right)$, we get

$$S_n(E) = 2(2 - \sqrt{2})^{\frac{n}{2}} + 2(2 + \sqrt{2})^{\frac{n}{2}}.$$

Since $S_n(P)$ is constant with respect to P, it follows that $S_n(A) = S_n(E)$, or $2^{\frac{n+2}{2}} + 2^n = 2(2 - \sqrt{2})^{\frac{n}{2}} + 2(2 + \sqrt{2})^{\frac{n}{2}}$.

It is obvious that $2^{\frac{n+2}{2}} 2(2 - \sqrt{2})^{\frac{n}{2}}$ for all $n \geq 1$. We also have $2^n > 2(2 + \sqrt{2})^{\frac{n}{2}}$ for all $n \geq 9$. The last inequality is equivalent to

$$\frac{1}{4} > \left(\frac{2 + \sqrt{2}}{4} \right)^n \quad \text{for } n \geq 9.$$

The left-hand side of the inequality decreases with n, so it suffices to observe that

$$\frac{1}{4} > \left(\frac{2 + \sqrt{2}}{4} \right)^9.$$

Therefore the inequality $S_n(A) = S_n(E)$ can hold only for $n \leq 8$. Now it is not difficult to verify that $S_n(P)$ is constant only for $n \in \{2, 4, 6\}$.

Problem 9. *A function $f : \mathbb{R}^2 \to \mathbb{R}$ is called Olympic if it has the following property: given $n \geq 3$ distinct points A_1, A_2, $\ldots$, $A_n \in \mathbb{R}^2$, if $f(A_1) = f(A_2) = \cdots = f(A_n)$, then the points A_1, A_2, $\ldots$, A_n are the vertices of a convex polygon. Let $P \in \mathbb{C}[X]$ be a nonconstant polynomial. Prove that the function $f : \mathbb{R}^2 \to \mathbb{R}$ defined by $f(x, y) = |P(x + iy)|$ is Olympic if and only if all the roots of P are equal.*

(Romanian Mathematical Olympiad—Final Round, 2000)

Solution. First suppose that all the roots of P are equal, and write $P(x) = a(z - z_0)^n$ for some a, $z_0 \in \mathbb{C}$ and $n \in \mathbb{N}$. If A_1, A_2, $\ldots$, A_n are distinct points in $\mathbb{R}^2$ such that $f(A_1) = f(A_2) = \cdots = f(A_n)$, then $A_1, \ldots, A_n$ are situated on a circle with center $(\text{Re}(z_0), \text{Im}(z_0))$ and radius $\sqrt[n]{|f(A_1)|/a|}$, implying that the points are the vertices of a convex polygon.

Conversely, suppose that not all the roots of P are equal, and write $P(x) = (z - z_1)(z - z_2)Q(z)$, where z_1 and z_2 are distinct roots of $P(x)$ such that $|z_1 - z_2|$ is minimal. Let l be the line containing $Z_1 = (\text{Re}(z_1), \text{Im}(z_1))$ and

$Z_2 = (\text{Re}(z_2), \text{Im}(z_2))$, and let $z_3 = \frac{1}{2}(z_1+z_2)$, so that $Z_3 = (\text{Re}(z_3), \text{Im}(z_3))$ is the midpoint of $[Z_1 Z_2]$. Also, let s_1, s_2 denote the rays $Z_3 Z_1$ and $Z_3 Z_2$, and let $d = f(Z_3) \geq 0$. We must have $r > 0$, because otherwise, z_3 would be a root of P such that $|z_1 - z_3| < |z_1 - z_2|$, which is impossible. Because $f(Z_3) = 0$,

$$\lim_{\substack{Z_3 \to \infty \\ Z \in s_1}} f(Z) = +\infty,$$

and f is continuous, there exists a point $Z_4 \in s_1$ on the side of Z_1 opposite Z_3 such that $f(Z_4) = r$. Similarly, there exists $Z_5 \in s_2$ on the side of Z_2 opposite Z_3 such that $f(Z_5) = r$. Thus, $f(Z_3) = f(Z_4) = f(Z_5)$ and Z_3, Z_4, Z_5 are not vertices of a convex polygon. Hence f is not Olympic.

Problem 10. *In a convex hexagon $ABCDEF$, $\hat{A} + \hat{C} + \hat{E} = 360°$ and*

$$AB \cdot CD \cdot EF = BC \cdot DE \cdot FA.$$

Prove that $AB \cdot FC \cdot EC = BF \cdot DE \cdot CA$.

(1999 Polish Mathematical Olympiad)

Solution. Position the hexagon in the complex plane and let $a = z_B - z_A$, $b = z_C - z_B, \ldots, f = z_A - z_F$. The product identity implies that $|ace| = |bdf|$, and the angle equality implies that $\dfrac{-b}{a} \cdot \dfrac{-d}{c} \cdot \dfrac{-f}{e}$ is real and positive. Hence, $ace = -bdf$. Also, $a+b+c+d+e+f = 0$. Multiplying this by ad and adding $ace + bdf = 0$ gives $a^2 d + abd + acd + ad^2 + ade + adf + ace + bdf = 0$, which factors as $a(d + e)(c + d) + d(a + b)(f + a) = 0$. Thus

$$|a(d + e)(c + d)| = |d(a + b)(f + a)|,$$

which is what we wanted.

Problem 11. *Let $n > 2$ be an integer and $f : \mathbb{R}^2 \to \mathbb{R}$ a function such that for every regular n-gon $A_1 A_2 \cdots A_n$,*

$$f(A_1) + f(A_2) + \cdots + f(A_n) = 0.$$

Prove that f is identically zero.

(Romanian Mathematical Olympiad—Final Round, 1996)

Solution. We identify $\mathbb{R}^2$ with the complex plane and let $\zeta = \cos\dfrac{2\pi}{n} + i\sin\dfrac{2\pi}{n}$. Then the condition is that for every $z \in \mathbb{C}$ and positive real number t,

$$\sum_{j=1}^{n} f(z + t\zeta^j) = 0.$$

In particular, for each of $k = 1, \ldots, n$, we have

$$\sum_{j=1}^{n} f(z - \zeta^k + \zeta^j) = 0.$$

Summing over k yields

$$\sum_{m=1}^{n} \sum_{k=1}^{n} f\left(z - (1 - \zeta^m)\zeta^k\right) = 0.$$

For $m = n$, the inner sum is $nf(z)$; for other m, the inner sum again runs over a regular polygon, hence is 0. Thus $f(z) = 0$ for all $z \in \mathbb{C}$.

Here are some proposed problems.

Problem 12. *Prove that there exists a convex 1990-gon with the following two properties:*

(a) all angles are equal;
(b) the lengths of the sides are the numbers $1^2, 2^2, 3^2, \ldots, 1989^2, 1990^2$ in some order.

(31st IMO)

Problem 13. *Let A and E be opposite vertices of a regular octagon. Let a_n be the number of paths of length n of the form $(P_0, P_1, \ldots, P_n)$, where P_i are vertices of the octagon and the paths are constructed using the following rule: $P_0 = A$, $P_n = E$, P_i, and P_{i+1} are adjacent vertices for $i = 0, \ldots, n-1$, and $P_i \neq E$ for $i = 0, \ldots, n-1$.*

Prove that $a_{2n-1} = 0$ and $a_{2n} = \dfrac{1}{\sqrt{2}}(x^{n-1} - y^{n-1})$, for all $n = 1, 2, 3, \ldots$, where $x = 2 + \sqrt{2}$ and $y = 2 - \sqrt{2}$.

(21st IMO)

Problem 14. *Let A, B, C be three consecutive vertices of a regular polygon and let us consider a point M on the major arc AC of the circumcircle.*
Prove that
$$MA \cdot MC = MB^2 - AB^2.$$

Problem 15. *Let $A_1 A_2 \cdots A_n$ be a regular polygon inscribed in a circle $\mathcal{C}$ of radius 1. Find the maximum value of $\prod_{j=1}^{n} PA_j$, where P is an arbitrary point on circle $\mathcal{C}$.*

(Romanian Mathematical Regional Contest "Grigore Moisil," 1992)

Problem 16. *Let $A_1 A_2 \cdots A_{2n}$ be a regular polygon with circumradius equal to 1 and consider a point P on the circumcircle. Prove that*

$$\sum_{k=0}^{n-1} PA_{k+1}^2 \cdot PA_{n+k+1}^2 = 2n.$$

Problem 17. *Let $A_1 A_2 \ldots A_n$ be a regular n-gon inscribed in a circle with center O and radius R. Prove that for each point M in the plane of the n-gon, the following inequality holds:*

$$\prod_{k=1}^{n} MA_k \leq (OM^2 + R^2)^{\frac{n}{2}}.$$

(Mathematical Reflections, 2009)

5.8 Complex Numbers and Combinatorics

Problem 1. *Compute the sum*

$$\sum_{k=0}^{3n-1} (-1)^k \binom{6n}{2k+1} 3^k.$$

Solution. We have

$$\sum_{k=0}^{3n-1} (-1)^k \binom{6n}{2k+1} 3^k = \sum_{k=0}^{3n-1} \binom{6n}{2k+1} (-3)^k$$

$$= \sum_{k=0}^{3n-1} \binom{6n}{2k+1} (i\sqrt{3})^{2k} = \frac{1}{i\sqrt{3}} \sum_{k=0}^{3n-1} \binom{6n}{2k+1} (i\sqrt{3})^{2k+1}$$

$$= \frac{1}{i\sqrt{3}} \text{Im}(1 + i\sqrt{3})^{6n} = \frac{1}{i\sqrt{3}} \text{Im} \left[2 \left(\cos \frac{\pi}{3} + i \sin \frac{\pi}{3} \right) \right]^{6n}$$

$$= \frac{1}{i\sqrt{3}} \text{Im}[2^{6n}(\cos 2\pi n + i \sin 2\pi n)] = 0.$$

Problem 2. *Calculate the sum $S_n = \sum_{k=0}^{n} \binom{n}{k} \cos k\alpha$, where $\alpha \in [0, \pi]$.*

Solution. Consider the complex number $z = \cos \alpha + i \sin \alpha$ and the sum $T_n = \sum_{k=0}^{n} \binom{n}{k} \sin k\alpha$. We have

$$S_n + iT_n = \sum_{k=0}^{n} \binom{n}{k} (\cos k\alpha + i \sin k\alpha) = \sum_{k=0}^{n} \binom{n}{k} (\cos \alpha + i \sin \alpha)^k$$

$$= \sum_{k=0}^{n} \binom{n}{k} z^k = (1+z)^n. \tag{1}$$

The polar form of the complex number $1 + z$ is

$$1 + \cos \alpha + i \sin \alpha = 2 \cos^2 \frac{\alpha}{2} + 2i \sin \frac{\alpha}{2} \cos \frac{\alpha}{2}$$

$$= 2 \cos \frac{\alpha}{2} \left(\cos \frac{\alpha}{2} + i \sin \frac{\alpha}{2} \right),$$

since $\alpha \in [0, \pi]$. From (1), it follows that

$$S_n + iT_n = \left(2 \cos \frac{\alpha}{2} \right)^n \left(\cos \frac{n\alpha}{2} + i \sin \frac{n\alpha}{2} \right),$$

i.e.,

$$S_n = \left(2 \cos \frac{\alpha}{2} \right)^n \cos \frac{n\alpha}{2} \text{ and } T_n = \left(2 \cos \frac{\alpha}{2} \right)^n \sin \frac{n\alpha}{2}.$$

Problem 3. *Prove the identity*

$$\left(\binom{n}{0} - \binom{n}{2} + \binom{n}{4} - \cdots \right)^2 + \left(\binom{n}{1} - \binom{n}{3} + \binom{n}{5} - \cdots \right)^2 = 2^n.$$

Solution. Set

$$x_n = \binom{n}{0} - \binom{n}{2} + \binom{n}{4} - \cdots \text{ and } y_n = \binom{n}{1} - \binom{n}{3} + \binom{n}{5} - \cdots$$

and observe that

$$(1+i)^n = x_n + y_n i. \tag{1}$$

Passing to the absolute value, it follows that

$$|x_n + y_n i| = |(1+i)^n| = |1+i|^n = 2^{\frac{n}{2}}.$$

This is equivalent to $x_n^2 + y_n^2 = 2^n$.

Remark. We can write the explicit formulas for x_n and y_n as follows. Observe that

$$(1+i)^n = \left(\sqrt{2} \left(\cos \frac{\pi}{4} + i \sin \frac{\pi}{4} \right) \right)^n = 2^{\frac{n}{2}} \left(\cos \frac{n\pi}{4} + i \sin \frac{n\pi}{4} \right).$$

From relation (1), we get

$$x_n = 2^{\frac{n}{2}} \cos \frac{n\pi}{4} \text{ and } y_n = 2^{\frac{n}{2}} \sin \frac{n\pi}{4}.$$

Problem 4. *If m and p are positive integers and $m > p$, then*

$$\binom{m}{0} + \binom{m}{p} + \binom{m}{2p} + \binom{m}{3p} + \cdots$$

$$= \frac{2^m}{p}\left(1 + 2\sum_{k=1}^{\left[\frac{p-1}{2}\right]}\left(\cos\frac{k\pi}{p}\right)^m \cos\frac{mk\pi}{p}\right).$$

Solution. We begin with the following simple but useful remark: If $f \in \mathbb{R}[X]$ is a polynomial, $f = a_0 + a_1X + \cdots + a_mX^m$, and $\varepsilon = \cos\dfrac{2\pi}{p} + i\sin\dfrac{2\pi}{p}$ is a primitive pth root of unity, then for all real numbers n, the following relation holds:

$$a_0 + a_px^p + a_{2p}x^{2p} + \cdots = \frac{1}{p}(f(x) + f(\varepsilon x) + \cdots + f(\varepsilon^{p-1}x)). \quad (1)$$

To prove (1), we use the relation

$$1 + \varepsilon^k + \varepsilon^{2k} + \cdots + \varepsilon^{(p-1)k} = \begin{cases} p, & \text{if } p|k, \\ 0, & \text{otherwise,} \end{cases}$$

on the right-hand side.

Consider the case that p is odd. Using relation (1) for the polynomial $f = (1+X)^m = \binom{m}{0} + \binom{m}{1}X + \cdots + \binom{m}{m}X^m$, we obtain

$$\binom{m}{0} + \binom{m}{p}x^p + \binom{m}{2p}x^{2p} + \cdots = \frac{1}{p}((1+x)^m + (1+\varepsilon x)^m + \cdots + (1+\varepsilon^{p-1}x)^m) \quad (2)$$

Substituting $x = 1$ in relation (2) we obtain

$$S_p = \binom{m}{0} + \binom{m}{p} + \binom{m}{2p} + \cdots = \frac{1}{p}(2^m + (1+\varepsilon)^m + \cdots + (1+\varepsilon^{p-1})^m). \quad (3)$$

From $\varepsilon^k = \cos\dfrac{2k\pi}{p} + i\sin\dfrac{2k\pi}{p}$, it follows that for all $k = 0, 1, \ldots, p-1$,

$$(1+\varepsilon^k)^m = 2^m\left(\cos\frac{k\pi}{p}\right)^m\left(\cos\frac{mk\pi}{p} + i\sin\frac{mk\pi}{p}\right).$$

Using the relation $\varepsilon^{p-k} = \overline{\varepsilon^k}$, we obtain

$$(1+\varepsilon^{p-k})^m = (1+\overline{\varepsilon^k})^m = \overline{(1+\varepsilon^k)^m}$$

$$= 2^m\left(\cos\frac{k\pi}{p}\right)^m\left(\cos\frac{mk\pi}{p} - i\sin\frac{mk\pi}{p}\right).$$

Substituting into (3), we obtain

$$S_p = \frac{1}{p} \sum_{k=0}^{p-1} (1 + \varepsilon^k)^m = \frac{1}{p} \left[\sum_{k=0}^{\frac{p-1}{2}} (1 + \varepsilon^k)^m + \sum_{k=1}^{\frac{p-1}{2}} (1 + \varepsilon^{p-k})^m \right]$$

$$= \frac{1}{p} \left[2^m + 2^m \sum_{k=1}^{\frac{p-1}{2}} \left(\cos \frac{k\pi}{p} \right)^m \left(\cos \frac{mk\pi}{p} + i \sin \frac{mk\pi}{p} \right) \right.$$

$$\left. + 2^m \sum_{k=1}^{\frac{p-1}{2}} \left(\cos \frac{k\pi}{p} \right)^m \left(\cos \frac{mk\pi}{p} - i \sin \frac{mk\pi}{p} \right) \right]$$

$$= \frac{2^m}{p} \left(1 + 2 \sum_{k=1}^{\frac{p-1}{2}} \left(\cos \frac{k\pi}{p} \right)^m \cos \frac{mk\pi}{p} \right).$$

Consider now the case that p is an even positive integer. Because $\varepsilon^{\frac{p}{2}} = -1$, we have

$$S_p = \frac{1}{p} \sum_{k=0}^{p-1} (1 + \varepsilon^k)^m = \frac{1}{p} \left[2^m + \sum_{k=1}^{\frac{p}{2}-1} (1 + \varepsilon^k)^m + \sum_{k=\frac{p}{2}+1}^{p-1} (1 + \varepsilon^k)^m \right]$$

$$= \frac{1}{p} \left[2^m + \sum_{k=1}^{\frac{p}{2}-1} 2^m \left(\cos \frac{k\pi}{p} \right)^m \left(\cos \frac{mk\pi}{p} + i \sin \frac{mk\pi}{p} \right) + \right.$$

$$\left. + \sum_{k=1}^{\frac{p}{2}-1} 2^m \left(\cos \frac{k\pi}{p} \right)^m \left(\cos \frac{mk\pi}{p} - i \sin \frac{mk\pi}{p} \right) \right]$$

$$= \frac{2^m}{p} \left(1 + 2 \sum_{k=1}^{\frac{p}{2}-1} \left(\cos \frac{k\pi}{p} \right)^m \cos \frac{mk\pi}{p} \right).$$

Problem 5. *The following identity holds:*

$$\binom{n}{m} + \binom{n}{m+p} + \binom{n}{m+2p} + \cdots = \frac{2^n}{p} \sum_{k=0}^{p-1} \left(\cos \frac{k\pi}{p} \right)^n \cos \frac{(n-2m)k\pi}{p}.$$

Solution. Let ε_0, ε_1, $\ldots$, ε_{p-1} be the pth roots of unity. Then

$$\sum_{k=0}^{p-1} \varepsilon_k^{-m} (1 + \varepsilon_k)^n = \sum_{k=0}^{n} \binom{n}{k} (\varepsilon_0^{k-m} + \cdots + \varepsilon_{p-1}^{k-m}). \tag{1}$$

Using the result in Proposition 3, Sect. 2.2.2, it follows that

$$\varepsilon_0^{k-m} + \cdots + \varepsilon_{p-1}^{k-m} = \begin{cases} p, & \text{if } p|(k-m), \\ 0, & \text{otherwise.} \end{cases} \tag{2}$$

Taking into account that

$$\varepsilon_k^{-m}(1+\varepsilon_k)^m$$

$$= \left(\cos\frac{2mk\pi}{p} - i\sin\frac{2mk\pi}{p}\right)\left(2\cos\frac{k\pi}{p}\right)^n\left(\cos\frac{nk\pi}{p} + i\sin\frac{nk\pi}{p}\right)$$

$$= 2^n\left(\cos\frac{k\pi}{p}\right)^n\left(\cos\frac{(n-2m)k\pi}{p} + i\sin\frac{(n-2m)k\pi}{p}\right)$$

and using (1) and (2), we obtain the desired identity.

Remark. The following interesting trigonometric relation holds:

$$\sum_{k=0}^{p-1}\left(\cos\frac{k\pi}{p}\right)^n \sin\frac{(n-2m)k\pi}{p} = 0. \tag{1}$$

Problem 6. *Consider the integers* a_n, b_n, c_n, *where*

$$a_n = \binom{n}{0} + \binom{n}{3} + \binom{n}{6} + \cdots,$$

$$b_n = \binom{n}{1} + \binom{n}{4} + \binom{n}{7} + \cdots,$$

$$c_n = \binom{n}{2} + \binom{n}{5} + \binom{n}{8} + \cdots.$$

Show the following:

(1) $a_n^3 + b_n^3 + c_n^3 - 3a_nb_nc_n = 2^n$.
(2) $a_n^2 + b_n^2 + c_n^2 - a_nb_n - b_nc_n - c_na_n = 1$.
(3) Two of the integers a_n, b_n, c_n *are equal, and the third differs by one.*

Solution.

(1) Let ε be a cube root of unity different from 1. We have

$$(1+1)^n = a_n+b_n+c_n, \quad (1+\varepsilon)^n = a_n+b_n\varepsilon+c_n\varepsilon^2, \quad (1+\varepsilon^2)^n = a_n+b_n\varepsilon^2+c_n\varepsilon.$$

Therefore,

$$a_n^3+b_n^3+c_n^3-3a_nb_nc_n = (a_n+b_n+c_n)(a_n+b_n\varepsilon + c_n\varepsilon^2)(a_n + b_n\varepsilon^2 + c_n\varepsilon)$$

$$= 2^n(1+\varepsilon)^n(1+\varepsilon^2)^n = 2^n(-\varepsilon^2)^n(-\varepsilon)^n = 2^n.$$

(2) Using the identity

$$x^3 + y^3 + z^3 - 3xyz = (x + y + z)(x^2 + y^2 + z^2 - xy - yz - zx)$$

and the above relation, it follows that

$$a_n^2 + b_n^2 + c_n^2 - a_n b_n - b_n c_n - c_n a_n = 1.$$

(3) Multiplying the above relation by 2 we obtain

$$(a_n - b_n)^2 + (b_n - c_n)^2 + (c_n - a_n)^2 = 2. \tag{1}$$

From (1), it follows that two of a_n, b_n, c_n are equal and the third differs by one.

Remark. From Problem 5, it follows that

$$a_n = \frac{1}{3}\left[2^n + \cos\frac{n\pi}{3} + (-1)^n \cos\frac{2n\pi}{3}\right] = \frac{1}{3}\left(2^n + 2\cos\frac{n\pi}{3}\right),$$

$$b_n = \frac{1}{3}\left[2^n + \cos\frac{(n-2)\pi}{3} + (-1)^n \cos\frac{(2n-4)\pi}{3}\right]$$

$$= \frac{1}{3}\left(2^n + 2\cos\frac{(n-2)\pi}{3}\right),$$

$$c_n = \frac{1}{3}\left[2^n + \cos\frac{(n-4)\pi}{3} + (-1)^n \cos\frac{(2n-8)\pi}{3}\right]$$

$$= \frac{1}{3}\left(2^n + 2\cos\frac{(n-4)\pi}{3}\right).$$

It is not difficult to see that

$$a_n = b_n \text{ if and only if } n \equiv 1 \pmod 3,$$
$$a_n = c_n \text{ if and only if } n \equiv 2 \pmod 3,$$
$$b_n = c_n \text{ if and only if } n \equiv 0 \pmod 3.$$

Problem 7. *How many positive integers of n digits chosen from the set $\{2, 3, 7, 9\}$ are divisible by 3?*

(Romanian Mathematical Regional Contest "Traian Lalescu," 2003)

Solution. Let x_n, y_n, z_n be the numbers of all positive n-digit integers whose digits are taken from the set $\{2, 3, 7, 9\}$ that are congruent to 0, 1, and 2 modulo 3, repsectively. We have to find x_n.

Consider $\varepsilon = \cos\dfrac{2\pi}{3} + i\sin\dfrac{2\pi}{3}$. It is clear that $x_n + y_n + z_n = 4^n$ and

$$x_n + \varepsilon y_n + \varepsilon^2 z_n = \sum_{j_1+j_2+j_3+j_4=n} \varepsilon^{2j_1+3j_2+7j_3+9j_4} = (\varepsilon^2 + \varepsilon^3 + \varepsilon^7 + \varepsilon^9)^n = 1.$$

It follows that $x_n - 1 + \varepsilon y_n + \varepsilon^2 z_n = 0$. Applying Proposition 4 in Sect. 2.2.2, we obtain $x_n - 1 = y_n = z_n = k$. Then $3k = x_n + y_n + z_n - 1 = 4^n - 1$, and we obtain $k = \frac{1}{3}(4^n - 1)$. Finally, $x_n = k + 1 = \frac{1}{3}(4^n + 2)$.

Problem 8. *Let n be a prime number and let a_1, a_2, ..., a_m be positive integers. Consider $f(k)$, the number of all m-tuples $(c_1, \ldots, c_m)$ satisfying $1 \le c_i \le a_i$ and $\sum_{i=1}^{m} c_i \equiv k \pmod{n}$. Show that $f(0) = f(1) = \cdots = f(n-1)$ if and only if $n \mid a_j$ for some $j \in \{1, \ldots, m\}$.*

(Rookie Contest, 1999)

Solution. Let $\varepsilon = \cos\dfrac{2\pi}{n} + i\sin\dfrac{2\pi}{n}$. Note that the following relations hold:

$$\prod_{i=1}^{m}(X + X^2 + \cdots + X^{a_i}) = \sum_{1 \le c_i \le a_i} X^{c_1 + \cdots + c_m}$$

and

$$f(0) + f(1)\varepsilon + \cdots + f(n-1)\varepsilon^{n-1} = \sum_{1 \le c_i \le a_i} \varepsilon^{c_1 + \cdots + c_m} = \prod_{i=1}^{m}(\varepsilon + \varepsilon^2 + \cdots + \varepsilon^{a_i}).$$

Applying the result in Proposition 4, Sect. 2.2.2, we have $f(0) = f(1) = \ldots = f(n-1)$ if and only if $f(0) + f(1)\varepsilon + \cdots + f(n-1)\varepsilon^{n-1} = 0$. This is equivalent to $\prod_{i=1}^{m}(\varepsilon + \varepsilon^2 + \cdots + \varepsilon^{a_i}) = 0$, i.e., $\varepsilon + \varepsilon^2 + \cdots + \varepsilon^{a_j} = 0$ for some $j \in \{1, \ldots, m\}$. It follows that $\varepsilon^{a_j} - 1 = 0$, i.e., $n \mid a_j$.

Problem 9. *For a finite set A of real numbers denote by $|A|$ the cardinal number of A and by $m(A)$ the sum of the elements of A.*

Let p be a prime and $A = \{1, 2, \ldots, 2p\}$. Find the number of all subsets $B \subset A$ such that $|B| = p$ and $p \mid m(B)$.

(36th IMO)

Solution. The case $p = 2$ is trivial. Consider $p \ge 3$ and $\varepsilon = \cos\dfrac{2\pi}{p} + i\sin\dfrac{2\pi}{p}$. Denote by x_j the number of all subsets $B \subset A$ with the properties $|B| = p$ and $m(B) \equiv j \pmod{p}$.

Then

$$\sum_{j=0}^{p-1} x_j \varepsilon^j = \sum_{B \subset A, |B| = p} \varepsilon^{mB} = \sum_{1 \le c_1 < \cdots < c_p \le 2p} \varepsilon^{c_1 + \cdots + c_p}.$$

The last sum is the coefficient of X^p in $(X + \varepsilon)(X + \varepsilon^2) \cdots (X + \varepsilon^{2p})$. Taking into account the relation $X^p - 1 = (X - 1)(X - \varepsilon) \cdots (X - \varepsilon^{p-1})$, we obtain

$(X + \varepsilon)(X + \varepsilon^2) \cdots (X + \varepsilon^{2p}) = (X^p + 1)^2$; hence the coefficient of X^p is 2. Therefore,

$$\sum_{j=0}^{p-1} x_j \varepsilon^j = 2,$$

i.e., $x_0 - 2 + x_1\varepsilon + \cdots + x_{p-1}\varepsilon^{p-1} = 0$. From Proposition 4, Sect. 2.2.2, it follows that $x_0 - 2 = x_1 = \cdots = x_{p-1} = k$. We obtain $pk = x_0 + \cdots + x_{p-1} - 2 = \binom{2p}{p} - 2$, and hence $k = \dfrac{1}{p}\left(\binom{2p}{p} - 2\right)$. Therefore, the desired number is

$$x_0 = 2 + k = 2 + \frac{1}{p}\left(\binom{2p}{p} - 2\right).$$

Problem 10. *Prove that the number* $\displaystyle\sum_{k=0}^{n} \binom{2n+1}{2k+1} 2^{3k}$ *is not divisible by 5 for any integer $n \geq 0$.*

(16th IMO)

Solution. Since $2^3 \equiv -2 \pmod 5$, an equivalent problem is to prove that $S_n = \displaystyle\sum_{k=0}^{n} \binom{2n+1}{2k+1} (-2)^k$ is not divisible by 5. Expanding $(1 + i\sqrt{2})^{2n+1}$ and then separating the even and odd terms, we get

$$(1 + i\sqrt{2})^{2n+1} = R_n + i\sqrt{2}S_n, \tag{1}$$

where $R_n = \displaystyle\sum_{k=0}^{n} \binom{2n+1}{2k} (-2)^k$.

Passing to the absolute value from (1), it follows that

$$3^{2n+1} = R_n^2 + 2S_n^2. \tag{2}$$

Since $3^2 \equiv -1 \pmod 5$, the relation (2) leads to

$$R_n^2 + 2S_n^2 \equiv \pm 3 \pmod 5. \tag{3}$$

Assume for the sake of a contradiction that $S_n \equiv 0 \pmod 5$ for some positive integer n. Then from (3), we obtain $R_n^2 \equiv \pm 3 \pmod 5$, a contradiction, since every square is congruent to 0, 1, or 4 modulo 5.

Here are some other problems concerning complex numbers and combinatorics.

Problem 11. *Calculate the sum* $s_n = \displaystyle\sum_{k=0}^{n} \binom{n}{k}^2 \cos kt$, *where $t \in [0, \pi]$.*

Problem 12. *Prove the following identities:*

(1) $\binom{n}{0} + \binom{n}{4} + \binom{n}{8} + \cdots = \frac{1}{4}\left(2^n + 2^{\frac{n}{2}+1}\cos\frac{n\pi}{4}\right).$

(Romanian Mathematical Olympiad—Second Round, 1981)

(2) $\binom{n}{0} + \binom{n}{5} + \binom{n}{10} + \cdots =$

$$= \frac{1}{5}\left[2^n + \frac{(\sqrt{5}+1)^n}{2^{n-1}}\cos\frac{n\pi}{5} + \frac{(\sqrt{5}-1)^n}{2^{n-1}}\cos\frac{2n\pi}{5}\right].$$

Problem 13. *Consider the integers* A_n, B_n, C_n *defined by*

$$A_n = \binom{n}{0} - \binom{n}{3} + \binom{n}{6} - \cdots,$$

$$B_n = -\binom{n}{1} + \binom{n}{4} - \binom{n}{7} + \cdots,$$

$$C_n = \binom{n}{2} - \binom{n}{5} + \binom{n}{8} - \cdots.$$

The following identities hold:

(1) $A_n^2 + B_n^2 + C_n^2 - A_n B_n - B_n C_n - C_n A_n = 3^n$;
(2) $A_n^2 + A_n B_n + B_n^2 = 3^{n-1}.$

Problem 14. *Let* $p \geq 3$ *be a prime and let* m, n *be positive integers divisible by* p *such that* n *is odd. For each* m-*tuple* $(c_1, \ldots, c_m)$, $c_i \in \{1, 2, \ldots, n\}$, *with the property that* $p \mid \sum_{i=1}^{m} c_i$, *let us consider the product* $c_1 \cdots c_m$. *Prove that the sum of all these products is divisible by* $\left(\dfrac{n}{p}\right)^m$.

Problem 15. *Let* k *be a positive integer and* $a = 4k - 1$. *Prove that for every positive integer* n, *the integer*

$$s_n = \binom{n}{0} - \binom{n}{2}a + \binom{n}{4}a^2 - \binom{n}{6}a^3 + \cdots \text{ is divisible by } 2^{n-1}.$$

(Romanian Mathematical Olympiad—Second Round, 1984)

Problem 16. *Let* m *and* n *be integers greater than 1. Prove that*

$$\sum_{\substack{k_1+k_2+\cdots+k_n=m \\ k_1,k_2,\ldots,k_n \geq 0}} \frac{1}{k_1!k_2!\ldots k_n!}\cos(k_1 + 2k_2 + \ldots + nk_n)\frac{2\pi}{n} = 0.$$

(Mathematical Reflections, 2009)

Problem 17. *Given an integer $n \geq 2$, let a_n, b_n, c_n be integers such that*

$$(\sqrt[3]{2} - 1)^n = a_n + b_n \sqrt[3]{2} + c_n \sqrt[3]{4}.$$

Show that $c_n \equiv 1 \pmod 3$ if and only if $n \equiv 2 \pmod 3$.

(Romanian IMO Team Selection Test, 2013)

5.9 Miscellaneous Problems

Problem 1. *Two unit squares K_1, K_2 with centers M, N are situated in the plane so that $MN = 4$. Two sides of K_1 are parallel to the line MN, and one of the diagonals of K_2 lies on MN. Find the locus of the midpoint of XY as X, Y vary over the interior of K_1, K_2, respectively.*

(1997 Bulgarian Mathematical Olympiad)

Solution. Introduce complex numbers with $M = -2$, $N = 2$. Then the locus is the set of points of the form $-(w + xi) + (y + zi)$, where $|w|, |x| < 1/2$ and $|x + y|, |x - y| < \sqrt{2}/2$. The result is an octagon with vertices $(1 + \sqrt{2})/2 + i/2, 1/2 + (1 + \sqrt{2})i/2$, and so on.

Problem 2. *Curves A, B, C, and D are defined in the plane as follows:*

$$A = \left\{ (x, y) : x^2 - y^2 = \frac{x}{x^2 + y^2} \right\},$$
$$B = \left\{ (x, y) : 2xy + \frac{y}{x^2 + y^2} = 3 \right\},$$
$$C = \left\{ (x, y) : x^3 - 3xy^2 + 3y = 1 \right\},$$
$$D = \left\{ (x, y) : 3x^2y - 3x - y^3 = 0 \right\}.$$

Prove that $A \cap B = C \cap D$.

(1987 Putnam Mathematical Competition)

Solution. Let $z = x + yi$. The equations defining A and B are the real and imaginary parts of the equation $z^2 = z^{-1} + 3i$, and similarly the equations defining C and D are the real and imaginary parts of $z^3 - 3iz = 1$. Hence for all real x and y, we have $(x, y) \in A \cap B$ if and only if $z^2 = z^{-1} + 3i$. This is equivalent to $z^3 - 3iz = 1$, i.e., $(x, y) \in C \cap D$.

Thus $A \cap B = C \cap D$.

Problem 3. *Determine with proof whether it is possible to consider 1975 points on the unit circle such that the distance between every pair of points is a rational number (the distances being taken along the chord).*

(17th IMO)

Solution. There are infinitely many points with rational coordinates on the unit circle. This is a well-known result arising from Pythagorean triangles and the corresponding equation

$$m^2 + n^2 = p^2.$$

Each such point $A(x_A, \, y_A)$ can be represented by a complex number

$$z_A = x_A + i y_A = \cos \alpha_A + i \sin \alpha_A,$$

where α_A is the argument of the complex number z_A and $\cos \alpha_A$, $\sin \alpha_A$ are rational numbers.

Taking on the unit circle complex numbers of the form

$$z_A^2 = \cos 2\alpha_A + i \sin 2\alpha_A,$$

we have for two such points

$$|z_A^2 - z_B^2| = \sqrt{(\cos 2\alpha_A - \cos 2\alpha_B)^2 + (\sin 2\alpha_A - \sin 2\alpha_B)^2}$$
$$= \sqrt{2[1 - \cos 2(\alpha_B - \alpha_A)]} = \sqrt{2 \cdot 2 \sin^2(\alpha_B - \alpha_A)} = 2|\sin(\alpha_B - \alpha_A)|$$
$$= 2|\sin \alpha_B \cos \alpha_A - \sin \alpha_A \cos \alpha_B| \in \mathbb{Q}.$$

Answer: Yes, it is possible.

Problem 4. *A tourist takes a trip through a city in stages. Each stage consists of three segments of length* 100 *m separated by right turns of* 60°. *Between the last segment of one stage and the first segment of the next stage, the tourist makes a left turn of* 60°. *At what distance will the tourist be from his initial position after* 1997 *stages?*

(1997 Rio Plata Mathematical Olympiad)

Solution. In one stage, the tourist traverses the complex number

$$x = 100 + 100\bar{\varepsilon} + 100\bar{\varepsilon}^2 = 100 - 100\sqrt{3}i,$$

where $\varepsilon = \cos \dfrac{\pi}{3} + i \sin \dfrac{\pi}{3}$.

Thus in 1997 stages, the tourist traverses the complex number

$$z = x + x\varepsilon + x\varepsilon^2 + \cdots + x\varepsilon^{1996} = x \frac{1 - \varepsilon^{1997}}{1 - \varepsilon} = x\varepsilon^2.$$

Hence, the tourist ends up $|z| = |x\varepsilon^2| = |x| = 200$ m away from his initial position.

Problem 5. *Let* A, B, C, *be fixed points in the plane. A man starts from a certain point* P_0 *and walks directly to* A. *At* A, *he turns by* 60° *to the left and* *walks to P_1 such that $P_0 A = A P_1$. After he performs the same action 1986 times successively around points A, B, C, A, B, C, ..., he returns to the starting point. Prove that ABC is an equilateral triangle and that the vertices A, B, C, are arranged counterclockwise.*

(27th IMO)

Solution. For convenience, let $A_1, A_2, A_3, A_4, A_5, \ldots$ be $A, B, C, A, B, \ldots$, respectively, and let P_0 be the origin. After the kth step, the position P_k will be $P_k = A_k + (P_{k-1} - A_k)\varepsilon$ for $k = 1, 2, \ldots$, where $\varepsilon = \cos \dfrac{4\pi}{3} + i \sin \dfrac{4\pi}{3}$. We easily obtain

$$P_k = (1 - \varepsilon)(A_k + \varepsilon A_{k-1} + \varepsilon^2 A_{k-2} + \cdots + \varepsilon^{k-1} A_1).$$

The condition $P = P_{1986}$ is equivalent to $A_{1986} + \varepsilon A_{1985} + \cdots + \varepsilon^{1984} A_2 + \varepsilon^{1985} A_1 = 0$, which as we see from keeping in mind that $A_1 = A_4 = A_7 = \cdots$, $A_2 = A_5 = A_8 = \cdots$, $A_3 = A_6 = A_9 = \cdots$, reduces to

$$662(A_3 + \varepsilon A_2 + \varepsilon^2 A_1) = (1 + \varepsilon^3 + \cdots + \varepsilon^{1983})(A_3 + \varepsilon A_2 + \varepsilon^2 A_1) = 0,$$

and the assertion follows from Proposition 2 in Sect. 3.4.

Problem 6. *Let a, n be integers and let p be prime such that $p > |a| + 1$. Prove that the polynomial $f(x) = x^n + ax + p$ cannot be represented as a product of two nonconstant polynomials with integer coefficients.*

(1999 Romanian Mathematical Olympiad)

Solution. Let z be a complex root of the polynomial. We shall prove that $|z| > 1$. Suppose $|z| \le 1$. Then $z^n + az = -p$, and we deduce that

$$p = |z^n + az| = |z||z^{n-1} + a| \le |z^{n-1}| + |a| \le 1 + |a|,$$

which contradicts the hypothesis.

Now, suppose $f = gh$ is a decomposition of f into nonconstant polynomials with integer coefficients. Then $p = f(0) = g(0)h(0)$, and either $|g(0)| = 1$ or $|h(0)| = 1$. Assume without loss generality that $|g(0)| = 1$. If $z_1, z_2, \ldots, z_k$ are the roots of g, then they are also roots of f. Therefore,

$$1 = |g(0)| = |z_1 z_2 \ldots z_k| = |z_1||z_2| \cdots |z_k| > 1,$$

a contradiction.

Problem 7. *Prove that if a, b, c are complex numbers such that*

$$\begin{cases} (a+b)(a+c) = b, \\ (b+c)(b+a) = c, \\ (c+a)(c+b) = a, \end{cases}$$

then a, b, c are real numbers.

(2001 Romanian IMO Team Selection Test)

Solution 1. Let $P(x) = x^3 - sx^2 + qx - p$ be the polynomial with roots a, b, c. We have $s = a+b+c$, $q = ab+bc+ca$, $p = abc$. The given equalities are equivalent to

$$\begin{cases} sa + bc = b, \\ sb + ca = c, \\ sc + ab = a. \end{cases} \tag{1}$$

Adding these equalities, we obtain $q = s - s^2$. Multiplying the equalities in (1) by a, b, c, respectively, and adding them, we obtain $s(a^2 + b^2 + c^2) + 3p = q$, or after a short computation,

$$3p = -3s^3 + s^2 + s. \tag{2}$$

If we write the given equations in the form

$$(s - c)(s - b) = b, \quad (s - a)(s - c) = c, \quad (s - b)(s - a) = a,$$

we obtain $((s - a)(s - b)(s - c))^2 = abc$, and by performing standard computations and using (2), we finally get

$$s(4s - 3)(s + 1)^2 = 0.$$

If $s = 0$, then $P(x) = x^3$, so $a = b = c = 0$. If $s = -1$, then $P(x) = x^3 + x^2 - 2x - 1$, which has the roots $2\cos\dfrac{2\pi}{7}$, $2\cos\dfrac{4\pi}{7}$, $2\cos\dfrac{6\pi}{7}$ (this is not obvious, but we can see that P changes sign on the intervals $(-2, -1)$, $(-1, 0)$, $(1, 2)$ of the real line; hence its roots are real). Finally, if $s = 3/4$, then

$$P(x) = x^3 - \frac{3}{4}x^2 + \frac{3}{16}x - \frac{1}{64},$$

which has roots $a = b = c = 1/4$.

Solution 2. Subtract the second equation from the first. We obtain $(a + b)(a - b) = b - c$. Analogously, $(b+c)(b - c) = c - a$ and $(c+a)(c - a) = a - b$. We can see that if two of the numbers are equal, then all three are equal, and the conclusion is obvious. Suppose that the numbers are distinct. Then after multiplying the equalities above, we obtain $(a+b)(b+c)(c+a) = 1$, and next, $b(b+c) = c(c+a) = a(a+b) = 1$. Now, if one of the numbers is real, it follows immediately that all three are real. Suppose none of the numbers are real. Then $\arg a$, $\arg b$, $\arg c \in (0, 2\pi)$. Two of the numbers $\arg a$, $\arg b$, $\arg c$ are contained in either $(0, \pi)$ or $[\pi, 2\pi)$. Suppose these are $\arg a$, $\arg b$, and that $\arg a \le \arg b$. Then $\arg a \le \arg(a+b) \le \arg b$ and $\arg a \le \arg a(a+b) \le \arg(a + b) \le \arg b$. This is a contradiction, since $a(a + b) = 1$.

Problem 8. *Find the smallest integer n such that an $n \times n$ square can be partitioned into 40×40 and 49×49 squares, with both types of squares present in the partition.*

(2000 Russian Mathematical Olympiad)

Solution. We can partition a 2000×2000 square into 40×40 and 49×49 squares: partition one 1960×1960 corner of the square into 49×49 squares and then partition the remaining portion into 40×40 squares.

We now show that n must be at least 2000. Suppose that an $n \times n$ square has been partitioned into 40×40 and 49×49 squares, using at least one of each type. Let $\zeta = \cos\dfrac{2\pi}{40} + i\sin\dfrac{2\pi}{40}$ and $\xi = \cos\dfrac{2\pi}{49} + i\sin\dfrac{2\pi}{49}$. Orient the $n \times n$ square so that two sides are horizontal, and number the rows and columns of unit squares from the top left: $0, 1, 2, \ldots, n-1$. For $0 \le j,\ k \le n-1$, write $\zeta^j \xi^k$ in square (j, k). If an $m \times m$ square has its top left-hand corner at (x, y), then the sum of the numbers written in it is

$$\sum_{j=x}^{x+m-1} \sum_{k=y}^{y+m-1} \zeta^j \xi^k = \zeta^x \xi^y \left(\frac{\zeta^m - 1}{\zeta - 1}\right)\left(\frac{\xi^m - 1}{\xi - 1}\right).$$

The first fraction in parentheses is 0 if $m = 40$, and the second fraction is 0 if $m = 49$. Thus, the sum of the numbers written inside each square in the partition is 0, so the sum of all the numbers must be 0. However, applying the above formula with $(m, x, y) = (n, 0, 0)$, we find that the sum of all the numbers equals 0 only if either $\zeta^n - 1$ or $\xi^n - 1$ equals 0. Thus, n must be either a multiple of 40 or a multiple of 49.

Let a and b be the number of 40×40 and 49×49 squares, respectively. The area of the square equals $40^2 \cdot a + 49^2 \cdot b = n^2$. If $40|n$, then $40^2|b$, and hence $b \ge 40^2$. Thus, $n^2 > 49^2 \cdot 40^2 = 1960^2$; because n is a multiple of 40, $n \ge 50 \cdot 40 = 2000$. If instead $49|n$, then $49^2|a$, $a \ge 49^2$, and again $n^2 > 1960^2$. Because n is a multiple of 49, $n \ge 41 \cdot 49 = 2009 > 2000$. In either case, $n \ge 2000$, and 2000 is the minimum possible value of n.

Problem 9. *The pair (z_1, z_2) of nonzero complex numbers has the following property: there is a real number $a \in [-2, 2]$ such that $z_1^2 - az_1z_2 + z_2^2 = 0$. Prove that all pairs (z_1^n, z_2^n), $n = 2, 3, \ldots$, have the same property.*

(Romanian Mathematical Olympiad—Second Round, 2001)

Solution 1. Set $t = \dfrac{z_1}{z_2}, t \in \mathbb{C}^*$. The relation $z_1^2 - az_1z_2 + z_2^2 = 0$ is equivalent to $t^2 - at + 1 = 0$. We have $\Delta = a^2 - 4 \le 0$, whence $t = \dfrac{a \pm i\sqrt{4 - a^2}}{2}$ and $|t| = \sqrt{\dfrac{a^2}{4} + \dfrac{4 - a^2}{4}} = 1$. If $t = \cos\alpha + i\sin\alpha$, then $\dfrac{z_1^n}{z_2^n} = t^n = \cos n\alpha + i\sin n\alpha$, and we can write $z_1^{2n} - a_n z_1^n z_2^n + z_2^{2n} = 0$, where $a_n = 2\cos n\alpha \in [-2, 2]$.

Solution 2. Because $a \in [-2, 2]$, we can write $a = 2\cos\alpha$. The relation $z_1^2 - az_1z_2 + z_2^2 = 0$ is equivalent to

$$\frac{z_1}{z_2} + \frac{z_2}{z_1} = 2\cos\alpha, \tag{1}$$

and by a simple inductive argument, it follows from (1) that

$$\frac{z_1^n}{z_2^n} + \frac{z_2^n}{z_1^n} = 2\cos n\alpha, \ n = 1, 2, \ \ldots.$$

Problem 10. *Find*

$$\min_{z \in \mathbb{C}\backslash\mathbb{R}} \frac{\mathrm{Im}z^5}{\mathrm{Im}^5 z}$$

and the values of z for which the minimum is reached.

Solution. Let a, b be real numbers such that $z = a + bi$, $b \neq 0$. Then $\mathrm{Im}(z)^5 = 5a^4b - 10a^2b^3 + b^5$ and

$$\frac{\mathrm{Im}z^5}{\mathrm{Im}^5 z} = 5\left(\frac{a}{b}\right)^4 - 10\left(\frac{a}{b}\right)^2 + 1.$$

Setting $x = \left(\dfrac{a}{b}\right)^2$ yields

$$\frac{\mathrm{Im}(z)^5}{\mathrm{Im}^5 z} = 5x^2 - 10x + 1 = 5(x-1)^2 - 4.$$

The minimum value is -4, and it is obtained for $x = 1$, i.e., for $z = a(1 \pm i)$, $a \neq 0$.

Problem 11. *Let z_1, z_2, z_3 be complex numbers, not all real, such that $|z_1| = |z_2| = |z_3| = 1$ and $2(z_1 + z_2 + z_3) - 3z_1z_2z_3 \in \mathbb{R}$.*
  Prove that

$$\max(\arg z_1, \arg z_2, \arg z_3) \geq \frac{\pi}{6}.$$

Solution. Let $z_k = \cos t_k + i\sin t_k$, $k \in \{1, 2, 3\}$.
  The condition $2(z_1 + z_2 + z_3) - 3z_1z_2 \in \mathbb{R}$ implies

$$(\sin t_1 + \sin t_2 + \sin t_3) = 3\sin(t_1 + t_2 + t_3). \tag{1}$$

Assume by way of contradiction that $\max(t_1, t_2, t_3) < \dfrac{\pi}{6}$; hence $t_1, t_2, t_3 < \dfrac{\pi}{6}$. Let $t = \dfrac{t_1 + t_2 + t_3}{3} \in \left(0, \dfrac{\pi}{6}\right)$. The sine function is concave on $\left[0, \dfrac{\pi}{6}\right)$, so

$$\frac{1}{3}(\sin t_1 + \sin t_2 + \sin t_3) \leq \sin\frac{t_1 + t_2 + t_3}{3}. \tag{2}$$

From the relations (1) and (2), we obtain

$$\frac{\sin(t_1 + t_2 + t_3)}{2} \leq \sin\frac{t_1 + t_2 + t_3}{3}.$$

Then

$$\sin 3t \leq 2\sin t.$$

It follows that

$$4\sin^3 t - \sin t \geq 0,$$

i.e., $\sin^2 t \geq \dfrac{1}{4}$. Hence $\sin t \geq \dfrac{1}{3}$, and then $t \geq \dfrac{\pi}{6}$, which contradicts that $t \in \left(0, \dfrac{\pi}{6}\right)$.

Therefore, $\max(t_1, t_2, t_3) \geq \dfrac{\pi}{6}$, as desired.

Here are some more problems.

Problem 12. *Solve in complex numbers the system of equations*

$$\begin{cases} x|y| + y|x| = 2z^2, \\ y|z| + z|y| = 2x^2, \\ z|x| + x|z| = 2y^2. \end{cases}$$

Problem 13. *Solve in complex numbers the following:*

$$\begin{cases} x(x - y)(x - z) = 3, \\ y(y - x)(y - z) = 3, \\ z(z - x)(z - y) = 3. \end{cases}$$

(Romanian Mathematical Olympiad—Second Round, 2002)

Problem 14. *Let X, Y, Z, T be four points in the plane. The segments [XY] and [ZT] are said to be connected if there is some point O in the plane such that the triangles OXY and OZT are right isosceles triangles in O.*

Let ABCDEF be a convex hexagon such that the pairs of segments [AB], [CE], and [BD], [EF] are connected. Show that the points A, C, D and F are the vertices of a parallelogram and that the segments [BC] and [EA] are connected.

(Romanian Mathematical Olympiad—Final Round, 2002)

Problem 15. *Let ABCDE be a cyclic pentagon inscribed in a circle with center O that has angles $\hat{B} = 120°, \hat{C} = 120°, \hat{D} = 130°, \hat{E} = 100°$. Show that the diagonals BD and CE meet at a point belonging to the diameter AO.*

(Romanian IMO, Team Selection Test, 2002)

Problem 16. *A function f is defined on the complex numbers by*

$$f(z) = (a + bi)z,$$

where a and b are positive numbers. This function has the property that the image of each point in the complex plane is equidistant from that point and *the origin. Given that $|a + bi| = 8$ and that $b^2 = m/n$, where m and n are relatively prime positive integers, find $m + n$.*

(1999 AIME, Problem 9)

Problem 17. *Let $F(z) = \dfrac{z+i}{z-i}$ for all complex numbers $z \neq i$, and let*

$$z_n = F(z_{n-1})$$

for all positive integers n. Given that $z_0 = \dfrac{1}{137} + i$ and $z_{2002} = a + i$, where a and b are real numbers, find $a + b$.

(2002 AIME I, Problem 12)

Problem 18. *Given a positive integer n, it can be shown that every complex number of the form $r + si$, where r and s are integers, can be uniquely expressed in the base $-n + i$ using the integers $1, 2, \ldots, n^2$ as digits. That is, the equation*

$$r + si = a_m(-n+i)^m + a_{m-1}(-n+i)^{m-1} + \ldots + a_1(-n+i) + a_0$$

is valid for a unique choice of nonnegative integer m and digits $a_0, a_1, \ldots, a_m$ chosen from the set $\{0, 1, 2, \ldots, n^2\}$, with $a_m \neq 0$. We write

$$r + si = (a_m a_{m-1} \ldots a_1 a_0)_{-n+i}$$

to denote the base-$(-n+i)$ expansion of $r + si$. There are only finitely many integers $k + 0i$ that have four-digit expansions

$$k = (a_3 a_2 a_1 a_0)_{-3+i}, \ a_3 \neq 0.$$

Find the sum of all such k.

(1989 AIME, Problem 14)

Problem 19. *There is a complex number z with imaginary part 164 and a positive integer n such that*

$$\frac{z}{z+n} = 4i.$$

Find n.

(2009 AIME, Problem 2)

Problem 20. *Let u, v, w be complex numbers of modulus 1. Prove that one can choose signs $+$ and $-$ such that*

$$|\pm u \pm v \pm w| \leq 1.$$

(Romanian Mathematical Olympiad—District Round, 2007)

Problem 21. *Consider a complex number z, $z \neq 0$, and the real sequence*

$$a_n = \left| z^n + \frac{1}{z^n} \right|, \quad n \geq 1.$$

(a) Show that if $a_1 > 2$, then

$$a_{n+1} < \frac{a_n + a_{n+2}}{2}, \quad \text{for all } n \in \mathbb{N}^*.$$

(b) Prove that if there exists $k \in \mathbb{N}^*$ such that $a_k \leq 2$, then $a_1 \leq 2$.

(Romanian Mathematical Olympiad—District Round, 2010)

Problem 22. *Consider the set $M = \{z \in \mathbb{C} \mid |z| = 1, \ \text{Re}z \in \mathbb{Q}\}$. Prove that the complex plane contains an infinity of equilateral triangles with vertices in M.*

(Romanian Mathematical Olympiad—Final Round, 2012)

Problem 23. *Let $(a_n)_{n \geq 1}$ be a sequence of nonnegative integers such that $a_n \leq n$, for all $n \geq 1$, and $\displaystyle\sum_{k=1}^{n-1} \cos \frac{\pi a_k}{n} = 0$, for all $n \geq 2$. Find a closed formula for the general term of the sequence.*

(Romanian Mathematical Olympiad—District Round, 2012)

Problem 24. *Let a and b be two rational numbers such that the absolute value of the complex number $z = a + bi$ is equal to 1. Prove that the absolute value of the complex number $z_n = 1 + z + z^2 + \ldots + z^{n-1}$ is a rational number for all odd integers n.*

(Romanian Mathematical Olympiad—District Round, 2012)

Chapter 6
Answers, Hints, and Solutions to Proposed Problems

In what follows, answers and solutions are presented to problems posed in previous chapters. We have preserved the title of the subsection containing the problem and the number of the proposed problem. Taking into account the complexity of some problems in the chapter on Olympiad-caliber problems, we have included the statements of these problems before the solutions.

6.1 Answers, Hints, and Solutions to Routine Problems

6.1.1 Problems (p. 19) from Section 1.1: Algebraic Representation of Complex Numbers

1.(a) $z_1 + z_2 + z_3 = (0, 4)$; (b)$z_1 z_2 + z_2 z_3 + z_3 z_1 = (-4, 5)$;

(c) $z_1 z_2 z_3 = (-9, 7)$; (d)$z_1^2 + z_2^2 + z_3^2 = (-8, -10)$;

(e) $\dfrac{z_1}{z_2} + \dfrac{z_2}{z_3} + \dfrac{z_3}{z_1} = \left(-\dfrac{311}{130}, \dfrac{65}{83}\right)$; (f)$\dfrac{z_1^2 + z_2^2}{z_2^2 + z_3^2} = \left(\dfrac{152}{221}, -\dfrac{72}{221}\right)$.

2.(a) $z = (7, -8)$; (b)$z = (-7, -4)$;

(c) $z = \left(\dfrac{23}{13}, -\dfrac{2}{13}\right)$; (d)$z = (-9, 7)$.

3.(a) $z_1 = \left(-\dfrac{1}{2}, \dfrac{\sqrt{3}}{2}\right)$, $z_2 = \left(-\dfrac{1}{2}, -\dfrac{\sqrt{3}}{2}\right)$;

(b) $z_1 = (-1, 0)$, $z_2 = \left(\dfrac{1}{2}, \dfrac{\sqrt{3}}{2}\right)$, $z_3 = \left(\dfrac{1}{2}, -\dfrac{\sqrt{3}}{2}\right)$.

T. Andreescu and D. Andrica, *Complex Numbers from A to ... Z*,
DOI 10.1007/978-0-8176-8415-0_6, © Springer Science+Business Media New York 2014

4. $\displaystyle\sum_{k=0}^{n} z^k = \begin{cases} (1,0) & \text{for } n = 4k; \\ (1,1) & \text{for } n = 4k+1; \\ (0,1) & \text{for } n = 4k+2; \\ (0,0) & \text{for } n = 4k+3. \end{cases}$

5.(a) $z = (1,1)$;   (b)$z_1 = (2,1)$, $z_2 = (-2,-1)$.

6. $z^2 = (a^2 - b^2, 2ab)$; $z^3 = (a^3 - 3ab^2, 3a^2b - b^3)$;
$z^4 = (a^4 - 6a^2b^2 + b^4, 4a^3b - 4ab^3)$.

7. $z_1 = \left(\sqrt{\dfrac{a + \sqrt{a^2 + b^2}}{2}}, \operatorname{sgn} b\sqrt{\dfrac{-a + \sqrt{a^2 + b^2}}{2}} \right)$,

$z_2 = \left(-\sqrt{\dfrac{a + \sqrt{a^2 + b^2}}{2}}, -\operatorname{sgn} b\sqrt{\dfrac{-a + \sqrt{a^2 + b^2}}{2}} \right)$.

8. For all nonnegative integers k, we have

$$z^{4k} = ((-4)^k, 0); \quad z^{4k+1} = ((-4)^k, -(-4)^k); \quad z^{4k+2} = (0, -2(-4)^k);$$
$$z^{4k+3} = (-2(-4)^k, -2(-4)^k); \text{ for } k \geq 0.$$

9.(a) $x = \dfrac{1}{4}$, $y = \dfrac{3}{4}$;   (b)$x = -2$, $y = 8$;   (c)$x = 0$, $y = 0$.

10.(a) $8 + 51i$;   (b)$4 - 43i$;   (c)2;   (d)$\dfrac{11}{4} - \dfrac{5\sqrt{7}}{2}i$;   (e)$\dfrac{61}{13} + \dfrac{4}{13}i$.

11.(a) $-i$;   (b)$E_{4k} = 1, E_{4k+1} = 1+i, E_{4k+2} = i, E_{4k+3} = 0$;   (c)$1$;   (d)$-3i$.

12.(a) $z_1 = \dfrac{\sqrt{2}}{2} + i\dfrac{\sqrt{2}}{2}$, $z_2 = -\dfrac{\sqrt{2}}{2} - i\dfrac{\sqrt{2}}{2}$;

(b) $z_1 = \dfrac{\sqrt{2}}{2} - i\dfrac{\sqrt{2}}{2}$, $z_2 = -\dfrac{\sqrt{2}}{2} + i\dfrac{\sqrt{2}}{2}$;

(c) $z_{1,2} = \pm\left(\dfrac{\sqrt{1 + \sqrt{3}}}{2} - \dfrac{\sqrt{\sqrt{3} - 1}}{2}i \right)$.

13. $z \in \mathbb{R}$ or $z = x + iy$ with $x^2 + y^2 = 1$.

14.(a) $\overline{E}_1 = E_1$;
(b) $\overline{E}_2 = E_2$.

15. Use substitute the formula that defines the modulus.

16. From the identity

$$\left(z + \frac{1}{z} \right)^3 = z^3 + \frac{1}{z^3} + 3\left(z + \frac{1}{z} \right),$$

we obtain

$$\left| z + \frac{1}{z} \right|^3 \leq 2 + 3\left| z + \frac{1}{z} \right|, \quad \text{or} \quad a^3 - 3a - 2 \leq 0,$$

where

$$a = \left| z + \frac{1}{z} \right|, a \geq 0.$$

Since

$$a^3 - 3a - 2 = (a - 2)(a^2 + 2a + 1) = (a - 2)(a + 1)^2,$$

we have $a \leq 2$, as desired.

17. The equation $|z^2 + \overline{z}^2| = 1$ is equivalent to $|z^2 + \overline{z}^2|^2 = 1$. That is, $(z^2 + \overline{z}^2)(\overline{z}^2 + z^2) = 1$. We obtain $(z^2 + \overline{z}^2)^2 = 1$, or $(z^2 + \frac{1}{z^2})^2 = 1$. The last equation is equivalent to $(z^4 + 1)^2 = z^4$, or $(z^4 - z^2 + 1)(z^4 + z^2 + 1) = 0$. The solutions are $\pm \frac{1}{2}i \pm \frac{\sqrt{3}}{2}$ and $\pm \frac{\sqrt{3}}{2} \pm \frac{1}{2}i$.

18. $z \in \left\{ \pm \sqrt{\frac{2}{3}}, \pm i\sqrt{2} \right\}$.

19. $z \in \{0, 1, -1, i, -i\}$.

20. Observe that $\left| \frac{1}{z} - \frac{1}{2} \right| < \frac{1}{2}$ is equivalent to $|2 - z| < |z|$, and consequently, $(2 - z)(2 - \overline{z}) < z \cdot \overline{z}$. It follows that $4 < 2(z + \overline{z}) = 4\mathrm{Re}(z)$, as needed.

21. $a^2 + b^2 + c^2 - ab - bc - ca$.

22.(a) $z = \frac{-6 + \sqrt{21}}{3} + 2i$; (b) $z = -\frac{7}{6} + 4i$; (c) $z = 2 + i$;

(d) $z_{1,2} = \frac{-2 \pm \sqrt{3}}{2} + \frac{1}{2}i$; (e) $z^2 = -1$, $z^2 = -5 - 6i$; (f) $z^2 = -\frac{13}{2} - \frac{9}{2}i$.

23. $m \in \{1, 5\}$

24. $z = -2y + 2 + iy, y \in \mathbb{R}$.

25. $z = x + iy$ with $x^2 + y^2 = 1$.

26. From $|z_1 + z_2| = \sqrt{3}$ it follows that $|z_1 + z_2|^2 = 3$, i.e., $(z_1 + z_2)(\overline{z_1 + z_2}) = 3$. We obtain $|z_1|^2 + (z_1\overline{z}_2 + \overline{z}_1 z_2) + |z_2|^2 = 3$. That is, $z_1\overline{z}_2 + \overline{z}_1 z_2 = 1$. On the other hand, we have $|z_1 - z_2|^2 = |z_1|^2 - (z_1\overline{z}_2 + \overline{z}_1 z_2) + |z_2|^2 = 2 - 1 = 1$, and hence $|z_1 - z_2| = 1$.

27. Letting $\varepsilon = -\frac{1}{2} + i\frac{\sqrt{3}}{2}$ and noticing that $\varepsilon^3 = 1$, we obtain $n = 3k, k \in \mathbb{Z}$.

28. Note that $z = 0$ is a solution. For $z \neq 0$, passing to the absolute value, we obtain $|z|^{n-1} = |z|$, i.e., $|z| = 1$. The equation is equivalent to $z^n = i\overline{z} \cdot z$, which reduces to $z^n = i$. The total number of solutions is $n + 1$.

29. Let

$$\alpha = |z_2 - z_3|, \quad \beta = |z_3 - z_1|, \quad \gamma = |z_1 - z_2|.$$

Since the inequality

$$\alpha\beta + \beta\gamma + \gamma\alpha \le \alpha^2 + \beta^2 + \gamma^2$$

holds and

$$\alpha^2 + \beta^2 + \gamma^2 = 3(|z_1|^2 + |z_2|^2 + |z_2|^2) - |z_1 + z_2 + z_3|^2$$

$$\le 3(|z_1|^2 + |z_2|^2 + |z_2|^2 = 9R^2,$$

it follows that

$$\alpha\beta + \beta\gamma + \gamma\alpha \le 9R^2.$$

30. Observe that

$$|w| = |v| \cdot \frac{|u - z|}{|\overline{u}z - 1|} = \frac{|u - z|}{|\overline{u}z - 1|} \le 1$$

if and only if

$$|u - z| \le |\overline{u}z - 1|.$$

This is equivalent to

$$|u - z|^2 \le |\overline{u}z - 1|^2.$$

We obtain

$$(u - z)(\overline{u} - \overline{z}) \le (\overline{u}z - 1)(u\overline{z} - 1),$$

i.e.,

$$|u|^2 + |z|^2 - |u|^2|z|^2 - 1 \le 0.$$

Finally,

$$(|u^2| - 1)(|z|^2 - 1) \ge 0.$$

Since $|u| \le 1$, it follows that $|w| \le 1$ if and only if $|z| \le 1$, as desired.

31. $z_1^2 + z_2^2 + z_3^2 = (z_1 + z_2 + z_3)^2 - 2(z_1z_2 + z_2z_3 + z_3z_1)$

$$= -2z_1z_2z_3 \left(\frac{1}{z_1} + \frac{1}{z_2} + \frac{1}{z_3} \right) = -2z_1z_2z_3(\overline{z_1} + \overline{z_2} + \overline{z_3}) = 0.$$

32. The relation $|z_k| = r$ implies $\overline{z_k} = \dfrac{r^2}{z_k}$ for $k \in \{1, 2, \dots, n\}$. Then

$$\overline{E} = \frac{\left(\dfrac{r^2}{z_1} + \dfrac{r^2}{z_2} \right) \left(\dfrac{r^2}{z_2} + \dfrac{r^2}{z_3} \right) \cdots \left(\dfrac{r^2}{z_n} + \dfrac{r^2}{z_1} \right)}{\dfrac{r^2}{z_1} \cdot \dfrac{r^2}{z_2} \cdots \dfrac{r^2}{z_n}}$$

$$= \frac{r^{2n} \cdot \dfrac{z_1 + z_2}{z_1z_2} \cdot \dfrac{z_2 + z_3}{z_2z_3} \cdots \dfrac{z_n + z_1}{z_nz_1}}{r^{2n} \cdot \dfrac{1}{z_1z_2 \cdots z_n}} = E;$$

hence $E \in \mathbb{R}$.

33. Observe that

$$z_1 \cdot \overline{z_1} = z_2 \cdot \overline{z_2} = z_3 \cdot \overline{z_3} = r^2$$

and

$$z_1 z_2 + z_3 \in \mathbb{R} \text{ if and only if } z_1 z_2 + z_3 = \overline{z_1} \cdot \overline{z_2} + \overline{z_3}.$$

Then

$$\frac{r^2}{z_1 z_2 z_3} = \frac{z_1 z_2 + z_3}{z_1 z_2 + r^2 z_3} = \frac{z_1 z_3 + z_2}{z_1 z_3 + r^2 z_2} = \frac{z_2 z_3 + z_1}{z_2 z_3 + r^2 z_1}$$

$$\frac{(z_1 - 1)(z_2 - z_3)}{(z_2 - z_3)(z_1 - r^2)} = \frac{z_1 - 1}{z_1 - r^2} = \frac{z_2 - 1}{z_2 - r^2} = \frac{z_3 - 1}{z_3 - r^2} = \frac{z_1 - z_2}{z_1 - z_2} = 1.$$

Hence $z_1 z_2 z_3 = r^2$, and consequently, $r^3 = r^2$. Therefore, $r = 1$ and $z_1 z_2 z_3 = 1$, as desired.

34. Note that $x_1^3 = x_2^3 = -1$.

  (a) -1;   (b) 1;   (c) Consider $n \in \{6k, 6k \pm 1, 6k \pm 2, 6k \pm 3\}$.

35.(a) $x^4 + 16 = x^4 + 2^4 = (x^2 + 4i)(x^2 - 4i)$

$$= [x^2 + (\sqrt{2}(1 + i))^2][x^2 - (\sqrt{2}(1 + i))^2]$$

$$= (x + \sqrt{2}(-1 + i))(x + \sqrt{2}(1 - i))(x - \sqrt{2}(1 + i))(x + \sqrt{2}(1 + i)).$$

  (b) $x^3 - 27 = x^3 - 3^3 = (x - 3)(x - 3\varepsilon)(x - 3\varepsilon^2)$, where $\varepsilon = -\dfrac{1}{2} + \dfrac{\sqrt{3}}{2}i$.

  (c) $x^3 + 8 = x^3 + 2^3 = (x + 2)(x + 1 + i\sqrt{3})(x + 1 - i\sqrt{3})$.

  (d) $x^4 + x^2 + 1 = (x^2 - \varepsilon)(x^2 - \varepsilon^2) = (x^2 - \varepsilon^{-2})(x^2 - \varepsilon^2)$

$$= (x - \varepsilon)(x + \varepsilon)(x - \overline{\varepsilon})(x + \overline{\varepsilon}), \text{ where } \varepsilon = -\dfrac{1}{2} + \dfrac{\sqrt{3}}{2}i.$$

36.(a) $x^2 - 14x + 50 = 0$;   (b) $x^2 - \dfrac{18}{5}x + \dfrac{26}{5} = 0$;   (c) $x^2 + 4x + 8 = 0$.

37. We have

$$2|z_1 + z_2| \cdot |z_2 + z_3| = 2|z_2(z_1 + z_2 + z_3) + z_1 z_3| \le 2|z_2| \cdot |z_1 + z_2 + z_3| + 2|z_1||z_3|,$$

and likewise,

$$2|z_2 + z_3| \cdot |z_3 + z_1| \le 2|z_3||z_1 + z_2 + z_3| + 2|z_2||z_1|,$$
$$2|z_3 + z_1| \cdot |z_1 + z_2| \le 2|z_1||z_1 + z_2 + z_3| + 2|z_2||z_3|.$$

Summing up these inequalities with

$$|z_1 + z_2|^2 + |z_2 + z_3|^2 + |z_3 + z_1|^2 = |z_1|^2 + |z_2|^2 + |z_3|^2 + |z_1 + z_2 + z_3|^2$$

yields

$$(|z_1 + z_2| + |z_2 + z_3| + |z_3 + z_1|)^2 \le (|z_1| + |z_2| + |z_3| + |z_1 + z_2 + z_3|)^2.$$

The conclusion is now obvious.

38. Note that $|z_i| = |(x - x_i)(y - y_i) - xy| \le |x - x_i||y - y_i| + |x||y|$. Hence by the Cauchy–Schwarz inequality,

$$\sum |z_i| \le \sum |x - x_i||y - y_i| + n|x||y|$$
$$\le \sqrt{\sum |x - x_i|^2} \sqrt{\sum |y - y_i|^2} + n|x||y|.$$

Now we have

$$\sum |x - x_i|^2 = \sum (|x|^2 + |x_i|^2 + 2\mathrm{Re}(x\overline{x_i}))$$
$$= n|x|^2 + n = 2\mathrm{Re}\left(x \sum \overline{x_i}\right)$$
$$= n|x|^2 + n - 2n\mathrm{Re}(x\overline{x})$$
$$= n(1 - |x|^2),$$

the last equality following from $\mathrm{Re}(x\overline{x}) = x\overline{x} = |x|^2$. Thus

$$\sqrt{\sum |x - x_i|^2} \sqrt{\sum |y - y_i|^2} + n|x||y| = n(\sqrt{1 - |x|^2}\sqrt{1 - |y|^2} + |x||y|) \le n,$$

where the last inequality is also proven by Cauchy–Schwarz, and we are done.

6.1.2 Problems (p. 29) from Section 1.2: Geometric Interpretation of the Algebraic Operations

1.(a) The circle of center $(2, 0)$ and radius 3.
 (b) The disk of center $(0, -1)$ and radius 1.
 (c) The exterior of the circle of center $(1, -2)$ and radius 3.
 (d) $M = \left\{(x, y) \in \mathbb{R}^2 | x \ge -\dfrac{1}{2}\right\} \cup \left\{(x, y) \in \mathbb{R}^2 | x < -\dfrac{1}{2}, 3x^2 - y^2 - 3 < 0\right\}$.
 (e) $M = \{(x, y) \in \mathbb{R}^2 | -1 < y < 0\}$.
 (f) $M = \{(x, y) \in \mathbb{R}^2 | -1 < y < 1\}$.
 (g) $M = \{(x, y) \in \mathbb{R}^2 | x^2 + y^2 - 3x + 2 = 0\}4$.
 (h) The union of the lines with equations $x = -\dfrac{1}{2}$ and $y = 0$.

2. $M = \{(x, y) \in \mathbb{R}^2 | y = 10 - x^2, y \ge 4\}$.
3. $z_3 = \sqrt{3}(1 - i)$ and $z'_3 = \sqrt{3}(1 + i)$.
4. $M = \{(x, y) \in \mathbb{R}^2 | x^2 + y^2 + x = 0, x \ne 0, x \ne -1\}$
   $\cup \{(0, y) \in \mathbb{R}^2 | y \ne 0\} \cup \{(-1, y) \in \mathbb{R}^2 | y \ne 0\}$.

5. The union of the circles with equations

$$x^2 + y^2 - 2y - 1 = 0 \text{ and } x^2 + y^2 + 2y - 1 = 0.$$

6.1.3 Polar Representation of Complex Numbers

1. (a) $r = 3\sqrt{2}, \ t^* = \dfrac{3\pi}{4}$; (b) $r = 8, \ t^* = \dfrac{7\pi}{6}$; (c) $r = 5, \ t^* = \pi$;

(d) $r = \sqrt{5}, \ t^* = \arctan\dfrac{1}{2} + \pi$; (e) $r = 2\sqrt{5}, t^* = \arctan\left(-\dfrac{1}{2}\right) + 2\pi$.

2. (a) $x = 1, \ y = \sqrt{3}$; (b) $x = \dfrac{16}{5}, \ y = -\dfrac{12}{5}$; (c) $x = -2, \ y = 0$;

(d) $x = -3, y = 0$ (e)$x = 0, y = 1$ (f)$x = 0, y = -4$.

3. $\arg(\overline{z}) = \begin{cases} 2\pi - \arg z, & \text{if } \arg z \neq 0, \\ 0, & \text{if } \arg z = 0; \end{cases}$

$\arg(-z) = \begin{cases} \pi + \arg z, & \text{if } \arg z \in [0, \pi), \\ -\pi + \arg z, & \text{if } \arg z \in [\pi, 2\pi). \end{cases}$

4. (a) The circle of radius 2 with center at the origin.
(b) The circle with center $(0, -1)$ and radius 2 and its exterior.
(c) The disk with center $(0, 1)$ and radius 3.
(d) The interior of the angle determined by the rays $y = 0, x \leq 0$ and $y = x, x \leq 0$.
(e) The fourth quadrant and the ray $(OY'$.
(f) The first quadrant and the ray $(OX$.
(g) The interior of the angle determined by the rays $y = \frac{\sqrt{3}}{3}x, x \leq 0$ and $y = \sqrt{3}x, x < 0$.
(h) The intersection of the disk with center $(-1, -1)$ and radius 3 with the interior of the angle determined by the rays $y = 0, x \geq 0$ and $y = \frac{\sqrt{3}}{3}x, x > 0$.

5. (a) $z_1 = 12\left(\cos\frac{\pi}{3} + i\sin\frac{\pi}{3}\right)$; (b)$z_2 = \frac{1}{2}\left(\cos\frac{2\pi}{3} + i\sin\frac{2\pi}{3}\right)$;

(c) $z_3 = \cos\frac{4\pi}{3} + i\sin\frac{4\pi}{3}$; (d)$z_4 = 18\left(\cos\frac{5\pi}{3} + i\sin\frac{5\pi}{3}\right)$;

(e) $z_5 = \sqrt{13}\left[\cos(2\pi - \arctan\frac{2}{3}) + i\sin\left(2\pi - \arctan\frac{2}{3}\right)\right]$;

(f) $z_6 = 4\left(\cos\frac{3\pi}{2} + i\sin\frac{3\pi}{2}\right)$.

6. (a) $z_1 = \cos(2\pi - a) + i\sin(2\pi - a), a \in [0, 2\pi)$;

(b) $z_2 = 2|\cos\frac{a}{2}| \cdot \left[\cos\left(\frac{\pi}{2} - \frac{a}{2}\right) + i\sin\left(\frac{\pi}{2} - \frac{a}{2}\right)\right]$ if $a \in [0, \pi)$;

$z_2 = 2|\cos\frac{a}{2}| \cdot \left[\cos\left(\frac{3\pi}{2} - \frac{a}{2}\right) + i\sin\left(\frac{3\pi}{2} - \frac{a}{2}\right)\right]$ if $a \in (\pi, 2\pi)$;

(c) $z_3 = \sqrt{2}\left[\cos\left(a + \frac{7\pi}{4}\right) + i\sin\left(a + \frac{7\pi}{4}\right)\right]$ if $a \in \left[0, \frac{\pi}{4}\right]$;

$z_3 = \sqrt{2}\left[\cos\left(a - \frac{\pi}{4}\right) + i\sin\left(a - \frac{\pi}{4}\right)\right]$ if $a \in \left(\frac{\pi}{4}, 2\pi\right)$;

(d) $z_4 = 2\sin\frac{a}{2}\left[\cos\left(\frac{\pi}{2} - \frac{a}{2}\right) + i\sin\left(\frac{\pi}{2} - \frac{a}{2}\right)\right]$ if $a \in [0, \pi)$;

$z_4 = 2\sin\frac{a}{2}\left[\cos\left(\frac{5\pi}{2} - \frac{a}{2}\right) + i\sin\left(\frac{5\pi}{2} - \frac{a}{2}\right)\right]$ if $a \in [\pi, 2\pi)$.

7. (a) $12\sqrt{2}\left(\cos\frac{7\pi}{4} + i\sin\frac{7\pi}{4}\right)$; (b)$4(\cos 0 + i \sin 0)$;
(c) $48\sqrt{2}\left(\cos\frac{5\pi}{12} + i\sin\frac{5\pi}{12}\right)$; (d)$30\left(\cos\frac{\pi}{2} + i \sin\frac{\pi}{2}\right)$.

8. (a) $|z| = 12, \arg z = 0, \operatorname{Arg} z = 2k\pi, \arg\overline{z} = 0, \arg(-z) = \pi$;
(b) $|z| = 14\sqrt{2}, \arg z = \frac{11\pi}{12}, \operatorname{Arg}z = \frac{11\pi}{12} + 2k\pi, \arg\overline{z} = \frac{13\pi}{12}, \arg(-z) = \frac{\pi}{12}$.

9. (a) $|z| = 2^{13} + \frac{1}{2^{13}}, \arg z = \frac{5\pi}{6}$; (b)$|z| = \frac{1}{2^9}, \arg z = \pi$;
(c) $|z| = 2^{n+1}|\cos\frac{5n\pi}{3}|, \arg z \in \{0, \pi\}$.

10. If $z = r(\cos t + i\sin t)$ and $n = -m$, where m is a positive integer, then

$$z^n = z^{-m} = \frac{1}{z^m} = \frac{1}{r^m(\cos mt + i\sin mt)} = \frac{1}{r^m} \cdot \frac{\cos 0 + i\sin 0}{\cos mt + i\sin mt}$$

$$= \frac{1}{r^m}[\cos(0 - m)t + i\sin(0 - m)t] = r^{-m}(\cos(-mt) + i\sin(-mt))$$

$$= r^n(\cos nt + i\sin nt).$$

11. (a) $2^n \sin^n\frac{a}{2}\left[\cos\frac{n(\pi-a)}{2} + i\sin\frac{n(\pi-a)}{2}\right]$ if $a \in [0, \pi)$;
$2^n \sin^n\frac{a}{2}\left[\cos\frac{n(5\pi-a)}{2} + i\sin\frac{n(5\pi-a)}{2}\right]$ if $a \in [\pi, 2\pi]$;
(b) $z^n + \frac{1}{z^n} = 2\cos\frac{n\pi}{6}$.

12. Applying the quadratic formula to $z^2 - (2\cos 3°)z + 1 = 0$, we have

$$z = \frac{2\cos 3° \pm \sqrt{4\cos^2 3° - 4}}{2} = \cos 3° \pm i\sin 3°.$$

Using de Moivre theorem, we have

$$z^{2000} = \cos 6000° + i\sin 6000°,$$

$$6000 = 16(360) + 240,$$

so

$$z^{2000} = \cos 240° + i\sin 240°.$$

We want $z^{2000} + \frac{1}{z^{2000}} = 2\cos 240° = -1$.

Finally, the least integer greater than -1 is 0.

13. We know by de Moivre's theorem that

$$(\cos t + i\sin t)^n = \cos nt + i\sin nt$$

for all real numbers t and all integers n. We would like, therefore, somehow to convert our given expression into a form from which we can apply de Moivre's theorem.

Recall the trigonometric identities

$$\cos\frac{\pi}{2} - u = \sin u \text{ and } \sin\frac{\pi}{2} - u = \cos u,$$

which hold for all real u. If our original equation holds for all t, it must certainly hold for $t = \dfrac{\pi}{2} - u$. Thus, the question is equivalent to asking for how many positive integers $n \le 1000$ we have such that

$$\left(\sin\left(\frac{\pi}{2} - u\right) + i\cos\left(\frac{\pi}{2} - u\right)\right)^n = \sin n\left(\frac{\pi}{2} - u\right) + i\cos n\left(\frac{\pi}{2} - u\right)$$

holds for all real u. From the de Moivre's theorem we have

$$\left(\sin\left(\frac{\pi}{2} - u\right) + i\cos\left(\frac{\pi}{2} - u\right)\right)^n = (\cos u + i\sin u)^n = \cos nu + i\sin nu.$$

We know that two complex numbers are equal if and only if both their real parts and imaginary parts are equal. Thus, we need to find all n such that

$$\cos nu = \sin n\left(\frac{\pi}{2} - u\right) \quad \text{and} \quad \sin nu = \cos n\left(\frac{\pi}{2} - u\right)$$

hold for all real u. Now, $\sin x = \cos y$ if and only if either $x + y = \dfrac{\pi}{2} + 2k\pi$ or $x - y = \dfrac{\pi}{2} + 2k\pi$ for some integer k. So from the equality of the real parts, we need either

$$nu + n\left(\frac{\pi}{2} - u\right) = \frac{\pi}{2} + 2k\pi,$$

in which case $n = 1 + 4k$, or

$$-nu + n\left(\frac{\pi}{2} - u\right) = \frac{\pi}{2} + 2k\pi,$$

in which case n will depend on u, and so the equation will not hold for all real values of u. Checking $n = 1 + 4k$ in the equation for the imaginary parts, we see that it works there as well, so exactly those values of n congruent to 1 modulo 4 work. There are 250 of them in the given range.

14. Let $R(x) = \cos x + i\sin x$. We have

$$(1 - \sqrt{3}i)^n = \left(2R\left(-\frac{\pi}{3}\right)\right)^n = 2^n R\left(-\frac{n\pi}{3}\right),$$

and thus by de Moivre's theorem, we get

$$x_n = 2^n \cos\left(-\frac{n\pi}{3}\right) = 2^n \cos n\theta,$$

$$y_n = 2^n \sin\left(-\frac{n\pi}{3}\right) = -2^n \sin n\theta,$$

where $\theta = \dfrac{\pi}{3}$.

Substituting these into the given expressions, we obtain

(a) $x_n y_{n-1} - x_{n-1} y_n = 2^{2n-1} \sin(n\theta - (n-1)\theta) = 2^{2n-1} \sin\theta = \sqrt{3} \cdot 4^{n-1}$,

(b) $x_n x_{n-1} + y_n y_{n-1} = 2^{2n-1} \cos(n\theta - (n-1)\theta) = 2^{2n-1} \cos\theta = 4^{n-1}$.

6.1.4 The nth Roots of Unity

1.(a) $z_k = \sqrt[4]{2}\left(\cos\frac{\frac{\pi}{4}+2k\pi}{2} + i\sin\frac{\frac{\pi}{4}+2k\pi}{2}\right), k \in \{0,1\};$

(b) $z_k = \cos\frac{\frac{\pi}{2}+2k\pi}{2} + i\sin\frac{\frac{\pi}{2}+2k\pi}{2}, k \in \{0,1\};$

(c) $z_k = \cos\frac{\frac{\pi}{4}+2k\pi}{2} + i\sin\frac{\frac{\pi}{4}+2k\pi}{2}, k \in \{0,1\};$

(d) $z_k = 2\left(\cos\frac{\frac{4\pi}{3}+2k\pi}{2} + i\sin\frac{\frac{4\pi}{3}+2k\pi}{2}\right), k \in \{0,1\};$

(e) $z_0 = 4 - 3i, z_1 = -4 + 3i.$

2.(a) $z_k = \cos\frac{\frac{3\pi}{2}+2k\pi}{3} + i\sin\frac{\frac{3\pi}{2}+2k\pi}{2}, k \in \{0,1,2\};$

(b) $z_k = 3\left(\cos\frac{\pi+2k\pi}{3} + i\sin\frac{\pi+2k\pi}{3}\right), k \in \{0,1,2\};$

(c) $z_k = \sqrt{2}\left(\cos\frac{\frac{\pi}{4}+2k\pi}{3} + i\sin\frac{\frac{\pi}{4}+2k\pi}{3}\right), k \in \{0,1,2\};$

(d) $z_k = \cos\frac{\frac{5\pi}{3}+2k\pi}{3} + i\sin\frac{\frac{5\pi}{3}+2k\pi}{3}, k \in \{0,1,2\};$

(e) $z_0 = 3+i, z_1 = (3+i)\varepsilon, z_2 = (3+i)\varepsilon^2$, where $1, \varepsilon, \varepsilon^2$ are the cube roots of 1.

3.(a) $z_k = \sqrt{2}\left(\cos\frac{\frac{5\pi}{4}+2k\pi}{4} + i\sin\frac{\frac{5\pi}{4}+2k\pi}{4}\right), k \in \{0,1,2,3\};$

(b) $z_k = \sqrt[4]{2}\left(\cos\frac{\frac{\pi}{6}+2k\pi}{4} + i\sin\frac{\frac{\pi}{6}+2k\pi}{4}\right), k \in \{0,1,2,3\};$

(c) $z_k = \cos\frac{\frac{\pi}{2}+2k\pi}{4} + i\sin\frac{\frac{\pi}{2}+2k\pi}{4}, k \in \{0,1,2,3\};$

(d) $z_k = \sqrt[4]{2}\left(\cos\frac{\frac{3\pi}{2}+2k\pi}{4} + i\sin\frac{\frac{3\pi}{2}+2k\pi}{4}\right), k \in \{0,1,2,3\};$

(e) $z_0 = 2+i, z_1 = -2-i, z_2 = -1+2i, z_3 = 1-2i.$

4. $z_k = \cos\frac{2k\pi}{n} + i\sin\frac{2k\pi}{n}, k \in \{0,1,\ldots,n-1\}, n \in \{5,6,7,8,12\}.$

5.(a) Consider $\varepsilon_j = \varepsilon^j, \varepsilon_k = \varepsilon^k$, where $\varepsilon = \cos\frac{2\pi}{n} + i\sin\frac{2\pi}{n}$. Then $\varepsilon_j \cdot \varepsilon_k = \varepsilon^{j+k}$. Let r be the remainder modulo n of $j + k$. We have $j + k = p \cdot n + r, r \in \{0,1,\ldots,n-1\}$ and $\varepsilon_j \cdot \varepsilon_k = \varepsilon^{p \cdot n+r} = (\varepsilon^n)^p \cdot \varepsilon^r = \varepsilon^r = \varepsilon_r \in U_n.$

(b) We can write $\varepsilon_j^{-1} = \frac{1}{\varepsilon_j} = \frac{1}{\varepsilon^j} = \frac{\varepsilon^n}{\varepsilon^j} = \varepsilon^{n-j} \in U_n.$

6.(a) $z_k = 5\left(\cos\frac{2k\pi}{3} + i\sin\frac{2k\pi}{3}\right), k \in \{0,1,2\};$

(b) $z_k = 2\left(\cos\frac{\pi+2k\pi}{4} + i\sin\frac{\pi+2k\pi}{4}\right), k \in \{0,1,2,3\};$

(c) $z_k = 4\left(\cos\frac{\frac{3\pi}{2}+2k\pi}{3} + i\sin\frac{\frac{3\pi}{2}+2k\pi}{3}\right), k \in \{0,1,2\};$

(d) $z_k = 3\left(\cos\frac{\frac{\pi}{2}+2k\pi}{3} + i\sin\frac{\frac{\pi}{2}+2k\pi}{3}\right), k \in \{0,1,2\}.$

7.(a) The equation is equivalent to $(z^4 - i)(z^3 - 2i) = 0.$

(b) We can write the equation as $(z^3 + 1)(z^3 + i - 1) = 0.$

(c) The equation is equivalent to $z^6 = -1 + i.$

(d) We can write the equation equivalently as $(z^5 - 2)(z^5 + i) = 0.$

8. It is clear that every solution is different from zero. Multiplying by z, we see that the equation is equivalent to $z^5 - 5z^4 + 10z^3 - 10z^2 + 5z - 1 = -1, z \neq 0$. We obtain the binomial equation $(z - 1)^5 = -1, z \neq 0$. The solutions are $z_k = 1 + \cos\frac{(2k+1)\pi}{5} + i\sin\frac{(2k+1)\pi}{5}, k = 0,1,3,4.$

9. It suffices to prove that

$$nz^n + \ldots + 2z^2 + z = \frac{n+1}{z-1}.$$

Clearly, $z^{n+1} - 1 = 0$, with $z \neq 1$. We have

$$z^n + z^{n-1} + \ldots + z = \frac{z(z^n - 1)}{z-1} = \frac{z^{n+1} - z}{z-1} = \frac{1-z}{z-1},$$

$$z^n + z^{n-1} + \ldots + z^2 = \frac{z^2(z^{n-1} - 1)}{z-1} = \frac{z^{n+1} - z^2}{z-1} = \frac{1-z^2}{z-1},$$

$$\cdots\cdots\cdots\cdots\cdots$$

$$z^n + z^{n-1} = \frac{z^{n-1}(z^2 - 1)}{z-1} = \frac{z^{n+1} - z^{n-1}}{z-1} = \frac{1-z^{n-1}}{z-1},$$

$$z^n = \frac{z^n(z-1)}{z-1} = \frac{z^{n+1} - z^n}{z-1} = \frac{1-z^n}{z-1}.$$

Hence

$$nz^n + \ldots + 2z^2 + z = \frac{n - (z + z^2 + \ldots + z^n)}{z-1} = \frac{n - \dfrac{z^{n+1} - z}{z-1}}{z-1}$$

$$= \frac{n - \dfrac{1-z}{z-1}}{z-1} = \frac{n+1}{z-1}.$$

10. The given condition is equivalent to

$$z^2 + z + 1 + \frac{1}{z} + \frac{1}{z^2} = 0,$$

that is, $\dfrac{z^5 - 1}{z^2(z-1)} = 0$, because $z \neq 1$.

For $n \equiv 0 \pmod 5$, our product is equal to $(1+1)(1+1+1) = 6$.
Otherwise,

$$\left(z^n + \frac{1}{z^n}\right)\left(z^n + \frac{1}{z^n} + 1\right) - 1 = z^{2n} + z^n + 1 + \frac{1}{z^n} + \frac{1}{z^{2n}} = \frac{(z^n)^5 - 1}{z^{2n}(z^n - 1)}$$

$$= \frac{(z^5)^n - 1}{z^{2n}(z^n - 1)} = 0,$$

so the answer to the problem is

$$\begin{cases} 6, \text{if } n \equiv 0 \pmod 5, \\ 1, \text{otherwise.} \end{cases}$$

11. Solution 1. We have $z^{1997} = 1 = 1(\cos 0 + i \sin 0)$.

By de Moivre's theorem, we find that $(k \in \{0, 1, \ldots, 1996\})$

$$z = \cos\left(\frac{2k\pi}{1997}\right) + i \sin\left(\frac{2k\pi}{1997}\right).$$

Now let v be the root corresponding to $\theta = \dfrac{2m\pi}{1997}$, and let w be the root corresponding to $\theta = \dfrac{2n\pi}{1997}$. The magnitude of $v + w$ is therefore

$$\sqrt{\left(\cos\left(\frac{2m\pi}{1997}\right) + \cos\left(\frac{2n\pi}{1997}\right)\right)^2 + \left(\sin\left(\frac{2m\pi}{1997}\right) + \sin\left(\frac{2n\pi}{1997}\right)\right)^2}$$

$$= \sqrt{2 + 2\cos\left(\frac{2m\pi}{1997}\right)\cos\left(\frac{2n\pi}{1997}\right) + 2\sin\left(\frac{2m\pi}{1997}\right)\sin\left(\frac{2n\pi}{1997}\right)}.$$

We need

$$\cos\left(\frac{2m\pi}{1997}\right)\cos\left(\frac{2n\pi}{1997}\right) + \sin\left(\frac{2m\pi}{1997}\right)\sin\left(\frac{2n\pi}{1997}\right) \geq \frac{\sqrt{3}}{2}.$$

The cosine difference identity simplifies the above to

$$\cos\left(\frac{2m\pi}{1997} - \frac{2n\pi}{1997}\right) \geq \frac{\sqrt{3}}{2}.$$

Thus

$$|m - n| \leq \frac{\pi}{6} \cdot \frac{1997}{2\pi} = \left\lfloor \frac{1997}{12} \right\rfloor = 166.$$

Therefore, m and n cannot be more than 166 away from each other. This means that for a given value of m, there are 332 values for n that satisfy the inequality; 166 of them greater than m, and 166 of them less than m. Since m and n must be distinct, n has 1996 possible values. Therefore, the probability is $\dfrac{332}{1996} = \dfrac{83}{499}$. The answer is then $499 + 83 = 582$.

Solution 2. The solutions of the equation $z^{1997} = 1$ are the 1997th roots of unity and are equal to

$$\cos\left(\frac{2k\pi}{1997}\right) + i \sin\left(\frac{2k\pi}{1997}\right) \quad \text{for } k = 0, 1, \ldots, 1996.$$

They are also located at the vertices of a regular 1997-gon that is centered at the origin in the complex plane.

Without loss of generality, let $v = 1$. Then

$$|v + w|^2 = \left| \cos\left(\frac{2k\pi}{1997}\right) + i\sin\left(\frac{2k\pi}{1997}\right) + 1 \right|^2$$

$$= \left| \left[\cos\left(\frac{2k\pi}{1997}\right) + 1 \right] + i\sin\left(\frac{2k\pi}{1997}\right) \right|^2$$

$$= \cos^2\left(\frac{2k\pi}{1997}\right) + 2\cos\left(\frac{2k\pi}{1997}\right) + 1 + \sin^2\left(\frac{2k\pi}{1997}\right)$$

$$= 2 + 2\cos\left(\frac{2k\pi}{1997}\right).$$

We want $|v + w|^2 \geq 2 + \sqrt{3}$. From what we just obtained, this is equivalent to

$$\cos\left(\frac{2k\pi}{1997}\right) \geq \frac{\sqrt{3}}{2}.$$

This occurs when $\frac{\pi}{6} \geq \frac{2k\pi}{1997} \geq -\frac{\pi}{6}$, which is satisfied by

$$k = 166, 165, \ldots, -165, -166$$

(we do not include 0, because that corresponds to v). So out of the 1996 possible k, 332 work. Thus, $m/n = 332/1996 = 83/499$. So our answer is $83 + 499 = 582$.

12. **Solution 1.** We have the following equations:

$$1 - \frac{z}{2z - i} = 1,$$

$$1 - \frac{z}{2z - i} = -1,$$

$$1 - \frac{z}{2z - i} = i,$$

$$1 - \frac{z}{2z - i} = -i.$$

Let us work each equation separately. The first one gives $\frac{z}{2z - i} = 0$, or $z = 0$. The second one gives $2 = \frac{z}{2z - i}$, or $4z - 2i = z$ or $z = \frac{2i}{3}$. The third one gives $\frac{z - i}{2z - i} = i$ or $z - i = 2zi + 1$, or $z = \frac{i + 1}{-2i + 1}$, which is $\frac{3i - 1}{5}$. The fourth one gives $(1 + i)(2z - i) = z$, or $z + 2iz - i + 1 = 0$, which gives the solution $\frac{3i + 1}{5}$.

Thus our answers are

$$(0+1)\left(-\frac{4}{9}+1\right)\left(-\frac{8}{25}-\frac{6i}{25}+1\right)\left(-\frac{8}{25}+\frac{6i}{25}+1\right)=\left(\frac{5}{9}\right)\left(\frac{13}{25}\right)=\frac{13}{45}.$$

Solution 2. $\dfrac{z^4}{(2z+i)^4}=1$ has roots $z_1+i,\ z_2+i,\ z_3+i,\ z_4+i$.

From $z^4=16z^4+\ldots+1$, we get the product of roots $\dfrac{1}{15}$.

$\dfrac{(z-2i)^4}{(2z-3i)^4}=1$ has roots $z_1-i,\ z_2-i,\ z_3-i,\ z_4-i$.

From $z^4+\ldots+16=16z^4+\ldots+81$, we get the product of roots $\dfrac{65}{15}=\dfrac{13}{3}$.

Hence

$$\prod_{r=1}^{4}(z_r^2+1)=\prod_{r=1}^{4}(z_r+i)\prod_{r=1}^{4}(z_r-i)=\left(\frac{1}{15}\right)\left(\frac{13}{3}\right)=\frac{13}{45}.$$

Solution 3. The roots of the given equation are the same as the roots of the polynomial

$$P(z)=(z-i)^4-(2z-i)^4.$$

If a is the leading coefficient of $P(z)$, in this case $a=1-2^4=-15$, then the desired expression is simply $\dfrac{P(i)}{a}\cdot\dfrac{P(-i)}{a}$. Since $P(i)=-i^4=-1$ and $P(-i)=(-2i)^4-(-3i)^4=16-81=-65$, we get

$$E=\frac{P(i)P(-i)}{a^2}=\frac{65}{225}=\frac{13}{45}.$$

13. Let $t=1/x$. After multiplying the equation by t^{10}, we have

$$1+(13-t)^{10}=0\Rightarrow(13-t)^{10}=-1.$$

Using de Moivre's theorem, we obtain $13-t=\text{cis}\left(\dfrac{(2k+1)\pi}{10}\right)$, where k is an integer between 0 and 9, and $\text{cis}\,\theta=\cos\theta+i\sin\theta$. We have

$$t=13-\text{cis}\left(\frac{(2k+1)\pi}{10}\right)\Rightarrow\bar{t}=13-\text{cis}\left(-\frac{(2k+1)\pi}{10}\right).$$

Since $\text{cis}\,\theta+\text{cis}(-\theta)=2\cos\theta$, we have

$$t\bar{t}=170-26\cos\left(\frac{(2k+1)\pi}{10}\right)$$

after expanding. Here k ranges from 0 to 4, because two angles that sum to 2π are involved in the product.

The expression to find is

$$\sum t\bar{t} = 850 - 26 \sum_{k=0}^{4} \cos \frac{(2k+1)\pi}{10}.$$

But

$$\cos \frac{\pi}{10} + \cos \frac{9\pi}{10} = \cos \frac{3\pi}{10} + \cos \frac{7\pi}{10} = \cos \frac{\pi}{2} = 0,$$

so the sum is 850.

14. Since the coefficients of the polynomial are real, it follows that the nonreal roots must come in complex-conjugate pairs. Let the first two roots be m, n. Since $m + n$ is not real, m, n are not conjugate, so the other pair of roots must be the conjugates of m, n. Let m' be the conjugate of m, and n' be the conjugate of n. Then

$$mn = 13 + i, \ m' + n' = 3 + 4i \Rightarrow m'n' = 13 - i, \ m + n = 3 - 4i.$$

By Viète's formulas, we have that

$$b = mm' + nn' + mn' + nm' + mn + m'n' = (m+n)(m'+n') + mn + m'n' = 51.$$

6.1.5 Some Geometric Transformations of the Complex Plane

1. Suppose that f, g are isometries. Then for all complex numbers a, b, we have $|f(g(a)) - f(g(b))| = |g(a) - g(b)| = |a - b|$, so $f \circ g$ is also an isometry.
2. Suppose that f is an isometry and let C be any point on the line AB. Let $f(C) = M$. Then $MA = f(C)f(A) = AC$, and similarly, $MB = BC$. Thus $|MA - MB| = AB$. Hence A, M, B are collinear. Now, from $MA = AC$ and $MB = BC$, we conclude that $M = C$. Hence $f(M) = M$, and the conclusion follows.
3. This follows immediately from the fact that every isometry f is of the form $f(z) = az + b$ or $f(z) = a\bar{z} + b$, with $|a| = 1$.
4. The function f is the product of the rotation $z \rightarrow iz$, the translation $z \rightarrow z + 4 - i$, and a reflection in the real axis. It is clear that f is an isometry.
5. The function f is the product of the rotation $z \rightarrow -iz$ with the translation $z \rightarrow z + 1 + 2i$.

6.2 Solutions to the Olympiad-Caliber Problems

6.2.1 Problems Involving Moduli and Conjugates

Problem 21. *Consider the set*

$$A = \{z \in \mathbb{C} : |z| < 1\},$$

a real number a with $|a| > 1$, and the function

$$f : A \to A, \ f(z) = \frac{1 + az}{z + a}.$$

Prove that f is bijective.

Solution. First, we prove that the function f is well defined, i.e., that $|f(z)| < 1$ for all z with $|z| < 1$.

Indeed, we have $|f(z)| < 1$ if and only if $\left|\frac{1+az}{z+a}\right| < 1$, i.e., $|1+az|^2 < |z+a|^2$. The last relation is equivalent to $(1 + az)(1 + \overline{az}) < (z + a)(\overline{z} + \overline{a})$. That is, $1 + |a|^2|z|^2 < |a|^2 + |z|^2$, or equivalently, $(|a|^2 - 1)(|z|^2 - 1) < 0$. The last inequality is obvious, since $|z| < 1$ and $|a| > 1$.

To prove that f is bijective, it suffices to observe that for every $y \in A$, there is a unique $z \in A$ such that

$$f(z) = \frac{1 + az}{z + a} = y.$$

We obtain

$$z = \frac{ay - 1}{a - y} = -f(-y),$$

and hence $|z| = |f(-y)| < 1$, as desired.

Problem 22. *Let z be a complex number such that $|z| = 1$ and both $\mathrm{Re}(z)$ and $\mathrm{Im}(z)$ are rational numbers. Prove that $|z^{2n} - 1|$ is rational for all integers $n \geq 1$.*

Solution. Let $z = \cos\varphi + i\sin\varphi$ with $\cos\varphi, \sin\varphi \in \mathbb{Q}$. Then

$$z^{2n} - 1 = \cos 2n\varphi + i\sin 2n\varphi - 1 = 1 - 2\sin^2 n\varphi + 2i\sin n\varphi \cos n\varphi - 1$$

$$= -2\sin n\varphi(\sin n\varphi - i\cos n\varphi)$$

and

$$|z^{2n} - 1| = 2|\sin n\varphi|.$$

It suffices to prove that $\sin n\varphi \in \mathbb{Q}$. We prove by induction on n that both $\sin n\varphi$ and $\cos n\varphi$ are rational numbers. The claim is obvious for $n = 1$.

Assume that $\sin n\varphi, \cos n\varphi \in \mathbb{Q}$. Then

$$\sin(n+1)\varphi = \sin n\varphi \cos \varphi + \cos n\varphi \cos \varphi \in \mathbb{Q}$$

and

$$\cos(n+1)\varphi = \cos n\varphi \cos \varphi - \sin n\varphi \sin \varphi \in \mathbb{Q},$$

as desired.

Problem 23. *Consider the function*

$$f : \mathbb{R} \to \mathbb{C}, \quad f(t) = \frac{1+ti}{1-ti}.$$

Prove that f is injective and determine its range.

Solution. To prove that the function f is injective, let $f(a) = f(b)$. Then

$$\frac{1+ai}{1-ai} = \frac{1+bi}{1-bi}.$$

This is equivalent to $1+ab+(a-b)i = 1+ab+(b-a)i$, i.e., $a = b$, as needed.

The image of the function f is the set of numbers $z \in \mathbb{C}$ such that there is $t \in \mathbb{R}$ with

$$z = f(t) = \frac{1+ti}{1-ti}.$$

From $z = \frac{1+ti}{1-ti}$, we obtain $t = \frac{z-1}{i(1+z)}$ if $z \neq 1$. Then $t \in \mathbb{R}$ if and only if $t = \bar{t}$. The last relation is equivalent to

$$\frac{z-1}{i(1+z)} = \frac{\bar{z}-1}{-i(1+\bar{z})},$$

i.e.,

$$-(z-1)(\bar{z}+1) = (z+1)(\bar{z}-1).$$

It follows that $2z\bar{z} = 2$, i.e., $|z| = 1$; hence the image of the function f is the set $\{z \in \mathbb{R} \mid |z| = 1 \text{ and } z \neq -1\}$, the unit circle without the point with coordinate $z = -1$.

Problem 24. *Let $z_1, z_2 \in \mathbb{C}^*$ such that $|z_1 + z_2| = |z_1| = |z_2|$. Compute $\frac{z_1}{z_2}$.*

Solution 1. Let $\frac{z_2}{z_1} = t \in \mathbb{C}$. Then

$$|z_1 + z_1 t| = |z_1| = |z_1 t| \text{ or } |1+t| = |t| = 1.$$

It follows that $t\bar{t} = 1$ and

$$1 = |1+t|^2 = (1+t)(1+\bar{t}) = 1+t+\bar{t}+1,$$

whence $t^2 + t + 1 = 0$.

Therefore, t is a nonreal cube root of unity.

Solution 2. Let A, B, C be the geometric images of the complex numbers $z_1, z_2, z_1 + z_2$, respectively. In the parallelogram $OACB$ we have $OA = OB = OC$; hence $\widehat{AOB} = 120°$. Then

$$\frac{z_2}{z_1} = \cos 120° + i \sin 120° \text{ or } \frac{z_1}{z_2} = \cos 120° + i \sin 120°,$$

and therefore,

$$\frac{z_2}{z_1} = \cos \frac{2\pi}{3} \pm i \sin \frac{2\pi}{3}.$$

Problem 25. *Prove that for all complex numbers z_1, z_2, ..., z_n, the following inequality holds:*

$$(|z_1| + |z_2| + \cdots + |z_n| + |z_1 + z_2 + \cdots + z_n|)^2$$
$$\geq 2(|z_1|^2 + \cdots + |z_n|^2 + |z_1 + z_2 + \cdots + z_n|^2).$$

Solution. We prove first the inequality

$$|z_k| \leq |z_1| + |z_2| + \cdots + |z_{k-1}| + |z_{k+1}| + \cdots + |z_n| + |z_1 + z_2 + \cdots + z_n|$$

for all $k \in \{1, 2, \ldots, n\}$. Indeed,

$$|z_k| = |(z_1 + z_2 + \cdots + z_{k-1} + z_k + z_{k+1} + \cdots + z_n)$$
$$- (z_1 + z_2 + \cdots + z_{k-1} + z_{k+1} + \cdots + z_n)|$$
$$\leq |z_1 + z_2 + \cdots + z_n| + |z_1| + \cdots + |z_{k-1}| + |z_{k+1}| + \cdots + |z_n|,$$

as claimed.

Set $S_k = |z_1| + \cdots + |z_{k-1}| + |z_{k+1}| + \cdots + |z_n|$ for all k. Then

$$|z_k| \leq S_k + |z_1 + z_2 + \cdots + z_n|, \text{ for all } k. \tag{1}$$

Moreover,

$$|z_1 + z_2 + \cdots + z_n| \leq |z_1| + |z_2| + \cdots + |z_n|. \tag{2}$$

Multiplying the inequalities (1) by $|z_k|$ and the inequalities (2) by $|z_1 + z_2 + \cdots + z_n|$, we obtain by summation

$$|z_1|^2 + |z_2|^2 + \cdots + |z_n|^2 + |z_1 + z_2 + \cdots + z_n|^2$$
$$\leq 2|z_1 + z_2 + \cdots + z_n| \sum_{k=1}^{n} |z_k| + \sum_{k=1}^{n} |z_k| S_k.$$

Adding the expression

$$|z_1|^2 + |z_2|^2 + \cdots + |z_n|^2 + |z_1 + z_2 + \cdots + z_n|^2$$

to both sides of the inequality yields

$$2(|z_1|^2 + |z_2|^2 + \cdots + |z_n|^2 + |z_1 + z_2 + \cdots + z_n|^2)$$
$$\leq (|z_1| + \cdots + |z_n| + |z_1 + z_2 + \cdots + z_n|)^2,$$

as desired.

Problem 26. *Let z_1, z_2, ..., z_{2n} be complex numbers such that $|z_1| = |z_2| = \cdots = |z_{2n}|$ and $\arg z_1 \leq \arg z_2 \leq \cdots \leq \arg z_{2n} \leq \pi$. Prove that*

$$|z_1 + z_{2n}| \leq |z_2 + z_{2n-1}| \leq \cdots \leq |z_n + z_{n+1}|.$$

Solution 1. Let $M_1, M_2, \ldots, M_{2n}$ be the points with coordinates $z_1, z_2, \ldots, z_{2n}$, and let $A_1, A_2, \ldots, A_n$ be the midpoints of the segments $M_1 M_{2n}, M_2 M_{2n-1}, \ldots, M_n M_{n+1}$.

The points $M_i, i = \overline{1, 2n}$, lie on the upper semicircle centered at the origin with radius 1. Moreover, the lengths of the chords $M_1 M_{2n}$, $M_2 M_{2n-1}, \ldots, M_n M_{n+1}$ are in decreasing order; hence $OA_1, OA_2, \ldots, OA_n$ are increasing. Thus

$$\left| \frac{z_1 + z_{2n}}{2} \right| \leq \left| \frac{z_2 + z_{2n-1}}{2} \right| \leq \cdots \leq \left| \frac{z_n + z_{n+1}}{2} \right|,$$

and the conclusion follows (Fig. 6.1).

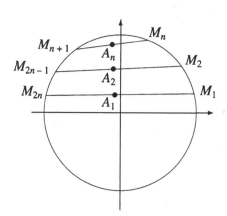

Figure 6.1.

Solution 2. Consider $z_k = r(\cos t_k + i \sin t_k), k = 1, 2, \ldots, 2n$ and observe that for every $j = 1, 2, \ldots, n$, we have

$$|z_j + z_{2n-j+1}|^2 = |r[(\cos t_j + \cos t_{2n-j+1}) + i(\sin t_j + \sin t_{2n-j+1})]|^2$$

$$= r^2[(\cos t_j + \cos t_{2n-j+1})^2 + (\sin t_j + \sin t_{2n-j+1})^2]$$

$$= r^2[2 + 2(\cos t_j \cos t_{2n-j+1} + \sin t_j \sin t_{2n-j+1})]$$

$$= 2r^2[1 + \cos(t_{2n-j+1} - t_j)] = 4r^2 \cos^2 \frac{t_{2n-j+1} - t_j}{2}.$$

Therefore, $|z_j + z_{2n-j+1}| = 2r \cos \frac{t_{2n-j+1}-t_j}{2}$, and the inequalities

$$|z_1 + z_{2n}| \le |z_2 + z_{2n-1}| \le \cdots \le |z_n + z_{n+1}|$$

are equivalent to $t_{2n} - t_1 \ge t_{2n-1} - t_2 \ge \cdots \ge t_{n+1} - t_n$. Because $0 \le t_1 \le t_2 \le \cdots \le t_{2n} \le \pi$, the last inequalities are obviously satisfied.

Problem 27. *Find all positive real numbers x and y satisfying the system of equations*

$$\sqrt{3x}\left(1 + \frac{1}{x+y}\right) = 2,$$

$$\sqrt{7y}\left(1 - \frac{1}{x+y}\right) = 4\sqrt{2}.$$

(1996 Vietnamese Mathematical Olympiad)

Solution. It is natural to make the substitution $\sqrt{x} = u, \sqrt{y} = v$. The system becomes

$$u\left(1 + \frac{1}{u^2+v^2}\right) = \frac{2}{\sqrt{3}},$$

$$v\left(1 - \frac{1}{u^2+v^2}\right) = \frac{4\sqrt{2}}{\sqrt{7}}.$$

But $u^2 + v^2$ is the square of the absolute value of the complex number $z = u + iv$. This suggests that we add the second equation multiplied by i to the first one. We obtain

$$u + iv + \frac{u-iv}{u^2+v^2} = \left(\frac{2}{\sqrt{3}} + i\frac{4\sqrt{2}}{\sqrt{7}}\right).$$

The quotient $(u-iv)/(u^2+v^2)$ is equal to $\bar{z}/|z|^2 = \bar{z}/(z\bar{z}) = 1/z$, so the above equation becomes

$$z + \frac{1}{z} = \left(\frac{2}{\sqrt{3}} + i\frac{4\sqrt{2}}{\sqrt{7}}\right).$$

Hence z satisfies the quadratic equation

$$z^2 - \left(\frac{2}{\sqrt{3}} + i\frac{4\sqrt{2}}{\sqrt{7}}\right)z + 1 = 0$$

with solutions

$$\left(\frac{1}{\sqrt{3}} \pm \frac{2}{\sqrt{21}}\right) + i\left(\frac{2\sqrt{2}}{\sqrt{7}} \pm \sqrt{2}\right),$$

where the signs $+$ and $-$ correspond.

This shows that the initial system has the solutions

$$x = \left(\frac{1}{\sqrt{3}} \pm \frac{2}{\sqrt{21}}\right)^2, \quad y = \left(\frac{2\sqrt{2}}{\sqrt{7}} \pm \sqrt{2}\right)^2,$$

where the signs $+$ and $-$ correspond.

Problem 28. *Let z_1, z_2, z_3 be complex numbers. Prove that $z_1 + z_2 + z_3 = 0$ if and only if $|z_1| = |z_2 + z_3|$, $|z_2| = |z_3 + z_1|$, and $|z_3| = |z_1 + z_2|$.*

Solution. The direct implication is obvious.

Conversely, let $|z_1| = |z_2 + z_3|, |z_2| = |z_1 + z_3|, |z_3| = |z_1 + z_2|$. It follows that

$$|z_1|^2 + |z_2|^2 + |z_3|^2 = |z_2 + z_3|^2 + |z_3 + z_1|^2 + |z_1 + z_2|^2.$$

This is equivalent to

$$z_1\overline{z_1} + z_2\overline{z_2} + z_3\overline{z_3} = z_2\overline{z_2} + z_2\overline{z_3} + \overline{z_2}z_3 + z_3\overline{z_3}$$
$$+ z_3\overline{z_1} + z_1\overline{z_3} + z_1\overline{z_1} + z_1\overline{z_1} + z_1\overline{z_2} + z_2\overline{z_1} + z_2\overline{z_2}, \quad \text{i.e.,}$$
$$z_1\overline{z_1} + z_2\overline{z_2} + z_3\overline{z_3} + z_1\overline{z_2} + z_2\overline{z_1} + z_1\overline{z_3} + \overline{z_1}z_3 + z_2\overline{z_3} + \overline{z_3}z_2 = 0.$$

We write the last relation as

$$(z_1 + z_2 + z_3)(\overline{z_1} + \overline{z_2} + \overline{z_3}) = 0,$$

and we obtain

$$|z_1 + z_2 + z_3|^2 = 0, \text{ i.e., } z_1 + z_2 + z_3 = 0,$$

as desired.

Problem 29. *Let z_1, z_2, $\ldots$, z_n be distinct complex numbers with the same modulus such that*

$$z_3z_4 \ldots z_{n-1}z_n + z_1z_4 \ldots z_{n-1}z_n + \cdots + z_1z_2 \ldots z_{n-2} = 0.$$

Prove that

$$z_1z_2 + z_2z_3 + \cdots + z_{n-1}z_n = 0.$$

Solution. Let $a = |z_1| = |z_2| = \cdots = |z_n|$. Then

$$\overline{z_k} = \frac{a^2}{z_k}, \quad k = \overline{1, n},$$

and

$$\overline{z_1 z_2 + z_2 z_3 + \cdots + z_{n-1} z_n} = \sum_{k=1}^{n-1} \overline{z_k z_{k+1}} = \sum_{k=1}^{n-1} \frac{a^4}{z_k z_{k+1}}$$

$$= \frac{a^4}{z_1 z_2 \cdots z_n} (z_3 z_4 \cdots z_n + z_1 z_4 \cdots z_n + \cdots + z_1 z_2 \cdots z_{n-2}) = 0;$$

hence

$$z_1 z_2 + z_2 z_3 + \cdots + z_{n-1} z_n = 0,$$

as desired.

Problem 30. *Let a and z be complex numbers such that $|z + a| = 1$. Prove that*

$$|z^2 + a^2| \geq \frac{|1 - 2|a|^2|}{\sqrt{2}}.$$

Solution. Let

$$z = r_1(\cos t_1 + i \sin t_1)$$

and

$$a = r_2(\cos t_2 + i \sin t_2).$$

We have

$$1 = |z + a| = \sqrt{(r_1 \cos t_1 + r_2 \cos t_2)^2 + (r_1 \sin t_1 + r_2 \sin t_2)^2}$$

$$= \sqrt{r_1^2 + r_2^2 + 2 r_1 r_2 \cos(t_1 - t_2)},$$

so

$$\cos(t_1 - t_2) = \frac{1 - r_1^2 - r_2^2}{2 r_1 r_2}.$$

Then

$$|z^2 + a^2| = |r_1^2(\cos 2t_1 + i \sin 2t_1) + r_2^2(\cos 2t_2 + i \sin 2t_2)|$$

$$= \sqrt{(r_1^2 \cos 2t_1 + r_2^2 \cos 2t_2)^2 + (r_1^2 \sin 2t_1 + r_2^2 \sin 2t_2)^2}$$

$$= \sqrt{r_1^4 + r_2^4 + 2 r_1^2 r_2^2 \cos 2(t_1 - t_2)}$$

$$= \sqrt{r_1^4 + r_2^4 + 2 r_1^2 r_2^2 (2 \cos^2(t_1 - t_2) - 1)}$$

$$= \sqrt{r_1^4 + r_2^4 + 2 r_1^2 r_2^2 \cdot \left(2 \left(\frac{1 - r_1^2 - r_2^2}{2 r_1 r_2}\right)^2 - 1\right)}$$

$$= \sqrt{2r_1^4 + 2r_2^4 + 1 - 2r_1^2 - 2r_2^2}.$$

The inequality

$$|z^2 + a^2| \geq \frac{|1 - 2|a|^2|}{\sqrt{2}}$$

is equivalent to

$$2r_1^4 + 2r_2^4 + 1 - 2r_1^2 - 2r_2^2 \geq \frac{(1 - 2r_2^2)^2}{2}, \text{i.e.,}$$

$$4r_1^4 + 4r_2^4 - 4r_1^2 - 4r_2^2 + 2 \geq 1 - 4r_2^2 + 4r_2^4.$$

We

$$(2r_1^2 - 1)^2 \geq 0,$$

and we are done.

Problem 31. *Find the geometric images of the complex numbers z for which*

$$z^n \cdot \mathrm{Re}(z) = \overline{z}^n \cdot \mathrm{Im}(z),$$

where n is an integer.

Solution. It is easy to see that $z = 0$ is a root of the equation. Consider $z = a + ib \neq 0, a, b \in \mathbb{R}$.

Observe that if $a = 0$, then $b = 0$, and if $b = 0$, then $a = 0$. Therefore, we may assume that $a, b \neq 0$.

Taking the modulus of both members of the equation

$$az^n = b\overline{z}^n \tag{1}$$

yields $|a| = |b|$, or $a = \pm b$.

Case 1. If $a = b$, (1) becomes

$$(a + ia)^n = (a - ia)^n.$$

This is equivalent to

$$\left(\frac{1+i}{1-i}\right)^n = 1, \text{i.e., } i^n = 1,$$

which has solutions only for $n = 4k, k \in \mathbb{Z}$. In that case, the solutions are

$$z = a(1 + i), \quad a \neq 0.$$

Case 2. If $a = -b$, (1) may be rewritten as

$$(a - ia)^n = -(a + ia)^n.$$

That is,

$$\left(\frac{1-i}{1+i}\right)^n = -1, \text{i.e.}, (-i)^n = -1,$$

which has solutions only for $n = 4k + 2, k \in \mathbb{Z}$. We obtain

$$z = a(1-i), \quad a \neq 0.$$

To conclude,

(a) if n is odd, then $z = 0$;
(b) if $n = 4k, k \in \mathbb{Z}$, then $z = \{a(1+i)|a \in \mathbb{R}\}$, i.e., a line through the origin;
(c) if $n = 4k + 2, k \in \mathbb{Z}$, then $z = \{a(1-i)|a \in \mathbb{R}\}$, i.e., a line through the origin.

Problem 32. *Let a, b be real numbers with $a + b = 1$ and let z_1, z_2 be complex numbers with $|z_1| = |z_2| = 1$.*
     Prove that

$$|az_1 + bz_2| \geq \frac{|z_1 + z_2|}{2}.$$

Solution. Let $z_1 = \cos t_1 + i \sin t_1$ and $z_2 = \cos t_2 + i \sin t_2$. The inequality

$$|az_1 + bz_2| \geq \frac{|z_1 + z_2|}{2}$$

is equivalent to

$$\sqrt{(a \cos t_1 + b \cos t_2)^2 + (a \sin t_1 + b \sin t_2)^2}$$

$$\geq \frac{1}{2}\sqrt{(\cos t_1 + \cos t_2)^2 + (\sin t_1 + \sin t_2)^2}.$$

That is,

$$2\sqrt{a^2 + b^2 + 2ab \cos(t_1 - t_2)} \geq \sqrt{2 + 2\cos(t_1 - t_2)}, \text{i.e.,}$$

$$4a^2 + 4(1-a)^2 + 8a(1-a) \cos(t_1 - t_2) \geq 2 + 2\cos(t_1 - t_2).$$

We obtain

$$8a^2 - 8a + 2 \geq (8a^2 - 8a + 2) \cos(t_1 - t_2), \text{i.e.}, 1 \geq \cos(t_1 - t_2),$$

which is obvious.
     Equality holds if and only if $t_1 = t_2$, i.e., $z_1 = z_2$ or $a = b = \frac{1}{2}$.

Problem 33. *Let k, n be positive integers and let z_1, z_2, $\ldots$, z_n be nonzero complex numbers with the same modulus such that*

$$z_1^k + z_2^k + \cdots + z_n^k = 0.$$

Prove that

$$\frac{1}{z_1^k} + \frac{1}{z_2^k} + \cdots + \frac{1}{z_n^k} = 0.$$

Solution. Let $r = |z_1| = |z_2| = \cdots = |z_n| > 0$. Then

$$\frac{1}{z_1^k} + \frac{1}{z_2^k} + \cdots + \frac{1}{z_n^k} = \frac{\overline{z_1}^k}{r^{2k}} + \frac{\overline{z_2}^k}{r^{2k}} + \cdots + \frac{\overline{z_n}^k}{r^{2k}}$$

$$= \frac{1}{r^{2k}} \left(\overline{z_1}^k + \overline{z_2}^k + \cdots + \overline{z_n}^k \right) = 0,$$

as desired.

Problem 34. *Find all pairs (a, b) of real numbers such that*

$$(a + bi)^5 = b + ai.$$

Solution. Let

$$a + bi = r(\cos \theta + i \sin \theta).$$

Then by taking the magnitude of both sides of the equation, we obtain

$$r^5 = r, \quad r \geq 0,$$

that is, $r = 0$ or $r = 1$.

If $r = 0$, then $a = b = 0$, and so

$$(a, b) = (0, 0).$$

Now if $r = 1$, then by de Moivre's law,

$$(a + bi)^5 = (\cos \theta + i \sin \theta)^5 = \cos(5\theta) + i \sin(5\theta).$$

By trigonometric identities we also have

$$b + ai = \cos(90° - \theta) + i \sin(90° - \theta).$$

It follows that

$$5\theta \equiv 90° - \theta \pmod{360}.$$

In other words,

$$5\theta = 90° - \theta + 360°k,$$

where k is an integer.

Hence

$$\theta = 15° + 70k°,$$

where $k \in \mathbb{Z}$. So $k = 15, 75, 135, 195, 255, 315$.

Now

$$(a, b) = \left(\pm\frac{\sqrt{2}}{2}, \mp\frac{\sqrt{2}}{2}\right), \left(\pm\frac{\sqrt{6} \pm \sqrt{2}}{4}, \pm\frac{\sqrt{6} \mp \sqrt{2}}{4}\right).$$

Combining the $r = 0$ and $r = 1$ cases, we get seven solutions:

$$(a, b) = (0, 0), \left(\pm\frac{\sqrt{2}}{2}, \mp\frac{\sqrt{2}}{2}\right), \left(\pm\frac{\sqrt{6} \pm \sqrt{2}}{4}, \pm\frac{\sqrt{6} \mp \sqrt{2}}{4}\right),$$

where the signs $+$ and $-$ correspond.

Problem 35. *For every value of $a \in \mathbb{R}$, find $\min |z^2 - az + a|$, where $z \in \mathbb{C}$ and $|z| \leq 1$.*

Solution. This was Problem 2, grade 10, at the 2009 Romanian National Olympiad, asked for just $a \in [2 + \sqrt{2}, 4]$.

The equation $z^2 - az + a = 0$ has roots $z_{1,2} = \frac{a}{2} \pm \frac{\sqrt{a^2 - 4a}}{2}$, and it is natural to consider various cases with respect to the values taken by a, since

$$|z^2 - az + a| = |(z - z_1)(z - z_2)|,$$

and if we denote by M, A_1, A_2 the points of coordinates z, z_1, z_2, the expression becomes $MA_1 \cdot MA_2$.

(1) $a \leq 1$. Then at least one of the roots z_1, z_2 is of magnitude at most 1; therefore,

$$\min |z^2 - az + a| = 0,$$

which is realized for the root.

(2) $a \geq 4$. Then the roots are real and greater than 1. Setting $T(1, 0)$ and $B(0, 1) = \{z \in \mathbb{C} \mid |z| \leq 1\}$, we have

$$MA_j \geq d(A_j, B(0, 1)) = TA_j,$$

and thus

$$|z^2 - az + a| = MA_1 \cdot MA_2 \geq TA_1 \cdot TA_2 = 1,$$

a minimum realized for $z = 1$.

(3) $1 \leq a \leq 4$. Then $a^2 - 4a \leq 0$, thus the roots are complex conjugates.

The following result will be useful. Consider the points $A_1(\alpha, \beta), A_2(\alpha, -\beta)$, and $M(x, y)$. Then

$$(MA_1 \cdot MA_2)^2 = ((x - \alpha)^2 + (y - \beta)^2)((x - \alpha)^2 + (y + \beta)^2)$$
$$= ((x - \alpha)^2 + y^2 + \beta^2)^2 - 4\beta^2 y^2,$$

having least value when x is closest to α.

Let us now compute $|z^2 - az + a|$ for $|z| = 1$. Thus $z = \cos\theta + i\sin\theta$. We have

$$|z^2 - az + a|^2 = |(\cos^2\theta - \sin^2\theta - a\cos\theta + a) + i(2\cos\theta\sin\theta - a\sin\theta)|^2$$

$$= (\cos\theta(2\cos\theta - a) + a - 1)^2 + (\sin\theta(2\cos\theta - a))^2$$

$$= (2\cos\theta - a)^2 + 2(a-1)\cos\theta(2\cos\theta - a) + (a-1)^2$$

$$= (2a^1 - 2a + 1) - 4a\left(\frac{a+1}{4}\right)^2 + 4a\left(\cos\theta - \frac{a+1}{4}\right)^2.$$

For $a \in [3, 4]$, the minimum of the expression above is obtained for $\cos\theta = 1$, thus again for $z = 1$ and is

$$\min|z^2 - az + a| = 1.$$

For $a \in [1, 3]$, the minimum of the expression above is obtained for $\cos\theta = \frac{a+1}{4}$ and is

$$\min|z^2 - az + a| = (2a^2 - 2a + 1) - 4a\left(\frac{a+1}{4}\right)^2 = 1 - \frac{a(a-3)^2}{4}.$$

The value $a = 2 + \sqrt{2} \in [3, 4]$ is of interest. The circle γ of center $(a/2, 0)$ and radius $\sqrt{4a - a^2}/2$, of diameter A_1A_2, is tangent at $T(1, 0)$ to the disk $B(0, 1)$. For $a \geq 2 + \sqrt{2}$, this allows for a special argumentation for the minimum of the expression $|z^2 - az + a|$ being 1, obtained for $z = 1$, and that was the approach of the official solution.

Problem 36. *Let a, b, c be three complex numbers such that*

$$a|bc| + b|ca| + c|ab| = 0.$$

Prove that

$$|(a-b)(b-c)(c-a)| \geq 3\sqrt{3}|abc|.$$

(Romanian Mathematical Olympiad—Final Round, 2008)

Solution. If one of the numbers is 0, then the conclusion is obvious. Otherwise, on dividing by $|abc|$ and setting $\alpha = \frac{a}{|a|}, \beta = \frac{b}{|b|}, \gamma = \frac{c}{|c|}$, the hypothesis becomes $\alpha + \beta + \gamma = 0$ and $|\alpha| = \|\beta| = |\gamma| = 1$. It is a well-known fact that in this case, the differences between the arguments of the numbers α, β, γ are $\pm\frac{2\pi}{3}$.

The law of cosines now gives $|a - b|^2 = |a|^2 + |b|^2 + |a||b| \geq 3|a||b|$ and two additional similar relations. Multiplication of the three inequalities yields the desired result.

Problem 37. *Let a and b be two complex numbers. Prove the inequality*

$$|1 + ab| + |a + b| \geq \sqrt{|a^2 - 1||b^2 - 1|}.$$

(Romanian Mathematical Olympiad—District Round, 2008)

Solution 1. By the triangle inequality,

$$|1 + ab| + |a + b| \geq |1 + ab + a + b|$$

and

$$|1 + ab| + |a + b| \geq |1 + ab - a - b|.$$

Multiply these two inequalities to get

$$(|1 + ab| + |a + b|)^2 \geq (|1 + ab)^2 - (a + b)^2|,$$

which is equivalent to $|1 + ab| + |a + b| \geq \sqrt{|a^2 - 1| \cdot |b^2 - 1|}$.

Solution 2. We have

$$|1 + 2ab + a^2b^2| + |a^2 + 2ab + b^2| \geq |a^2b^2 + 1 - a^2 - b^2| = |a^2 - 1| \cdot |b^2 - 1|,$$

which is equivalent to

$$(|1 + ab| + |a + b|)^2 \geq |(1 + ab)^2 - (a + b)^2|.$$

Problem 38. *Consider complex numbers a, b, and c such that $a + b + c = 0$ and $|a| = |b| = |c| = 1$. Prove that for every complex number z, $|z| \leq 1$, we have*

$$3 \leq |z - a| + |z - b| + |z - c| \leq 4.$$

(Romanian Mathematical Olympiad—Final Round, 2012)

Solution. Consider points A, B, C, and M having complex coordinates a, b, c, and z, respectively. The triangle ABC is equilateral and inscribed in the unit circle centered at the origin of the complex plane O.

To prove the left-hand inequality, we have successively

$$\sum |z - a| = \sum |\bar{a}||z - a| = \sum |\bar{a}z - \bar{a}a| \geq \left|\sum (\bar{a}z - 1)\right| = \left|z \left(\sum \bar{a}\right) - 3\right| = 3.$$

For the right-hand inequality, consider a chord containing M and denote by P, Q its points of intersection with the unit circle. Let p and q be the complex coordinates of P and Q. Let $\alpha \in [0, 1]$ be such that $m = \alpha p + (1 - \alpha)q$. We get

$$\sum |z - a| = \sum |\alpha p + (1 - \alpha)q - a| \leq \alpha \sum |p - a| + (1 - \alpha) \sum |q - a|,$$

so

$$\sum |z - a| \leq \max\left\{\sum |p - a|, \sum |q - a|\right\}.$$

We can suppose without loss of generality that

$$\max\left\{\sum |p - a|, \sum |q - a|\right\} = \sum |p - a|$$

and that P is on the arc from A to C. By Ptolemy's relation, $PA + PC = PB$, that is,

$$|p - a| + |p - c| = |p - b|.$$

Then

$$\sum |z - a| \leq \sum |p - a| = 2|p - b| \leq 4,$$

which concludes the proof.

Remark. Equality holds in the left-hand inequality when $z = 0$, and in the right-hand one when $z \in \{-a, -b, -c\}$.

6.2.2 Algebraic Equations and Polynomials

Problem 11. *Let* a, b, c *be complex numbers with* $a \neq 0$. *Prove that if the roots of the equation* $az^2 + bz + c = 0$ *have equal moduli, then* $\bar{a}b|c| = |a|\bar{b}c$.

Solution. Let $r = |z_1| = |z_2|$.
 The relation $\bar{a}b|c| = |a|\bar{b}c$ is equivalent to

$$\frac{\bar{a}b|c|}{\bar{a}a|a|} = \frac{|a|\bar{b}c}{\bar{a}a|a|}.$$

This relation can be written as

$$\frac{b}{a} \cdot \left|\frac{c}{a}\right| = -\left(\frac{\bar{b}}{a}\right) \cdot \frac{c}{a}.$$

That is,

$$-(x_1 + x_2) \cdot |x_1 x_2| = -(\overline{x_1} + \overline{x_2}) \cdot x_1 x_2, \text{ i.e.,}$$
$$(x_1 + x_2)r^2 = |x_1|^2 x_2 + x_1 |x_2|^2.$$

It follows that

$$(x_1 + x_2)r^2 = (x_1 + x_2)r^2,$$

which is certainly true.

Problem 12. *Let z_1, z_2 be the roots of the equation $z^2 + z + 1 = 0$, and let z_3, z_4 be the roots of the equation $z^2 - z + 1 = 0$. Find all integers n such that $z_1^n + z_2^n = z_3^n + z_4^n$.*

Solution. Observe that $z_1^3 = z_2^3 = 1$ and $z_3^3 = z_4^3 = -1$. If $n = 6k + r$, with $k \in \mathbb{Z}$ and $r \in \{0, 1, 2, 3, 4, 5\}$, then $z_1^n + z_2^n = z_1^r + z_2^r$ and $z_3^n + z_4^n = z_3^r + z_4^r$.

The equality $z_1^n + z_2^n = z_3^n + z_4^n$ is equivalent to $z_1^r + z_2^r = z_3^r + z_4^r$ and holds only for $r \in \{0, 2, 4\}$. Indeed,

(i) if $r = 0$, then $z_1^0 + z_2^0 = 2 = z_3^0 + z_4^0$;
(ii) if $r = 2$, then $z_1^2 + z_2^2 = (z_1 + z_2)^2 - 2z_1 z_2 = (-1)^2 - 2 \cdot 1 = -1$ and $z_3^2 + z_4^2 = (z_3 + z_4)^2 - 2z_3 z_4 = 1^2 - 2 \cdot 1 = -1$;
(iii) if $r = 4$, then $z_1^4 + z_2^4 = z_1 + z_2 = -1$ and $z_3^4 + z_4^4 = -(z_3 + z_4) = -(+1) = -1$.

The other cases are as follows:
(iv) $r = 1$: then $z_1 + z_2 = -1 \neq z_3 + z_4 = 1$;
(v) $r = 3$: then $z_1^3 + z_2^3 = 1 + 1 = 2 \neq z_3^3 + z_4^3 = -1 - 1 = -2$;
(vi) $r = 5$: then $z_1^5 + z_2^5 = z_1^2 + z_2^2 = -1 \neq z_3^5 + z_4^5 = -(z_3^2 + z_4^2) = 1$.

Therefore, the desired numbers are the even numbers.

Problem 13. *Consider the equation with real coefficients*

$$x^6 + ax^5 + bx^4 + cx^3 + bx^2 + ax + 1 = 0,$$

and denote by $x_1, x_2, \ldots, x_6$ the roots of the equation.
  Prove that

$$\prod_{k=1}^{6}(x_k^2 + 1) = (2a - c)^2.$$

Solution. Let

$$f(x) = x^6 + ax^5 + bx^4 + cx^3 + bx^2 + ax + 1$$

$$= \prod_{k=1}^{6}(x - x_k) = \prod_{k=1}^{6}(x_k - x), \text{ for all } x \in \mathbb{C}.$$

We have

$$\prod_{k=1}^{6}(x_k^2 + 1) = \prod_{k=1}^{6}(x_k + i) \cdot \prod_{k=1}^{6}(x_k - i) = f(-i) \cdot f(i)$$

$$= (i^6 + ai^5 + bi^4 + ci^3 + bi^2 + ai + 1) \cdot (i^6 - ai^5 + bi^4 - ci^3 + bi^2 - ai + 1)$$

$$= (2ai - ci)(-2ai + ci) = (2a - c)^2,$$

as desired.

Problem 14. *Let a and b be complex numbers and let $P(z) = az^2 + bz + i$. Prove that there exists $z_0 \in \mathbb{C}$ with $|z_0| = 1$ such that $|P(z_0)| \geq 1 + |a|$.*

Solution. For a complex number z with $|z| = 1$, observe that

$$P(z) + P(-z) = az^2 + bz + i + az^2 - bz + i = 2(az^2 + i).$$

It suffices to choose z_0 such that $az_0^2 = |a|i$. Let

$$a = |a|(\cos t + i \sin t), \quad t \in [0, 2\pi).$$

The equation $az^2 = |a|i$ is equivalent to

$$z_0^2 = \cos\left(\frac{\pi}{2} - t\right) + i \sin\left(\frac{\pi}{2} - t\right).$$

Set

$$z_0 = \cos\left(\frac{\pi}{4} - \frac{t}{2}\right) + i \sin\left(\frac{\pi}{4} - \frac{t}{2}\right),$$

and we are done.

Therefore, we have

$$P(z_0) + P(-z_0) = 2(|z|i + i) = 2i(1 + |a|).$$

Passing to absolute values, it follows that

$$|P(z_0)| + |P(-z_0)| \geq 2(1 + |a|).$$

That is, $|P(z_0)| \geq 1 + |a|$ or $|P(-z_0)| \geq 1 + |a|$.

Note that $|z_0| = |-z_0| = 1$, as needed.

Problem 15. *Find all polynomials f with real coefficients satisfying, for every real number x, the relation $f(x)f(2x^2) = f(2x^3 + x)$.*

(21st IMO—Shortlist)

Solution. Let z be a complex root of the polynomial f. From the given relation, it follows that $2z^3 + z$ is also a root of f. Observe that if $|z| > 1$, then

$$|2z^3 + z| = |z||2z^2 + 1| \geq |z|(2|z|^2 - 1) > |z|.$$

Hence, if f has a root z_1 with $|z_1| > 1$, then f has a root $z_2 = 2z_1^3 + z_1$ with $|z_2| > |z_1|$. We can continue this procedure and obtain an infinite number of roots of f, namely $z_1, z_2, \ldots$ with $|z_1| < |z_2| < \cdots$, a contradiction. Therefore, all roots of f satisfy $|z| \leq 1$.

We will show that f is not divisible by x. Assume, for the sake of a contradiction, the contrary and choose the greatest $k \geq 1$ with the property that x^k divides f. It follows that $f(x) = x^k(a + xg(x))$ with $a \neq 0$; hence

$$f(2x^2) = x^{2k}(a_1 + 2^{k+1}x^2 g(2x^2)) = x^{2k}(a_1 + xg_1(x))$$

and

$$f(2x^3 + x) = x^k(2x^2 + 1)^k(a + (2x^2 + 1)xg(x)) = x^k(a + xg_2(x)),$$

where g, g_1, g_2 are polynomials and $a_1 \neq 0$ is a real number. The relation $f(x)f(2x^2) = f(2x^3 + x)$ is equivalent to $x^k(a + xg(x))x^{2k}(a_1 + xg_1(x)) = x^k(a + xg_2(x))$, which is not possible for $a \neq 0$ and $k > 0$. Let m be the degree of polynomial f. The polynomials $f(2x^2)$ and $f(2x^3 + x)$ have degrees $2m$ and $3m$, respectively.

If $f(x) = b_m x^m + \cdots + b_0$, then $f(2x^2) = 2^m b_m x^{2m} + \cdots$ and $f(2x^3 + x) = 2^m b_m x^{3m} + \cdots$. From the given relation, we obtain $b_m \cdot 2^m \cdot b_m = 2^m b_m$, hence $b_m = 1$. Again using the given relation, it follows that $f^2(0) = f(0)$, i.e., $b_0^2 = b_0$; hence $b_0 = 1$.

The product of the roots of the polynomial f is ± 1. Taking into account that for every root z of f we have $|z| \leq 1$, it follows that the roots of f have modulus 1.

Consider z a root of f. Then $|z| = 1$ and $1 = |2z^3 + z| = |z||2z^2 + 1| = |2z^2 + 1| \geq |2z^2| - 1 = 2|z|^2 - 1 = 1$. Equality is possible if and only if the complex numbers $2z^2$ and -1 have the same argument; that is, $z = \pm i$.

Because f has real coefficients and its roots are $\pm i$, it follows that f is of the form $(x^2 + 1)^n$ for some positive integer n. Using the identity

$$(x^2 + 1)(4x^4 + 1) = (2x^3 + x)^2 + 1,$$

we obtain that the desired polynomials are $f(x) = (x^2 + 1)^n$, where n is an arbitrary positive integer.

Problem 16. *Find all complex numbers z such that*

$$(z - z^2)(1 + z + z^2)^2 = \frac{1}{7}.$$

(Mathematical Reflections, 2013)

Solution. From the well-known identity

$$(x + y)^7 = x^7 + y^7 + 7xy(x + y)(x^2 + xy + y^2)^2,$$

we deduce

$$(1 - z)^7 = 1 - z^7 - 7z(1 - z)(1 - z + z^2)^2.$$

Hence our equation is equivalent to $(1 - z)^7 = -z^7$, that is,

$$\left(-\frac{1}{z} + 1\right)^7 = 1.$$

It follows that

$$-\frac{1}{z_k} + 1 = \cos\frac{2k\pi}{7} + i\sin\frac{2k\pi}{7}, \quad k = 0, 1, \ldots, 6.$$

This reduces to

$$\frac{1}{z_k} = 1 - \cos\frac{2k\pi}{7} - i\sin\frac{2k\pi}{7} = 2\sin^2\frac{k\pi}{7} - 2i\sin\frac{k\pi}{7}\cos\frac{k\pi}{7},$$

which is equivalent to

$$z_k = \frac{1}{-2i\sin\frac{k\pi}{7}\left(\cos\frac{k\pi}{7} - i\sin\frac{k\pi}{7}\right)} = \frac{\cos\frac{k\pi}{7} + i\sin\frac{k\pi}{7}}{-2i\sin\frac{k\pi}{7}}$$

$$= \frac{1}{2}\left(-1 + i\cot\frac{k\pi}{7}\right), \quad k = 0, 1, \ldots, 6.$$

Problem 17. *Determine all pairs (z, n) such that*

$$z + z^2 + \ldots + z^n = n|z|,$$

where $z \in \mathbb{C}$ and $|z| \in \mathbb{Z}_+$.

(Mathematical Reflections, 2008)

Solution. For $n = 1$, we obtain $z = |z|$, and $(z, 1)$ is a solution iff $z \in \mathbb{Z}^+$. For $|z| = 1$, we obtain

$$n = |z + z^2 + \ldots + z^n| \leq |z| + |z|^2 + \ldots + |z|^n = n,$$

with equality iff all z^k are collinear, i.e., iff $z \in \mathbb{R}$, and $(1, n)$ is the only possible solution with $|z| = 1$, but it is valid for every positive integer n. These solutions may be considered "trivial."

Let us now look for nontrivial solutions. For $n = 2$, the equation becomes $z + z^2 = 2|z|$, which after expressing

$$z = |z|(\cos\theta + i\sin\theta)$$

and separating into real and imaginary parts yields

$$\cos\theta + |z|\cos(2\theta) = 2 \quad \text{and} \quad \sin\theta + |z|\sin(2\theta) = 0.$$

The latter results in either

$$\sin\theta = 0 \quad \text{or} \quad \cos\theta = -\frac{1}{2|z|},$$

the second option yielding

$$\cos(2\theta) = 2\cos^2\theta - 1 = \frac{1 - 2|z|^2}{2|z|^2}.$$

Insertion into the former yields $|z| = -2$, which is obviously absurd. So $\sin\theta = 0$, resulting in $|z| = 2 + 1 = 3$ when $\cos\theta = -1$, and $|z| = 2 - 1 = 1$ when $\cos\theta = 1$, for the trivial solution $(1, 2)$ and the additional solution $(-3, 2)$.

If $|z| > 1$, consider $n|z|(z - 1) = z^{n+1} - z$, which after expressing

$$z = |z|(\cos\theta + i\sin\theta)$$

and separating real and imaginary parts yields

$$|z|^n \cos((n+1)\theta) = (n|z| + 1)\cos\theta - n,$$

$$|z|^n \sin((n+1)\theta) = (n|z| + 1)\sin\theta.$$

Squaring both equations and adding them, we obtain

$$|z|^n = \sqrt{(n|z| + 1)^2 + n^2 - 2n(n|z| + 1)\cos\theta} \le n|z| + n + 1,$$

with equality iff $\cos\theta = -1$. The derivatives of $|z|^n$ and $n|z| + n + 1$ with respect to n are respectively $|z|^n \ln|z| > \dfrac{|z|^n}{2} \ge |z|^{n-1}$ and $|z| + 1$, where we have used that $|z| \ge 2$ and $\ln 2 > \dfrac{1}{2}$, since $4 > e$. Note that for $n \ge 3$ and $|z| \ge 2$, $|z|^{n-1} \ge 2|z| > |z| + 1$, and since for $n = 3$, $|z| = 3$, we obtain $|z|^n = 27 > 13 = n|z| + n + 1$, and for $n = 4$, $|z| = 2$, we obtain $|z|^n = 16 > 13 = n|z| + n + 1$, we cannot have solutions for $n \ge 4$ when $|z| = 2$, nor for $n \ge 3$ when $|z| \ge 3$. Since the cases $n = 1, 2$ for every $|z|$, and $|z| = 1$ for every n, have already been discussed, we need to find only whether solutions exist for $|z| = 2$ and $n = 3$. In this case, the equations become

$$8\cos(4\theta) = 7\cos\theta - 3 \text{ and } 8\sin(4\theta) = 7\sin\theta,$$

for $64 = 49 + 9 - 42\cos\theta$, or $\cos\theta = -\dfrac{1}{7}$. But this is absurd, since substitution in the previous equations yields $\cos(4\theta) = -\dfrac{1}{2}$, but direct calculation yields

$$\cos(2\theta) = 2\cos^2\theta - 1 = -\frac{47}{49},$$

and similarly,

$$\cos(4\theta) = \frac{2 \cdot 47^2 - 49^2}{49^2} \ne -\frac{1}{2},$$

and the potential solution found was actually artificially introduced in the squaring-and-adding process. Hence no solution exists for $n = 3$ and $|z| = 3$, and the only possible solutions are $(-3, 2)$, and the trivial solutions $(1, n)$ for all positive integers n and $(z, 1)$ for all positive integers z.

Problem 18. *Let a, b, c, d be nonzero complex numbers such that $ad - bc \neq 0$ and let n be a positive integer. Consider the equation*

$$(ax + b)^n + (cx + d)^n = 0.$$

(a) Prove that for $|a| = |c|$, the roots of the equation are situated on a line.
(b) Prove that for $|a| \neq |c|$, the roots of the equation are situated on a circle.
(c) Find the radius of the circle when $|a| \neq |c|$.

(Mathematical Reflections, 2010)

Solution. If there is a root x such that $cx + d = 0$, then we have $ax + b = 0$. It follows that $ad - bc = 0$, a relation that is contrary to the hypothesis. Therefore, we can assume that $cx + d \neq 0$. We can write the equation in the equivalent form

$$\left(\frac{ax + b}{cx + d}\right)^n = -1. \tag{1}$$

This is, in fact, the binomial equation $z^n = -1$, where $z = \dfrac{ax + b}{cx + d}$. The roots of this equation are

$$z_k = \cos\frac{(2k + 1)\pi}{n} + i\sin\frac{(2k + 1)\pi}{n}, \text{ where } k = 0, 1, \ldots, n - 1.$$

It is clear that the roots of our equation and the roots of the binomial equation $z^n = -1$ are related by $z_k = \dfrac{ax_k + b}{cx_k + d}$, $k = 0, 1, \ldots, n - 1$. Because $|z_k| = 1$, it follows that

$$\left|\frac{ax_k + b}{cx_k + d}\right| = 1 \text{ for } k = 0, 1, \ldots, n - 1.$$

The last relation is equivalent to

$$\frac{\left|x_k + \dfrac{b}{a}\right|}{\left|x_k + \dfrac{d}{c}\right|} = \frac{|c|}{|a|}. \tag{2}$$

If $|a| = |c|$, then

$$\left|x_k + \frac{b}{a}\right| = \left|x_k + \frac{d}{c}\right|,$$

i.e., the roots x_k are situated on the perpendicular bisector of the segment determined by the points of complex coordinates $-\dfrac{b}{a}$ and $-\dfrac{d}{c}$.

If $|a| \neq |c|$, then from (2), it follows that x_k belongs to the circle of Apollonius corresponding to the constant $\dfrac{|c|}{|a|}$.

In order to find the radius of this circle, we will use the following result, which can be obtained from Stewart's theorem: Let α, β, and $K \geq 0$ be fixed real numbers, and let A and B be fixed points in the plane. If

$$K > \frac{\alpha\beta}{\alpha + \beta} \cdot AB^2,$$

then the locus of points M in the plane with the property

$$\alpha \cdot MA^2 + \beta \cdot MB^2 = K, \tag{3}$$

is a circle of radius

$$R = \sqrt{\frac{K}{\alpha + \beta} - \frac{\alpha\beta}{(\alpha + \beta)^2} \cdot AB^2}.$$

In our case, we have just to take $K = 0$, $\alpha = |a|$, and $\beta = -|c|$ and the fixed points $A\left(-\dfrac{b}{a}\right)$ and $B\left(-\dfrac{d}{c}\right)$. We get

$$R = \frac{|b| \cdot |ad - bc|}{|c| \cdot ||a| - |b||}.$$

Problem 19. *Let n be a positive integer. Prove that a complex number of absolute value 1 is a solution to $z^n + z + 1 = 0$ if and only if $n = 3m + 2$ for some positive integer m.*

(Romanian Mathematical Olympiad—Final Round, 2007)

Solution 1. If $n = 3m + 2$ for some positive integer m, then the complex number $\cos(2\pi/3) + i\sin(2\pi/3)$ is clearly a solution of absolute value 1. Conversely, if z is a solution of absolute value 1, then so is $\overline{z} = 1/z$. Hence $z^n + z + 1 = 0 = z^n + z^{n-1} + 1$, which yields successively $z^{n-2} = 1$, $z^2 + z + 1 = 0$, $z^3 = 1$ with $z \neq 1$, so $n = 3m + 2$ for some positive integer m.

Solution 2. Let $P(z) = z^n + z + 1 = 0$. If $P(\omega) = 0$, $|\omega| = 1$, then

$$\omega = \cos\theta + i\sin\theta,$$

and so using de Moivre's formula, we obtain $\omega^n = \cos n\theta + i\sin n\theta$. Then

$$0 = (\cos n\theta + \cos\theta + 1) + i(\sin n\theta + \sin\theta),$$

whence $\sin^2 n\theta = \sin^2\theta$ and $\cos^2 n\theta = \cos^2\theta + 2\cos\theta + 1$, so $\cos\theta = -\dfrac{1}{2}$. It follows that $\omega^3 = 1$ and $\omega^2 + \omega + 1 = 0$, and therefore $\omega^n = \omega^2$, so $n \equiv 2 \pmod{3}$.

Conversely, if $n \equiv 2 \pmod 3$, then for ω a cube root of unity, $\omega \neq 1$, we have $P(\omega) = 0$. In fact, then $P(z) = z^n + z + 1 = (z^2 + z + 1)Q(z)$ for some Q (with integer coefficients).

Problem 20. *Let a and b be two complex numbers. Prove that the following statements are equivalent:*

(1) The absolute values of the roots of the equation $x^2 - ax + b = 0$ are equal to the absolute values of the roots of the equation

$$x^2 - bx + a = 0.$$

(2) $a^3 = b^3$ or $b = \overline{a}$.

(Romanian Mathematical Olympiad—District Round, 2011)

Solution. Let

$$|x_1| = |x_3|, \quad |x_2| = |x_4| \tag{1}$$

and observe that $|a| = |x_3 x_4| = |x_1 x_2| = |b|$ to derive

$$|x_1 + x_2| = |x_3 + x_4|. \tag{2}$$

Relations (1) and (2) show that there exists a number $k \in \mathbb{C}$ such that $x_2 = kx_1$, $x_4 = kx_3$ or $x_2 = kx_1$, $x_4 = \overline{k}x_3$.

In the first case, we have $a = kx_3^2 = (1+k)x_1$ and $b = kx_1^2 = (1+k)x^3$, so

$$a^3 = k(1+k)^2 x_1^2 x_3^2 = b^3.$$

In the latter case, we have $a = \overline{k}x_3^2 = (1+k)x_1$ and $b = kx_1^2 = (1+\overline{k})x_3$. It follows that $x_1^2 \overline{x}_1 = x_3 \overline{x}_2$, so $x_1 = \overline{x}_3$ or $a = b = 0$, and furthermore, $x_2 = \overline{x}_4$, and hence $a = \overline{b}$.

Conversely, if $b = \overline{a}$, then $x_1 + x_2 = \overline{x}_3 + \overline{x}_4$, $x_1 x_2 = \overline{x}_3 \overline{x}_4$, implying $\{x_1, x_2\} = \{\overline{x}_3, \overline{x}_4\}$. If $a^3 = b^3$, then $a = \varepsilon b$, $\varepsilon^3 = 1$. The roots satisfy the relations $x_1 + x_2 = \varepsilon(x_3 + x_4)$, $x_1 x_2 = \varepsilon^2 x_3 x_4$. Both cases lead to

$$\{|x_1|, |x_2|\} = \{|x_3|, |x_4|\},$$

as needed.

6.2.3 From Algebraic Identities to Geometric Properties

Problem 12. *Let a, b, c, d be distinct complex numbers with $|a| = |b| = |c| = |d|$ and $a + b + c + d = 0$.*
Then the geometric images of a, b, c, d are the vertices of a rectangle.

Solution 1. Let A, B, C, D be the points with coordinates a, b, c, d, respectively.

If $a + b = 0$, then $c + d = 0$. Hence $a + b = c + d$, i.e., $ABCD$ is a parallelogram inscribed in the circle of radius $R = |a|$, and we are done.

If $a + b \neq 0$, then the points M and N with coordinates $a + b$ and $c + d$, respectively, are symmetric with respect to the origin O of the complex plane. Since AB is a diagonal in the rhombus $OAMB$, it follows that AB is the perpendicular bisector of the segment OM. Likewise, CD is the perpendicular bisector of the segment ON. Therefore, A, B, C, D are the intersection points of the circle of radius R with the perpendicular bisector of the segments OM and ON, so A, B, C, D are the vertices of a rectangle.

Solution 2. First, let us note that it follows from $a + b + c + d = 0$ that $a + d = -(b + c)$, i.e., $|a + d| = |b + c|$. Hence $|a + d|^2 = |b + c|^2$, and using properties of the real product, we find that $(a + d) \cdot (a + d) = (b + c) \cdot (b + c)$. That is, $|a|^2 + |d|^2 + 2a \cdot d = |b|^2 + |c|^2 + 2b \cdot c$. Taking into account that $|a| = |b| = |c| = |d|$, one obtains $a \cdot d = b \cdot c$.

On the other hand, $AD^2 = |d - a|^2 = (d - a) \cdot (d - a) = |d|^2 + |a|^2 - 2a \cdot d = 2(R^2 - a \cdot d)$. Analogously, we have $BC^2 = 2(R^2 - b \cdot c)$. Since $a \cdot d = b \cdot c$, it follows that $AD = BC$, so $ABCD$ is a rectangle.

Problem 13. *The complex numbers z_i, $i = 1, 2, 3, 4, 5$, have the same nonzero modulus, and*

$$\sum_{i=1}^{5} z_i = \sum_{i=1}^{5} z_i^2 = 0.$$

Prove that $z_1, z_2, \ldots, z_5$ are the coordinates of the vertices of a regular pentagon.

(Romanian Mathematical Olympiad—Final Round, 2003)

Solution. Consider the polynomial

$$P(X) = X^5 + aX^4 + bX^3 + cX^2 + dX + e$$

with roots $z_k, k = \overline{1,5}$. Then

$$a = -\sum z_1 = 0 \text{ and } b = \sum z_1 z_2 = \frac{1}{2}\left(\sum z_1\right)^2 - \frac{1}{2}\sum z_1^2 = 0.$$

Denoting by r the common modulus and taking conjugates, we also get

$$0 = \sum \overline{z_1} = \sum \frac{r^2}{z_1} = \frac{r^2}{z_1 z_2 z_3 z_4 z_5} \sum z_1 z_2 z_3 z_4,$$

from which $d = 0$ and

$$0 = \sum \overline{z_1}\overline{z_2} = \sum \frac{r^4}{z_1 z_2} = \frac{r^4}{z_1 z_2 z_3 z_4 z_5} \sum z_1 z_2 z_3.$$

Therefore, $c = 0$. It follows that $P(X) = X^5 + e$, so $z_1, z_2, \ldots, z_5$ are the fifth roots of e, and the conclusion is proved.

Problem 14. *Let ABC be a triangle.*

(a) Prove that if M is any point in its plane, then

$$AM \sin A \le BM \sin B + CM \sin C.$$

(b) Let A_1, B_1, C_1 be points on the sides BC, AC, and AB, respectively, such that the angles of the triangle $A_1B_1C_1$ are in order α, β, γ. Prove that

$$\sum_{\text{cyc}} AA_1 \sin \alpha \le \sum_{\text{cyc}} BC \sin \alpha.$$

(Romanian Mathematical Olympiad—Second Round, 2003)

Solution.

(a) Consider a complex plane with origin at M. Denote by a, b, c the coordinates of A, B, C, respectively. Since

$$a(b - c) = b(a - c) + c(b - a),$$

we have

$$|a||b - a| = |b(a - c) + c(b - a)| \le |b||a - c| + |c||b - a|.$$

Thus

$$AM \cdot BC \le BM \cdot AC + CM \cdot AB,$$

or

$$2R \cdot AM \cdot \sin A \le 2R \cdot BM \cdot \sin B + 2R \cdot CM \cdot \sin C,$$

which gives

$$AM \cdot \sin A \le BM \cdot \sin B + CM \cdot \sin C.$$

(b) From (a), we have

$$AA_1 \cdot \sin \alpha \le AB_1 \cdot \sin \beta + AC_1 \cdot \sin \gamma,$$
$$BB_1 \cdot \sin \beta \le BA_1 \cdot \sin \alpha + BC_1 \cdot \sin \gamma,$$
$$CC_1 \cdot \sin \gamma \le CA_1 \cdot \sin \alpha + CB_1 \cdot \sin \beta,$$

which when summed give the desired conclusion.

Problem 15. *Let M and N be points inside triangle ABC such that*

$$\widehat{MAB} = \widehat{NAC} \text{ and } \widehat{MBA} = \widehat{NBC}.$$

Prove that

$$\frac{AM \cdot AN}{AB \cdot AC} + \frac{BM \cdot BN}{BA \cdot BC} + \frac{CM \cdot CN}{CA \cdot CB} = 1.$$

(39th IMO—Shortlist)

Solution. Let the coordinates of A, B, C, M, and N be a, b, c, m, and n, respectively. Since the lines AM, BM, and CM are concurrent, as well as the lines AN, BN, and CN, it follows from Ceva's theorem that

$$\frac{\sin \widehat{BAM}}{\sin \widehat{MAC}} \cdot \frac{\sin \widehat{CBM}}{\sin \widehat{MBA}} \cdot \frac{\sin \widehat{ACM}}{\sin \widehat{MCB}} = 1, \tag{1}$$

$$\frac{\sin \widehat{BAN}}{\sin \widehat{NAC}} \cdot \frac{\sin \widehat{CBN}}{\sin \widehat{NBA}} \cdot \frac{\sin \widehat{ACN}}{\sin \widehat{NCB}} = 1. \tag{2}$$

By hypothesis, $\widehat{BAM} = \widehat{NAC}$ and $\widehat{MBA} = \widehat{CBN}$. Hence $\widehat{BAN} = \widehat{MAC}$ and $\widehat{NBA} = \widehat{CBM}$. Combined with (1) and (2), these equalities imply

$$\sin \widehat{ACM} \cdot \sin \widehat{ACN} = \sin \widehat{MCB} \cdot \sin \widehat{NCB}.$$

Thus,

$$\cos(\widehat{NCM} + 2\widehat{ACM}) - \cos \widehat{NCM} = \cos(\widehat{NCM} + 2\widehat{NCB}) - \cos \widehat{NCM},$$

and hence $\widehat{ACM} = \widehat{NCB}$ (Fig. 6.2).

Since $\widehat{BAM} = \widehat{NAC}, \widehat{MBA} = \widehat{CBN}$ and $\widehat{ACN} = \widehat{MCB}$, the following complex ratios are all positive real numbers:

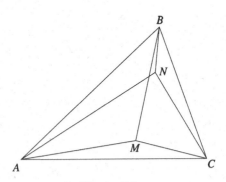

Figure 6.2.

$$\frac{m - a}{b - a} : \frac{c - a}{n - a}, \quad \frac{m - b}{a - b} : \frac{c - b}{n - b}, \quad \frac{m - c}{b - c} : \frac{a - c}{n - c}.$$

Hence each of these equals its absolute value, and so

$$\frac{AM \cdot AN}{AB \cdot AC} + \frac{BM \cdot BN}{BA \cdot BC} + \frac{CM \cdot CN}{CA \cdot CB}$$

$$= \frac{(m-a)(n-a)}{(b-a)(c-a)} + \frac{(m-b)(n-b)}{(a-b)(c-b)} + \frac{(m-c)(n-c)}{(b-c)(a-c)} = 1.$$

6.2.4 Solving Geometric Problems

Problem 26. Let ABC be a triangle such that $AC^2 + AB^2 = 5BC^2$. Prove that the medians from the vertices B and C are perpendicular.

Solution. Let a, b, c be the coordinates of the points A, B, C, respectively. Using the real product of complex numbers, we have

$$AC^2 + AB^2 = 5BC^2 \text{ if and only if } |c-a|^2 + |b-a|^2 = 5|c-b|^2, \text{ i.e.,}$$

$$(c-a) \cdot (c-a) + (b-a) \cdot (b-a) = 5(c-b) \cdot (c-b).$$

The last relation is equivalent to

$$c^2 - 2a \cdot c + a^2 + b^2 - 2a \cdot b + a^2 = 5c^2 - 10b \cdot c + 5b^2, \text{ i.e.,}$$

$$2a^2 - 4b^2 - 4c^2 - 2a \cdot b - 2a \cdot c + 10b \cdot c = 0.$$

It follows that

$$a^2 - 2b^2 - 2c^2 - a \cdot b - a \cdot c + 5b \cdot c = 0, \text{ i.e.,}$$

$$(a+c-2b) \cdot (a+b-2c) = 0, \text{ so } \left(\frac{a+c}{2} - b\right) \cdot \left(\frac{a+b}{2} - c\right) = 0.$$

The last relation shows that the medians from B and C are perpendicular, as desired.

Problem 27. On the sides BC, CA, AB of a triangle ABC the points A', B', C' are chosen such that

$$\frac{A'B}{A'C} = \frac{B'C}{B'A} = \frac{C'A}{C'B} = k.$$

Consider the points A'', B'', C'' on the segments $B'C'$, $C'A'$, $A'B'$ such that

$$\frac{A''C'}{A''B'} = \frac{C''B'}{C''A'} = \frac{B''A'}{B''C'} = k.$$

Prove that triangles ABC and $A''B''C''$ are similar.

Solution. Denoting by the corresponding lowercase letter the coordinates of a point denoted by an uppercase letter, we obtain

$$a' = \frac{b - kc}{1 - k}, \quad b' = \frac{c - ka}{1 - k}, \quad c' = \frac{a - kb}{1 - k},$$

and

$$a'' = \frac{c' - kb'}{1 - k} = \frac{(1 + k^2)a - k(b + c)}{(1 - k)^2},$$

$$b'' = \frac{a' - kc'}{1 - k} = \frac{(1 + k^2)b - k(a + c)}{(1 - k)^2},$$

$$c'' = \frac{b' - ka'}{1 - k} = \frac{(1 + k^2)c - k(b + a)}{(1 - k)^2}.$$

Then

$$\frac{c'' - a''}{b'' - a''} = \frac{(1 + k^2)(c - a) - k(a - c)}{(1 + k^2)(b - a) - k(a - b)} = \frac{c - a}{b - a},$$

which proves that triangles ABC and $A''B''C''$ are similar.

Problem 28. *Prove that the following inequality holds in every triangle:*

$$\frac{R}{2r} \geq \frac{m_\alpha}{h_\alpha}.$$

Equality holds only for equilateral triangles.

Solution. Consider the complex plane with origin at the circumcircle of triangle ABC and let z_1, z_2, z_3 be the coordinates of points A, B, C.

The inequality $\frac{R}{2r} \geq \frac{m_\alpha}{h_\alpha}$ is equivalent to

$$2rm_\alpha \leq Rh_\alpha, \text{ i.e., } 2\frac{K}{s}m_\alpha \leq R\frac{2K}{\alpha}.$$

Hence $\alpha m_\alpha \leq Rs$.

Using complex numbers, we have

$$2\alpha m_\alpha = 2|z_2 - z_3|\left|z_1 - \frac{z_2 + z_3}{2}\right| = |(z_2 - z_3)(2z_1 - z_2 - z_3)|$$

$$= |z_2(z_1 - z_2) + z_1(z_2 - z_3) + z_3(z_3 - z_1)|$$

$$\leq |z_2||z_1 - z_2| + |z_1||z_2 - z_3| + |z_3||z_3 - z_1| = R(\alpha + \beta + \gamma) = 2Rs.$$

Hence $\alpha m_\alpha \leq Rs$, as desired.

Problem 29. *Let $ABCD$ be a quadrilateral inscribed in the circle $\mathcal{C}(O;\ R)$. Prove that*

$$AB^2 + BC^2 + CD^2 + DA^2 = 8R^2$$

if and only if $AC \perp BD$ or one of the diagonals is a diameter of $\mathcal{C}$.

Solution. Consider the complex plane with origin at the circumcenter O, and let a, b, c, d be the coordinates of points A, B, C, D.

The midpoints E and F of the diagonals AC and BD have the coordinates $\dfrac{a+c}{2}$ and $\dfrac{b+d}{2}$.

Using the real product the complex numbers, we have

$$AB^2 + BC^2 + CD^2 + DA^2 = 8R^2$$

if and only if

$$(b-a) \cdot (b-a) + (c-b) \cdot (c-b) + (d-c) \cdot (d-c) + (a-d) \cdot (a-d) = 8R^2,$$

i.e.,

$$2a \cdot b + 2b \cdot c + 2c \cdot d + 2d \cdot a = 0.$$

The last relation is equivalent to

$$b \cdot (a+c) + d \cdot (a+c) = 0, \text{ i.e., } (b+d) \cdot (a+c) = 0.$$

We obtain

$$\frac{b+d}{2} \cdot \frac{a+c}{2} = 0, \text{ i.e., } OE \perp OF$$

or $E = O$ or $F = O$.

That is, $AC \perp BD$, or one of the diagonals AC and BD is a diameter of the circle $\mathcal{C}$.

Problem 30. *On the sides of the convex quadrilateral $ABCD$, equilateral triangles ABM, BCN, CDP, and DAQ are drawn external to the figure. Prove that quadrilaterals $ABCD$ and $MNPQ$ have the same centroid.*

Solution. Denote by the corresponding lowercase letter the coordinate of a point denoted by an uppercase letter, and let

$$\varepsilon = \cos 120° + i \sin 120°.$$

Since triangles ABM, BCN, COP, and DAQ are equilateral we have

$$m + b\varepsilon + a\varepsilon^2 = 0, \quad n + c\varepsilon + b\varepsilon^2 = 0, \quad p + d\varepsilon + c\varepsilon^2 = 0, \quad q + a\varepsilon + d\varepsilon^2 = 0.$$

Summing these equalities yields

$$(m + n + p + q) + (a + b + c + d)(\varepsilon + \varepsilon^2) = 0,$$

and since $\varepsilon + \varepsilon^2 = -1$, it follows that $m + n + p + q = a + b + c + d$. Therefore, the quadrilaterals $ABCD$ and $MNPQ$ have the same centroid.

Problem 31. *Let $ABCD$ be a quadrilateral and consider the rotations $\mathcal{R}_1$, $\mathcal{R}_2$, $\mathcal{R}_3$, $\mathcal{R}_4$ with centers A, B, C, D through the angle α and of the same orientation.*

Points M, N, P, Q are the images of points A, B, C, D under the rotations $\mathcal{R}_2$, $\mathcal{R}_3$, $\mathcal{R}_4$, $\mathcal{R}_1$, respectively.

Prove that the midpoints of the diagonals of the quadrilaterals $ABCD$ and $MNPQ$ are the vertices of a parallelogram.

Solution. Denote by the corresponding lowercase letter the coordinate of a point denoted by an uppercase letter. Using the rotation formula, we obtain

$$m = b + (a - b)\varepsilon, \quad n = c + (b - c)\varepsilon, \quad p = d + (c - d)\varepsilon, \quad q = a + (d - a)\varepsilon,$$

where $\varepsilon = \cos\alpha + i\sin\alpha$.

Let E, F, G, H be the midpoints of the diagonals BD, AC, MP, NQ, respectively; then

$$e = \frac{b+d}{2}, \quad f = \frac{a+c}{2}, \quad g = \frac{b + d + (a + c - b - d)\varepsilon}{2}$$

$$\text{and } h = \frac{a + c + (b + d - a - c)\varepsilon}{2}.$$

Since $e + f = g + h$, it follows that $EGFH$ is a parallelogram, as desired.

Problem 32. *Prove that in every cyclic quadrilateral $ABCD$, the following hold:*

(a) $AD + BC\cos(A + B) = AB\cos A + CD\cos D$;
(b) $BC\sin(A + B) = AB\sin A - CD\sin D$.

Solution. Consider the points E, F, G, H such that

$$OE \perp AB, \quad OE = CD, \quad OF \perp BC, \quad OF = AD,$$

$$OG \perp CD, \quad OG = AB, \quad OH \perp AD, \quad OH = BC,$$

where O is the circumcenter of $ABCD$.

We prove that $EFGH$ is a parallelogram. Since $OE = CD$, $OF = AD$, and $\widehat{EOF} = 180° - \widehat{ABC} = \widehat{ADC}$, it follows that triangles EOF and ADC are congruent; hence $EF = GH$. Likewise, $FG = EH$, and the claim is proved.

Consider the complex plane with origin at O such that F is on the positive real axis. Denote by the corresponding lowercase letter the coordinate of a point denoted by an uppercase letter. We have

$$|e| = CD, \quad |f| = AD, \quad |g| = AB, \quad |h| = BC.$$

Furthermore,

$$\widehat{FOG} = 180° - \hat{C} = \hat{A}, \quad \widehat{GOH} = \hat{B}, \quad \widehat{HOE} = \hat{C},$$

whence

$$f = |f| = AD, \quad g = |g|(\cos A + i \sin A) = AD(\cos A + i \sin A),$$

$$h = |h|[\cos(A + B) + i \sin(A + B)] = BC[\cos(A + B) + i \sin(A + B)],$$

$$e = |e|[\cos(A + B + C) + i \sin(A + B + C)] = CD(\cos D - i \sin D).$$

Since $e + g = f + h$, we obtain

$$AD + BC \cos(A + B) + iBC \sin(A + B)$$

$$= CD(\cos D - i \sin D) + AB(\cos A + i \sin A),$$

and the conclusion follows.

Problem 33. Let O_9, I, G be the 9-point center, the incenter, and the centroid, respectively, of a triangle ABC. Prove that lines O_9G and AI are perpendicular if and only if $\hat{A} = \frac{\pi}{3}$.

Solution. Consider the complex plane with origin at the circumcenter O of the triangle. Let a, b, c, ω, g, z_I be the coordinates of the points A, B, C, O_9, G, I, respectively.

Without loss of generality, we may assume that the circumradius of the triangle ABC is equal to 1, and hence $|a| = |b| = |c| = 1$.

We have

$$\omega = \frac{a+b+c}{2}, \quad g = \frac{a+b+c}{3}, \quad z_I = \frac{a|b-c| + b|a-c| + c|a-b|}{|a-b| + |b-c| + |a-c|}.$$

Using the properties of the real product of complex numbers, we have

$$O_9G \perp AI \text{ if and only if } (\omega - g) \cdot (a - z_I) = 0, \text{ i.e.,}$$

$$\frac{a+b+c}{6} \cdot \frac{(a-b)|a-c| + (a-c)|a-b|}{|a-b| + |b-c| + |a-c|} = 0.$$

This is equivalent to

$$(a+b+c) \cdot [(a-b)|a-c| + (a-c)|a-b|] = 0, \text{ i.e.,}$$

$$\operatorname{Re}\{(a+b+c)[(\bar{a}-\bar{b})|a-c| + (\bar{a}-\bar{c})|a-b|]\} = 0.$$

We find that

$$\operatorname{Re}\{|a-c|(a\bar{a} + b\bar{a} + c\bar{a} - a\bar{b} - b\bar{b} - c\bar{b})$$
$$+ |a-b|(a\bar{a} + b\bar{a} + c\bar{a} - a\bar{c} - b\bar{c} - c\bar{c})\} = 0. \tag{1}$$

Observe that

$$a\bar{a} = b\bar{b} = c\bar{c} = 1 \text{ and } \operatorname{Re}(b\bar{a} - \bar{a}b) = \operatorname{Re}(c\bar{a} - a\bar{c}) = 0;$$

hence the relation (1) is equivalent to

$$\text{Re}\{|a - c|(c\bar{a} - c\bar{b}) + |a - b|(b\bar{a} - b\bar{c})\} = 0, \text{ i.e.,}$$

$$|a - c|(c\bar{a} + \bar{c}a - \bar{c}b - c\bar{b}) + |a - b|(\bar{a}b + a\bar{b} - b\bar{c} - \bar{b}c) = 0.$$

It follows that

$$|a - c|[(b\bar{b} - \bar{b}c - \bar{c}b + c\bar{c}) - (a\bar{a} - c\bar{a} - \bar{c}a + c\bar{c})]$$

$$+|a - b|[(b\bar{b} - \bar{b}c - \bar{c}b + c\bar{c}) - (a\bar{a} - \bar{a}b - a\bar{b} + b\bar{b})] = 0, \text{ i.e.,}$$

$$|a - c|(|b - c|^2 - |a - c|^2) + |a - b|(|b - c|^2 - |a - b|^2) = 0.$$

This is equivalent to

$$AC \cdot BC^2 - AC^3 + AB \cdot BC^2 - AB^3 = 0.$$

The last relation can be written as

$$BC^2(AC + AB) = (AC + AB)(AC^2 - AC \cdot AB + AB^2),$$

so

$$AC \cdot AB = AC^2 + AB^2 - BC^2.$$

We obtain

$$\cos A = \frac{1}{2}, \text{ i.e., } \hat{A} = \frac{\pi}{3},$$

as desired.

Problem 34. *Two circles ω_1 and ω_2 are given in the plane, with centers O_1 and O_2, respectively. Let M'_1 and M'_2 be two points on ω_1 and ω_2, respectively, such that the lines $O_1M'_1$ and $O_2M'_2$ intersect. Let M_1 and M_2 be points on ω_1 and ω_2, respectively, such that when measured clockwise, the angles $\widehat{M'_1O_1M_1}$ and $\widehat{M'_2O_2M_2}$ are equal.*

(a) Determine the locus of the midpoint of $[M_1M_2]$.

(b) Let P be the point of intersection of lines O_1M_1 and O_2M_2. The circumcircle of triangle M_1PM_2 intersects the circumcircle of triangle O_1PO_2 at P and another point Q. Prove that Q is fixed, independent of the locations of M_1 and M_2.

(2000 Vietnamese Mathematical Olympiad)

Solution.

(a) Let a lowercase letter denote the complex number associated with the point labeled by the corresponding uppercase letter. Let M', M, and O denote the midpoints of segments $[M'_1M'_2], [M_1M_2]$, and $[O_1O_2]$, respectively. Also let $z = \dfrac{m_1 - o_1}{m'_1 - o_1} = \dfrac{m_2 - o_2}{m'_2 - o_2}$, so that multiplication by z is a rotation about the origin through some angle. Then $m = \dfrac{m_1 + m_2}{2}$ equals

$$\frac{1}{2}(o_1 + z(m_1' - o_1)) + \frac{1}{2}(o_2 + z(m_2' - o_2)) = o + z(m' - o),$$

i.e., the locus of M is the circle centered at O with radius OM'.

(b) We shall use directed angles modulo π. Observe that

$$\widehat{QM_1M_2} = \widehat{QPM_2} = \widehat{QPO_2} = \widehat{QO_1O_2}.$$

Similarly, $\widehat{QM_2M_1} = \widehat{QO_2O_1}$, implying that triangles QM_1M_2 and QO_1O_2 are similar with the same orientation. Hence,

$$\frac{q - o_1}{q - o_2} = \frac{q - m_1}{q - m_2},$$

or equivalently,

$$\frac{q - o_1}{q - o_2} = \frac{(q - m_1) - (q - o_1)}{(q - m_2) - (q - o_2)} = \frac{o_1 - m_1}{o_2 - m_2} = \frac{o_1 - m_1'}{o_2 - m_2'}.$$

Because lines O_1M_1' and O_2M_2' meet, we have $o_1 - m_1' \neq o_2 - m_2'$, and we can solve this equation to find a unique value for q.

Problem 35. *Isosceles triangles $A_3A_1O_2$ and $A_1A_2O_3$ are constructed externally along the sides of a triangle $A_1A_2A_3$ with $O_2A_3 = O_2A_1$ and $O_3A_1 = O_3A_2$. Let O_1 be a point on the opposite side of line A_2A_3 from A_1, with $\widehat{O_1A_3A_2} = \dfrac{1}{2}\widehat{A_1O_3A_2}$ and $\widehat{O_1A_2A_3} = \dfrac{1}{2}\widehat{A_1O_2A_3}$, and let T be the foot of the perpendicular from O_1 to A_2A_3. Prove that $A_1O_1 \perp O_2O_3$ and that*

$$\frac{A_1O_1}{O_2O_3} = 2\frac{O_1T}{A_2A_3}.$$

(2000 Iranian Mathematical Olympiad)

Solution. Without loss of generality, assume that triangle $A_1A_2A_3$ is oriented counterclockwise (i.e., angle $A_1A_2A_3$ is oriented clockwise). Let P be the reflection of O_1 across T.

We use the complex numbers with origin O_1, where each point denoted by an uppercase letter is represented by the complex number with the corresponding lowercase letter. Let $\zeta_k = a_k/p$ for $k = 1, 2$, so that $z \mapsto \zeta_k(z - z_0)$ is a similarity through angle $\widehat{PO_1A_k}$ with ratio O_1A_3/O_1P about the point corresponding to z_0.

Because O_1 and A_1 lie on opposite sides of line A_2A_3, angles $A_2A_3O_1$ and $A_2A_3A_1$ have opposite orientations, i.e., the former is oriented counterclockwise. Thus, angles PA_3O_1 and $A_2O_3A_1$ are both oriented counterclockwise. Because $\widehat{PA_3O_1} = 2\widehat{A_2A_3O_2} = \widehat{A_2O_3A_1}$, it follows that isosceles triangles PA_3O_1 and $A_2O_3A_1$ are similar and have the same orientation. Hence, $o_3 = a_1 + \zeta_3(a_2 - a_1)$.

Similarly, $o_2 = a_1 + \zeta_2(a_3 - a_1)$. Hence,

$$o_3 - o_2 = (\zeta_2 - \zeta_3)a_1 + \zeta_3 a_2 - \zeta_2 a_3$$

$$= \zeta_2(a_2 - a_3) + \zeta_3(\zeta_2 p) - \zeta_2(\zeta_3 p) = \zeta_2(a_2 - a_3),$$

or (recalling that $o_1 = 0$ and $t = 2p$)

$$\frac{o_3 - o_2}{a_1 - o_1} = \zeta_2 = \frac{a_2 - a_3}{p - o_1} = \frac{1}{2}\frac{a_2 - a_3}{t - o_1}.$$

Thus, the angle between $[O_1A_1]$ and $[O_2O_3]$ equals the angle between $[O_1T]$ and $[A_3A_2]$, which is $\pi/2$. Furthermore, $O_2O_3/O_1A_1 = \frac{1}{2}A_3A_2/O_1T$, or $O_1A_1/O_2O_3 = 2O_1T/A_2A_3$. This completes the proof.

Problem 36. *A triangle $A_1A_2A_3$ and a point P_0 are given in the plane. We define $A_s = A_{s-3}$ for all $s \geq 4$. We construct a sequence of points P_0, P_1, P_2, ... such that P_{k+1} is the image of P_k under the rotation with center A_{k+1} through the angle $120°$ clockwise $(k = 0, 1, 2, ...)$. Prove that if $P_{1986} = P_0$, then the triangle $A_1A_2A_3$ is equilateral.*

(27th IMO)

Solution. Assume that the origin O of the coordinate system in the complex plane is the center of the circumscribed circle. Then the vertices A_1, A_2, A_3 are represented by complex numbers w_1, w_2, w_3 such that

$$|w_1| = |w_2| = |w_3| = R.$$

Let $\varepsilon = \cos\frac{2\pi}{3} + i\sin\frac{2\pi}{3}$. Then $\varepsilon^2 + \varepsilon + 1 = 0$ and $\varepsilon^3 = 1$. Suppose that P_0 is represented by the complex number z_0. The point P_1 is represented by the complex number

$$z_1 = z_0\varepsilon + (1 - \varepsilon)w_1. \tag{1}$$

The point P_2 is represented by

$$z_2 = z_0\varepsilon^2 + (1 - \varepsilon)w_1\varepsilon + (1 - \varepsilon)w_2,$$

and P_3 by

$$z_3 = z_0\varepsilon^3 + (1 - \varepsilon)w_1\varepsilon^2 + (1 - \varepsilon)w_2\varepsilon + (1 - \varepsilon)w_3$$

$$= z_0 + (1 - \varepsilon)(w_1\varepsilon^2 + w_2\varepsilon + w_3).$$

An easy induction on n shows that after n cycles of three such rotations, we obtain that P_{3n} is represented by

$$z_{3n} = z_0 + n(1 - \varepsilon)(w_1 \varepsilon^2 + w_2 \varepsilon + w_3).$$

In our case, for $n = 662$ we obtain

$$z_{1996} = z_0 + 662(1 - \varepsilon)(w_1 \varepsilon^2 + w_2 \varepsilon + w_3) = z_0.$$

Thus, we have the equality

$$w_1 \varepsilon^2 + w_2 \varepsilon + w_3 = 0. \tag{2}$$

This can be written in the equivalent form

$$w_3 = w_1(1 + \varepsilon) + (-\varepsilon)w_2. \tag{3}$$

Taking into account that $1 + \varepsilon = \cos \frac{\pi}{3} + i \sin \frac{\pi}{3}$, the equality (3) can be translated, using the rotation formula, into the following: the point A_3 is obtained under the rotation of point A_1 about the center A_2 through the angle $\frac{\pi}{3}$. This proves that $A_1 A_2 A_3$ is an equilateral triangle.

Problem 37. *Two circles in a plane intersect. Let A be one of the points of intersection. Starting simultaneously from A, two points move with constant speed, each point traveling along its own circle in the same direction. After one revolution, the two points return simultaneously to A. Prove that there exists a fixed point P in the plane such that, at any time, the distances from P to the moving points are equal.*

(21st IMO)

Solution. Let $B(b,0), C(c,0)$ be the centers of the given circles and let $A(0,a), X(0,-a)$ be their intersection points. The complex numbers associated with these point are $z_B = b$, $z_C = c$, $z_A = ia$, and $z_X = -ia$, respectively (Fig. 6.3). After rotating A through angle t about B, we obtain a point M, and after rotating A about C, we obtain the point N. Their corresponding complex numbers are given by the formulas

$$z_M = (ia - b)w + b = iaw + (1 - w)b$$

and

$$z_N = iaw + (1 - w)c.$$

The required result is equivalent to the following: the bisector lines l_{MN} of the segments MN pass through a fixed point $P(x_0, y_0)$. Let R be the midpoint of the segment MN. Then $z_R = \frac{1}{2}(z_M + z_N)$. A point Z of the plane is a point of l_{MN} if and only if the lines RZ and MN are orthogonal. Using the real product of complex numbers, we obtain

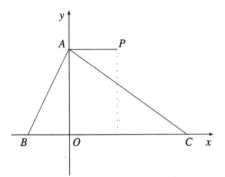

Figure 6.3.

$$\left(z - \frac{z_M + z_N}{2}\right) \cdot (z_N - z_M) = 0.$$

This is equivalent to

$$z \cdot (z_N - z_M) = \frac{1}{2}\left(|z_N|^2 - |z_M|^2\right).$$

By noting that $z = x + iy$, we obtain

$$x(c - b)(1 - \cos t) - y(c - b)\sin t = \frac{1}{2}\left(|z_N|^2 - |z_M|^2\right).$$

After an easy computation, we obtain

$$|z_M|^2 = 2b^2 + a^2 - 2b^2\cos t - 2ab\sin t$$

and

$$|z_N|^2 = 2c^2 + a^2 - 2c^2\cos t - 2ac\sin t.$$

Thus, the orthogonality condition yields

$$x(1 - \cos t) - y\sin t = (b + c) - (b + c)\cos t - a\sin t.$$

This can be written in the form

$$(x - b - c)(1 - \cos t) = (y - a)\sin t.$$

This equation shows that the point $P(x_0, y_0)$ where $x_0 = b + c, y_0 = a$ is a fixed point of the family of lines l_{MN}.

The point P belongs to the line through A parallel to BC, and it is the symmetric point of X with respect to the midpoint of the segment BC. This follows from the equality

$$z_P + z_X = \frac{b + c}{2}.$$

Problem 38. *Inside the square $ABCD$, the equilateral triangles ABK, BCL, CDM, DAN are inscribed. Prove that the midpoints of the segments KL, LM, MN, NK and the midpoints of the segments AK, BK, BL, CL, CM, DM, DN, AN are the vertices of a regular dodecagon.*

(19th IMO)

Solution. Let $A(1 + i), B(-1 + i), C(-1 - i), D(1 - i)$ be the vertices of the square. Using the symmetry of the configuration of points with respect to the axes and center O of the square, we will do computations for the points lying in the first quadrant. Then L, M are represented by the complex numbers $L(\sqrt{3} - 1), M((\sqrt{3} - 1)i)$. The midpoint of the segment LM is $P\left(\frac{\sqrt{3}-1}{2} + i\frac{\sqrt{3}-1}{2}\right)$. Since K is represented by $K(-i(\sqrt{3} - 1))$, the midpoint of AK is $Q\left(\frac{1}{2} + i\frac{2-\sqrt{3}}{2}\right)$. In the same way, the midpoint of AN is $R\left(\frac{2-\sqrt{3}}{2} + \frac{i}{2}\right)$, and the midpoint of BL is $S\left(\frac{-2+\sqrt{3}}{2} + \frac{i}{2}\right)$ (Fig. 6.4). It is sufficient to prove that $SR = RP = PQ$ and $\widehat{SRP} = \widehat{RPQ} = \frac{5\pi}{6}$. For a point X, we denote by z_X the corresponding complex number. We have

$$RS^2 = |z_S - z_R|^2 = (-2 + \sqrt{3})^2 = 7 - 4\sqrt{3},$$

$$RP^2 = |z_P - z_R|^2 = \left|\frac{\sqrt{3}-1}{2} + i\frac{\sqrt{3}-1}{2} - \frac{2-\sqrt{3}}{2} - \frac{i}{2}\right|^2$$

$$= \left|\frac{2\sqrt{3}-3}{2} + i\frac{\sqrt{3}-2}{2}\right|^2 = \frac{(2\sqrt{3}-3)^2 + (2\sqrt{3}-2)^2}{4}$$

$$= \frac{28 - 16\sqrt{3}}{4} = 7 - 4\sqrt{3}.$$

Using reflection in OA, we also have $PQ^2 = RP^2 = 7 - 4\sqrt{3}$ (Fig. 6.4). For angles, we have

$$\cos \widehat{SRP} = \frac{\dfrac{3-2\sqrt{3}}{2}(2 - \sqrt{3}) + \dfrac{2-2\sqrt{3}}{2} \cdot 0}{7 - 4\sqrt{3}}$$

$$= \frac{(12 - 7\sqrt{3})(7 + 4\sqrt{3})}{2(7 - 4\sqrt{3})(7 + 4\sqrt{3})} = -\frac{\sqrt{3}}{2}.$$

This proves that $\widehat{SRP} = \frac{5\pi}{6}$. In the same way, $\cos \widehat{RPQ} = -\frac{\sqrt{3}}{2}$ and $\widehat{RPQ} = \frac{5\pi}{6}$.

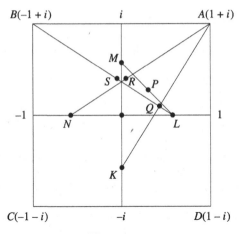

Figure 6.4.

Problem 39. *Let ABC be an equilateral triangle and let M be a point in the interior of angle $\widehat{BAC}$. Points D and E are the images of points B and C under the rotations with center M and angle 120°, counterclockwise and clockwise, respectively. Prove that the fourth vertex of the parallelogram with sides MD and ME is the reflection of point A across point M.*

Solution. Let $1, \varepsilon, \varepsilon^2$, be the coordinates of points A, B, C, M, respectively, where $\varepsilon = \cos 120° + i \sin 120°$ (Fig. 6.5).

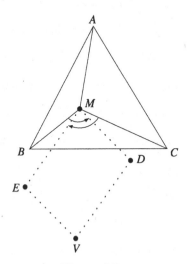

Figure 6.5.

Consider point V such that $MEVD$ is a parallelogram. If d, e, v are the coordinates of points D, E, V, respectively, then

$$v = e + d - m.$$

Using the rotation formula, we obtain

$$d = m + (\varepsilon - m)\varepsilon \text{ and } e = m + (\varepsilon^2 - m)\varepsilon^2;$$

hence

$$v = m + \varepsilon^2 - m\varepsilon + m + \varepsilon^4 - m\varepsilon^2 - m$$

$$= m + \varepsilon^2 + \varepsilon - m(\varepsilon^2 + \varepsilon) = m - 1 + m = 2m - 1.$$

This relation shows that M is the midpoint of the segment $[AV]$, and the conclusion follows.

Problem 40. *Prove that the following inequality holds for every point M inside parallelogram $ABCD$:*

$$MA \cdot MC + MB \cdot MD \geq AB \cdot BC.$$

Solution 1. Consider the complex plane with origin at the center of the parallelogram $ABCD$. Let a, b, c, d, m be the coordinates of points A, B, C, D, M, respectively. It follows that $c = -a$ and $d = -b$.
  It suffices to prove that

$$|m - a| \cdot |m + a| + |m - b||m + b| \geq |a - b||a + b|,$$

or

$$|m^2 - a^2| + |m^2 - b^2| \geq |a^2 - b^2|.$$

This follows immediately from the triangle inequality.

Solution 2. By a translation $t_{\overrightarrow{AB}}$ of vector $\overrightarrow{AB}$, the points in our configuration are transformed as follows: $A \to B$, $D \to C$, $B \to B'$, $C \to C'$, $M \to M'$. Now the desired relation is just Ptolemy's inequality in the quadrilateral $MBM'C$.

Problem 41. *Let ABC be a triangle, H its orthocenter, O its circumcenter, and R its circumradius. Let D be the reflection of A across BC, let E be that of B across CA, and F that of C across AB. Prove that D, E, and F are collinear if and only if $OH = 2R$.*

(39th IMO—Shortlist)

Solution. Let the coordinates of $A, B, C, H,$ and O be $a, b, c, h,$ and o, respectively. Consequently, $a\bar{a} = b\bar{b} = c\bar{c} = R^2$ and $h = a + b + c$. Since D is symmetric to A with respect to line BC, the coordinates d and a satisfy

$$\frac{d - b}{c - b} = \overline{\left(\frac{a - b}{c - b}\right)}, \quad \text{or} \quad (\bar{b} - \bar{c})d - (b - c)\bar{a} + (b\bar{c} - \bar{b}c) = 0. \qquad (1)$$

Since

$$\bar{b} - \bar{c} = -\frac{R^2(b-c)}{bc} \quad \text{and} \quad b\bar{c} - \bar{b}c = \frac{R^2(b^2 - c^2)}{bc},$$

by inserting these expressions in (1), we obtain that

$$d = \frac{-bc + ca + ab}{a} = \frac{k - 2bc}{a},$$

$$\bar{d} = \frac{R^2(-a + b + c)}{bc} = \frac{R^2(h - 2a)}{bc},$$

where $k = bc + c + ab$. Similarly, we have

$$e = \frac{k - 2ca}{b}, \quad \bar{e} = \frac{R^2(h - 2b)}{ca}, \quad f = \frac{k - 2ab}{c} \quad \text{and} \quad \bar{f} = \frac{R^2(h - 2c)}{ab}.$$

Since

$$\Delta = \begin{vmatrix} d & \bar{d} & 1 \\ e & \bar{e} & 1 \\ f & \bar{f} & 1 \end{vmatrix} = \begin{vmatrix} e - d & \bar{e} - \bar{d} \\ f - d & \bar{f} - \bar{d} \end{vmatrix}$$

$$= \begin{vmatrix} \frac{(b-a)(k-2ab)}{ab} & \frac{R^2(a-b)(h-2c)}{abc} \\ \frac{(c-a)(k-2ca)}{ca} & \frac{R^2(a-c)(h-2b)}{abc} \end{vmatrix}$$

$$= \frac{R^2(c-a)(a-b)}{a^2 b^2 c^2} \times \begin{vmatrix} -(ck - 2abc) & (h - 2c) \\ (bk - 2abc) & -(h - 2b) \end{vmatrix}$$

$$= \frac{-R^2(b-c)(c-a)(a-b)(hk - 4abc)}{a^2 b^2 c^2}$$

and $\bar{h} = R^2 k / abc$, it follows that $D, E,$ and F are collinear if and only if $\Delta = 0$. This is equivalent to $hk - 4abc = 0$, i.e., $h\bar{h} = 4R^2$. From the last relation, we obtain $OH = 2R$.

Problem 42. *Let ABC be a triangle such that $\widehat{ACB} = 2\widehat{ABC}$. Let D be the point on the side BC such that $CD = 2BD$. The segment AD is extended to E so that $AD = DE$. Prove that*

$$\widehat{ECB} + 180° = 2\widehat{EBC}.$$

(39th IMO—Shortlist)

Solution. Let the coordinates of $A, B, C, D,$ and E be $a, b, c, d,$ and e, respectively. Then $d = (2b + c)/3$ and $e = 2d - a$. Since $\widehat{ACB} = 2\widehat{ABC}$, the ratio

$$\left(\frac{a - b}{c - b}\right)^2 : \frac{b - c}{a - c}$$

is real and positive. It is equal to $(AB^2 \cdot AC)/BC^3$. On the other hand, a direct computation shows that the ratio

$$\frac{e-c}{b-c} : \left(\frac{c-b}{e-b}\right)^2$$

is equal to

$$\frac{1}{(b-c)^3} \times \left(\frac{(b-a)+2(c-a)}{3}\right)^2 \left(\frac{4(b-a)-(c-a)}{3}\right)$$

$$= \frac{4}{27} + \frac{(b-a)^2(c-a)}{(b-c)^3} = \frac{4}{27} - \frac{AB^2 \cdot AC}{BC^3},$$

which is a real number. Hence the arguments of $(e-c)/(b-c)$ and $(c-b)^2/(e-b)^2$, namely $\widehat{ECB}$ and $2\widehat{EBC}$, differ by an integer multiple of $180°$. We easily infer that either $\widehat{ECB} = 2\widehat{EBC}$ or $\widehat{ECB} = 2\widehat{EBC} - 180°$, according to whether the ratio is positive or negative. To prove that the latter holds, we have to show that $AB^2 \cdot AC/BC^3$ is greater than $4/27$. Choose a point F on the ray AC such that $CF = CB$ (Fig. 6.6).

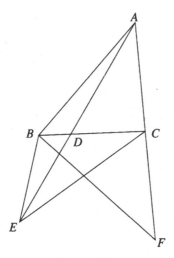

Figure 6.6.

Since $\triangle CBF$ is isosceles and $\widehat{ACB} = 2\widehat{ABC}$, we have $\widehat{CFB} = \widehat{ABC}$. Thus $\triangle ABF$ and $\triangle ACB$ are similar and $AB : AF = AC : AB$. Since

$AF = AC + BC$, $AB^2 = AC(AC + BC)$. Let $AC = u^2$ and $AC + BC = v^2$. Then $AB = uv$ and $BC = v^2 - u^2$. From $AB + AC > BC$, we obtain $u/v > 1/2$. Thus

$$\frac{AB^2 \cdot AC}{BC^3} = \frac{u^4 v^2}{(v^2 - u^2)^3} = \frac{(u/v)^4}{(1 - u^2/v^2)^3} > \frac{(1/2)^4}{(1 - 1/4)^3} = \frac{4}{27},$$

and the conclusion follows.

Problem 43. *Let P be a point situated in the interior of a circle. Two variable perpendicular lines through P intersect the circle at A and B. Find the locus of the midpoint of the segment AB.*

(Mathematical Reflections, 2010)

Solution 1. We can assume, without loss of generality, that $P = t \in [0, 1]$ and the circle C is given by $C = \{|z| = 1\}$. Let $A = z = x + iy \in C$. Then $B = w = si(z - P) + C \in C$ with some $s > 0$. Hence

$$1 = |w|^2 = (t - sy)^2 + s^2(x - t)^2. \tag{1}$$

The midpoint of the segment AB is given by $M = (A + B)/2$. Now we verify that

$$|M - P/2| = \sqrt{2 - |P|^2}/2.$$

In fact, by (1),

$$(2|M - P/2|)^2 = (x = sy)^2 + (s(x - t) + y)^2 = x^2 + y^2 + 1 - t^2 = 2 - t^2.$$

Hence the required locus is a circle with center $P/2$ and radius $\sqrt{2 - |P|^2}/2$. In the general setting, if the circle C has center at P_0 and radius R, then the locus is a circle with center $(P_0 + P)/2$ and radius $\sqrt{2R^2 - |P - P_0|^2}/2$.

Solution 2. Let $ABCD$ be a quadrilateral and let M and N be the midpoints of sides AB and CD, respectively. Using the median theorem or direct computation with complex coordinates, it is easy to prove that the following relation holds:

$$AC^2 + BD^2 + BC^2 + DA^2 = AB^2 + CD^2 + 4MN^2.$$

Let M be the midpoint of the segment AB and let N be the midpoint of the segment OP, where O is the center of the given circle. Applying the relation above in the quadrilateral $ABPO$, we obtain

$$AP^2 + R^2 + BP^2 + R^2 = AB^2 + OP^2 + 4MN^2.$$

It is clear that $AP^2 + BP^2 = AB^2$; hence we get

$$4MN^2 = 2R^2 - OP^2,$$

that is,

$$NM = \frac{1}{2}\sqrt{2R^2 - OP^2}.$$

Since the point N is fixed, it follows that the desired locus is the circle with center N and radius $\frac{1}{2}\sqrt{2R^2 - OP^2}$.

Problem 44. *Let ABC be a triangle and consider the points $M \in (BC)$, $N \in (CA)$, $P \in (AB)$ such that*

$$\frac{AP}{PB} = \frac{BM}{MC} = \frac{CN}{NA}.$$

Prove that if MNP is an equilateral triangle, then ABC is an equilateral triangle as well.

(Romanian Mathematical Olympiad—District Round, 2006)

Solution. Let $\lambda = \frac{AP}{AB} = \frac{BM}{BC} = \frac{CN}{CA}$. We use complex numbers, and we choose the point M the as origin. Furthermore, we can assume that the complex numbers corresponding to the points N and P are 1 and $\varepsilon = \cos\frac{\pi}{3} + i\sin\frac{\pi}{3}$, respectively.

Suppose that the complex numbers corresponding to the points A, B, C are a, b, c, respectively. We have then

$$\varepsilon = (1 - \lambda)a + \lambda b, \ 0 = (1 - \lambda)b + \lambda c, \text{ and } 1 = (1 - \lambda)c + \lambda a.$$

It follows that $\frac{c - a}{b - a} = \varepsilon$. Therefore, $AC = AB$ and $A = \frac{\pi}{3}$.

Problem 45. *Consider the triangle ABC and the points $D \in (BC)$, $E \in (CA)$, $F \in (AB)$, such that*

$$\frac{BD}{DC} = \frac{CE}{EA} = \frac{AF}{FB}.$$

Prove that if the circumcenter of triangles DEF and ABC coincide, then the triangle ABC is equilateral.

(Romanian Mathematical Olympiad—Final Round, 2008)

Solution. Consider complex coordinates, the origin being taken at the circumcenter of the triangle ABC, and use lowercase letters to denote the coordinates of the points. Then, if $\frac{BD}{DC} = k$, we have $d = \frac{b + kc}{1 + k}$, and so on.

The triangles DEF and ABC have the same circumcenter if and only if $|d| = |e| = |f|$, that is, $d\bar{d} = e\bar{e} = f\bar{f}$.

Since $a\bar{a} = b\bar{b} = c\bar{c}$, this amounts to $a\bar{b} + b\bar{a} = a\bar{c} + c\bar{a} = b\bar{c} + c\bar{b}$, which is equivalent to $|a - b|^2 = |a - c|^2 = |b - c|^2$, whence the conclusion.

Problem 46. *On the exterior of a nonequilateral triangle ABC, consider the similar triangles (in this order) ABM, BCN, and CAP such that the triangle MNP is equilateral. Find the angles of the triangles ABM, BCN and CAP.*

(Romanian Mathematical Olympiad—Final Round, 2010)

Solution. All angles are directly oriented. Denote by the corresponding lowercase letter the coordinate of a point denoted by an uppercase letter.

The given similarity can be rewritten as

$$\frac{m-b}{a-b} = \frac{n-c}{b-c} = \frac{p-a}{c-a} = k;$$

hence

$$m = ka + (1-k)b,$$
$$n = kb + (1-k)c,$$
$$p = kc + (1-k)a.$$

Since the triangle MNP is equilateral, we have

$$m + \varepsilon n + \varepsilon^2 p = 0,$$

where $\varepsilon = \cos\dfrac{2\pi}{3} + i\sin\dfrac{2\pi}{3}$. Substituting, we infer that

$$0 = k(a + b\varepsilon + c\varepsilon^2) + (1-k)(b + c\varepsilon + a\varepsilon^2)$$

$$= k(a + b\varepsilon + c\varepsilon^2) + \frac{1-k}{\varepsilon}(a + b\varepsilon + c\varepsilon^2)$$

$$= (a + b\varepsilon + c\varepsilon^2)\left(k + \frac{1-k}{\varepsilon}\right).$$

The triangle ABC is not equilateral, so $a + b\varepsilon + c\varepsilon^2 \neq 0$, and consequently

$$k = \frac{1}{1-\varepsilon}.$$

The equality $m = ka + (1-k)b$ yields $m - a = \varepsilon(m-b)$, showing that triangle AMB is isosceles, with an angle $\dfrac{2\pi}{3}$ and two angles $\dfrac{\pi}{6}$.

6.2.5 Solving Trigonometric Problems

Problem 11. *Sum the following two n-term series for $\theta = 30°$:*

(i) $1 + \dfrac{\cos\theta}{\cos\theta} + \dfrac{\cos(2\theta)}{\cos^2\theta} + \dfrac{\cos(3\theta)}{\cos^3\theta} + \cdots + \dfrac{\cos((n-1)\theta)}{\cos^{n-1}\theta}$, *and*

(ii) $\cos\theta\cos\theta + \cos^2\theta\cos(2\theta) + \cos^3\theta\cos(3\theta) + \cdots + \cos^n\theta\cos(n\theta)$.

(Crux Mathematicorum, 2003)

Solution.

(i) Consider the complex number

$$z = \frac{1}{\cos\theta}(\cos\theta + i\sin\theta).$$

From the identity

$$\sum_{k=0}^{n-1} z^k = \frac{1-z^n}{1-z}, \tag{1}$$

we derive

$$\sum_{k=0}^{n-1}\frac{1}{\cos^k\theta}(\cos k\theta + i\sin k\theta) = \frac{1 - \dfrac{1}{\cos^n\theta}(\cos n\theta + i\sin n\theta)}{1 - \dfrac{1}{\cos\theta}(\cos\theta + i\sin\theta)}$$

$$= \frac{\cos\theta - \dfrac{1}{\cos^{n-1}\theta}(\cos n\theta + i\sin n\theta)}{-i\sin\theta} = \frac{\sin n\theta}{\sin\theta\cos^{n-1}\theta} + i\frac{\cos^n\theta - \cos n\theta}{\sin\theta\cos^{n-1}\theta}.$$

It follows that

$$\sum_{k=0}^{n-1}\frac{\cos k\theta}{\cos^k\theta} = \frac{\sin n\theta}{\sin\theta\cos^{n-1}\theta},$$

and we have just to substitute $\theta = 30°$.

(ii) We proceed in an analogous way by considering the complex number $z = \cos\theta(\cos\theta + i\sin\theta)$. Using identity (1), we obtain

$$\sum_{k=1}^{n} z^k = \frac{z - z^{n+1}}{1-z}.$$

Hence

$$\sum_{k=1}^{n}\cos^k\theta(\cos k\theta + i\sin k\theta)$$

$$= \frac{\cos\theta(\cos\theta + i\sin\theta) - \cos^{n+1}\theta(\cos(n+1)\theta + i\sin(n+1)\theta)}{\sin^2\theta - i\cos\theta\sin\theta}$$

$$= i\frac{\cos\theta(\cos\theta + i\sin\theta) - \cos^{n+1}\theta(\cos(n+1)\theta + i\sin(n+1)\theta)}{\sin\theta(\cos\theta + i\sin\theta)}$$

$$= i\left[\cotan\theta - \frac{\cos^{n+1}\theta(\cos n\theta + i\sin\theta)}{\sin\theta}\right]$$

$$= \frac{\sin n\theta\cos^{n+1}\theta}{\sin\theta} + i\left(\cotan\theta - \frac{\cos^{n+1}\theta\cos n\theta}{\sin\theta}\right).$$

It follows that

$$\sum_{k=1}^{n} \cos^k \theta \cos k\theta = \frac{\sin n\theta \cos^{n+1} \theta}{\sin \theta}.$$

Finally, we let $\theta = 30°$ in the above sum.

Problem 12. *Prove that*

$$1 + \cos^{2n}\left(\frac{\pi}{n}\right) + \cos^{2n}\left(\frac{2\pi}{n}\right) + \cdots + \cos^{2n}\left(\frac{(n-1)\pi}{n}\right)$$

$$= n \cdot 4^{-n}\left(2 + \binom{2n}{n}\right),$$

for all integers $n \geq 2$.

Solution. Let

$$\omega = \cos\frac{2\pi}{n} + i\sin\frac{2\pi}{n}$$

for some integer n. Consider the sum

$$S_n = 4^n + (1+\omega)^{2n} + (1+\omega^2)^{2n} + \cdots + (1+\omega^{n-1})^{2n}.$$

For all $k = 1, \ldots, n-1$, we have

$$1 + \omega^k = 1 + \cos\frac{2k\pi}{n} + i\sin\frac{2k\pi}{n} = 2\cos\frac{k\pi}{n}\left(\cos\frac{k\pi}{n} + i\sin\frac{k\pi}{n}\right)$$

and

$$(1 + \omega^k)^{2n} = 2^{2n}\cos^{2n}\frac{k\pi}{n}(\cos 2k\pi + i\sin 2k\pi) = 4^n\cos^{2n}\frac{k\pi}{n}.$$

Hence

$$S_n = 4^n + \sum_{k=1}^{n-1}(1+\omega^k)^{2n}$$

$$= 4^n\left[1 + \cos^{2n}\left(\frac{\pi}{n}\right) + \cos^{2n}\left(\frac{2\pi}{n}\right) + \cdots + \cos^{2n}\left(\frac{(n-1)\pi}{n}\right)\right]. \quad (1)$$

On the other hand, using the binomial expansion, we have

$$S_n = \sum_{k=0}^{n-1}(1+\omega^k)^{2n} = \sum_{k=0}^{n-1}\left(\binom{2n}{0} + \binom{2n}{1}\omega^k + \right.$$

$$+ \binom{2n}{2}\omega^{2k} + \cdots + \binom{2n}{n}\omega^{nk} + \binom{2n}{2n-1}\omega^{(2n-1)k} + \left.\binom{2n}{2n}\right)$$

$$= n\binom{2n}{0} + n\binom{2n}{n} + n\binom{2n}{2n} + \sum_{\substack{j=1 \\ i\neq n}}^{2n-1}\binom{2n}{j} \cdot \sum_{k=0}^{n-1}\omega^{jk}$$

$$= 2n + n\binom{2n}{n} + \sum_{\substack{j=1 \\ i\neq n}}^{2n-1}\binom{2n}{j} \cdot \frac{1-\omega^{jn}}{1-\omega^j} = 2n + n\binom{2n}{n}. \qquad (2)$$

The relations (1) and (2) give the desired identity.

Problem 13. *For every integer $p \geq 0$, there are real numbers $a_0, a_1, \ldots, a_p$ with $a_p \neq 0$ such that*

$$\cos 2p\alpha = a_0 + a_1 \sin^2 \alpha + \cdots + a_p \cdot (\sin^2 \alpha)^p, \text{ for all } \alpha \in \mathbb{R}.$$

Solution. For $p = 0$, take $a_0 = 1$. If $p \geq 1$, let $z = \cos\alpha + i\sin\alpha$ and observe that

$$z^{2p} = \cos 2p\alpha + i\sin 2p\alpha,$$

$$z^{-2p} = \cos 2p\alpha - i\sin 2p\alpha,$$

and

$$\cos 2p\alpha = \frac{z^{2p} + z^{-2p}}{2} = \frac{1}{2}[(\cos\alpha + i\sin\alpha)^{2p} + (\cos\alpha - i\sin\alpha)^{2p}].$$

Using the binomial expansion, we obtain

$$\cos 2p\alpha = \binom{2p}{0}\cos^{2p}\alpha - \binom{2p}{2}\cos^{2p-2}\alpha\sin^2\alpha + \cdots + (-1)^p\binom{2p}{2p}\sin^{2p}\alpha.$$

Hence $\cos 2p\alpha$ is a polynomial of degree p in $\sin^2\alpha$, so there are $a_0, a_1, \ldots, a_p \in \mathbb{R}$ such that

$$\cos 2p\alpha = a_0 + a_1 \sin^2\alpha + \cdots + a_p \sin^{2p}\alpha \text{ for all } \alpha \in \mathbb{R},$$

with

$$a_p = \binom{2p}{0} - \binom{2p}{2}(-1)^{p-1} + \binom{2p}{4}(-1)^{p-2} + \cdots + \binom{2p}{2p}(-1)^p$$

$$= (-1)^p\left(\binom{2p}{0} + \binom{2p}{2} + \cdots + \binom{2p}{2p}\right) \neq 0.$$

Problem 14. *Let*

$$x = \frac{\displaystyle\sum_{n=1}^{44} \cos n^6}{\displaystyle\sum_{n=1}^{44} \sin n^6}.$$

What is the greatest integer that does not exceed $100x$?

(1997, AIME Problem 11)

Solution 1. We have

$$x = \frac{\displaystyle\sum_{n=1}^{44} \cos n^\circ}{\displaystyle\sum_{n=1}^{44} \sin n^\circ} = \frac{\cos 1 + \cos 2 + \ldots + \cos 44}{\sin 1 + \sin 2 + \ldots + \sin 44}$$

$$= \frac{\cos(45 - 1) + \cos(45 - 2) + \ldots + \cos(45 - 44)}{\sin 1 + \sin 2 + \ldots + \sin 44}.$$

Using the identity

$$\sin a + \sin b = 2 \sin \frac{a+b}{2} \cos \frac{a-b}{2} \Rightarrow$$

$$\sin x + \cos x = \sin x + \sin(90 - x)$$

$$= 2 \sin 45 \cos(45 - x) = \sqrt{2} \cos(45 - x),$$

the expression of x reduces to

$$x = \left(\frac{1}{\sqrt{2}}\right) \left(\frac{(\cos 1 + \cos 2 + \ldots + \cos 44) + (\sin 1 + \sin 2 + \ldots + \sin 44)}{\sin 1 + \sin 2 + \ldots + \sin 44}\right)$$

$$= \left(\frac{1}{\sqrt{2}}\right) \left(1 + \frac{\cos 1 + \cos 2 + \ldots + \cos 44}{\sin 1 + \sin 2 + \ldots + \sin 44}\right),$$

$$x = \left(\frac{1}{\sqrt{2}}\right)(1 + x),$$

$$\frac{1}{\sqrt{2}} = x \left(\frac{\sqrt{2} - 1}{\sqrt{2}}\right),$$

$$x = \frac{1}{\sqrt{2} - 1} = 1 + \sqrt{2}.$$

Therefore,

$$\lfloor 100x \rfloor = \lfloor 100(1 + \sqrt{2}) \rfloor = 241.$$

Solution 2. For a slight variant of the above solution, note that

$$\sum_{n=1}^{44} \cos n + \sum_{n=1}^{44} \sin n = \sum_{n=1}^{44} \sin n + \sin(90 - n)$$

$$= \sqrt{2} \sum_{n=1}^{44} \cos(45 - n) = \sqrt{2} \sum_{n=1}^{44} \cos n,$$

$$\sum_{n=1}^{44} \sin n = (\sqrt{2} - 1) \sum_{n=1}^{44} \cos n.$$

This is the ratio we are looking for. The number x reduces to $\dfrac{1}{\sqrt{2} - 1} = \sqrt{2} + 1$, and

$$\lfloor 100(\sqrt{2} + 1) \rfloor = 241.$$

Solution 3. Consider the sum $\displaystyle\sum_{n=1}^{44} \text{cis}\,n^\circ$, where cis $t = \cos t + i \sin t$. The fraction is given by the real part divided by the imaginary part. By de Moivre's theorem with geometric series, the sum can be written

$$-1 + \sum_{n=0}^{44} \text{cis}\,n^\circ = -1 + \frac{\text{cis}45^\circ - 1}{\text{cis}1^\circ - 1}$$

$$= -1 + \frac{\dfrac{\sqrt{2}}{2} - 1 + \dfrac{i\sqrt{2}}{2}}{\text{cis}1^\circ - 1} = -1 + \frac{\left(\dfrac{\sqrt{2}}{2} - 1 + \dfrac{i\sqrt{2}}{2}\right)(\text{cis}(-1^\circ) - 1)}{(\cos 1^\circ - 1)^2 + \sin^2 1^\circ}$$

$$= -1 + \frac{\left(\dfrac{\sqrt{2}}{2} - 1\right)(\cos 1^\circ - 1) + \dfrac{\sqrt{2}}{2}\sin 1^\circ + i\left(\left(1 - \dfrac{\sqrt{2}}{2}\right)\sin 1^\circ + \dfrac{\sqrt{2}}{2}(\cos 1^\circ - 1)\right)}{2(1 - \cos 1^\circ)}$$

$$= -\frac{1}{2} - \frac{\sqrt{2}}{4} - \frac{i\sqrt{2}}{4} + \frac{\sin 1^\circ\left(\dfrac{\sqrt{2}}{2} + i\left(1 - \dfrac{\sqrt{2}}{2}\right)\right)}{2(1 - \cos 1^\circ)}.$$

Using the tangent half-angle formula, this becomes

$$\left(-\frac{1}{2} + \frac{\sqrt{2}}{4}[\cot(1/2^\circ) - 1]\right) + i\left(\frac{1}{2}\cos(1/2^\circ) - \frac{\sqrt{2}}{4}[\cot(1/2^\circ) + 1]\right).$$

Dividing the two parts and multiplying each part by 4, we see that the fraction is

$$\frac{-2 + \sqrt{2}[\cot(1/2°) - 1]}{2\cot(1/2°) - \sqrt{2}[\cot(1/2°) + 1]}.$$

Although computing an exact value for $\cot(1/2°)$ in terms of radicals would be difficult, it is clear that the value, whatever it is, is really large!
So treat it as though it were ∞. The fraction is approximated by

$$\frac{\sqrt{2}}{2 - \sqrt{2}} = \frac{\sqrt{2}(2 + \sqrt{2})}{2} = 1 + \sqrt{2} \Rightarrow \lfloor 100(1 + \sqrt{2}) \rfloor = 241.$$

Problem 15. *Prove that*

$$\sum_{k=0}^{n} \binom{n}{k} \cos[(n-k)x + ky] = \left(2\cos\frac{x-y}{2}\right)^n \cos n\frac{x+y}{2}$$

for all positive integers n and all real numbers x and y.

(Mathematical Reflections, 2009)

Solution. The real number $\sum_{k=0}^{n} \binom{n}{k} \cos[(n-k)x + ky]$ is the real part of the complex number

$$Z = \sum_{k=0}^{n} \binom{n}{k} e^{i((n-k)x+ky)} = \sum_{k=0}^{n} \binom{n}{k} (e^{ix})^{n-k} (e^{iy})^k,$$

where $e^{it} = \cos t + i\sin t$.

From the binomial theorem, we have $Z = (e^{ix} + e^{iy})^n$ which can be rewritten as

$$Z = \left(e^{i\frac{x+y}{2}}\left(e^{i\frac{x-y}{2}} + e^{-i\frac{x-y}{2}}\right)\right)^n = \left(2\cos\frac{x-y}{2}\right)^n e^{ni\frac{x+y}{2}}.$$

Thus, the real part of Z is also

$$\left(2\cos\frac{x-y}{2}\right)^n \cos n\frac{x+y}{2},$$

and the result follows.

Problem 16. *Let k be a fixed positive integer and let*

$$S_n^{(j)} = \binom{n}{j} + \binom{n}{j+k} + \binom{n}{j+2k} + \cdots, \quad j = 0, 1, \ldots, k-1.$$

Prove that

$$\left(S_n^{(0)} + S_n^{(1)} \cos \frac{2\pi}{k} + \ldots + S_n^{(k-1)} \cos \frac{2(k-1)\pi}{k}\right)^2$$

$$+ \left(S_n^{(1)} \sin \frac{2\pi}{k} + S_n^{(2)} \sin \frac{4\pi}{k} + \ldots + S_n^{(k-1)} \sin \frac{2(k-1)\pi}{k}\right)^2 = \left(2 \cos \frac{\pi}{k}\right)^{2n}.$$

(Mathematical Reflections, 2010)

Solution. Let $\mathbb{Z}_+ = \mathbb{N} \cup \{0\}$ and

$$D_j = \{j + mk \mid m \in \mathbb{Z}_+ \text{ and } j + mk \le n\}.$$

Then $S_n^{(j)} = \sum_p \binom{n}{p}$ and $\bigcup_{j=0}^{k-1} D_j = \{0, 1, 2, \ldots, n\}$. Let

$$a = \sum_{j=0}^{k-1} S_n^{(j)} \cos \frac{2j\pi}{k},$$

$$b = \sum_{j=0}^{k-1} S_n^{(j)} \sin \frac{2j\pi}{k},$$

$$\varepsilon = \cos \frac{2\pi}{k} + i \sin \frac{2\pi}{k} \sum_{j=0}^{k-1} S_n^{(j)} \left(\cos \frac{2\pi}{k} + i \sin \frac{2\pi}{k}\right)^j.$$

Then $\varepsilon^k = 1$ and

$$a + ib = \sum_{j=0}^{k-1} S_n^{(j)} \cos \frac{2j\pi}{k} + i \sum_{j=0}^{k-1} S_n^{(j)} \cos \frac{2j\pi}{k}$$

$$= \sum_{j=0}^{k-1} S_n^{(j)} \left(\cos \frac{2j\pi}{k} + i \sin \frac{2j\pi}{k}\right) = \sum_{j=0}^{k-1} S_n^{(j)} \varepsilon^j = \sum_{j=0}^{k-1} \sum_{p \in D_j} \binom{n}{p} \varepsilon^p$$

$$= \sum_{p \in D_j} \binom{n}{p} \varepsilon^p = \sum_{p=1}^{n} \binom{n}{p} \varepsilon^p = (1 + \varepsilon)^n = \left(1 + \cos \frac{2\pi}{k} + i \sin \frac{2\pi}{k}\right)^n$$

$$= \left(2 \cos \frac{\pi}{k} \left(\cos \frac{\pi}{k} + i \sin \frac{\pi}{k}\right)\right)^n = \left(2 \cos \frac{\pi}{k}\right)^n \left(\cos \frac{\pi}{k} + i \sin \frac{\pi}{k}\right)^n.$$

Hence

$$|a+ib| = \left|\left(2\cos\frac{\pi}{k}\right)^n\left(\cos\frac{\pi}{k}+i\sin\frac{\pi}{k}\right)^n\right| = \left|\left(2\cos\frac{\pi}{k}\right)^n\right|\left|\left(\cos\frac{\pi}{k}+i\sin\frac{\pi}{k}\right)^n\right|$$

$$= \left|\left(2\cos\frac{\pi}{k}\right)\right|^n\left|\left(\frac{\pi}{k}+i\sin\frac{\pi}{k}\right)\right|^n = \left|\left(2\cos\frac{\pi}{k}\right)\right|^n.$$

Therefore,

$$a^2 + b^2 = \left(2\cos\frac{\pi}{k}\right)^{2n}.$$

Problem 17.

(a) Let z_1, z_2, z_3, z_4 be distinct complex numbers of zero sum, having equal absolute values. Prove that the points with complex coordinates z_1, z_2, z_3, z_4 are the vertices of a rectangle.

(b) Let x, y, z, t be real numbers such that $\sin x + \sin y + \sin z + \sin t = 0$ and $\cos x + \cos y + \cos z + \cos t = 0$. Prove that for every integer n,

$$\sin(2n+1)x + \sin(2n+1)y + \sin(2n+1)z + \sin(2n+1)t = 0.$$

(Romanian Mathematical Olympiad—District Round, 2011)

Solution.

(a) The equality $z_1 + z_2 + z_3 + z_4 = 0$ implies $\bar{z}_1 + \bar{z}_2 + \bar{z}_3 + \bar{z}_4 = 0$, and furthermore,

$$\frac{1}{z_1} + \frac{1}{z_2} + \frac{1}{z_3} + \frac{1}{z_4} = 0, \tag{1}$$

for $|z_1| = |z_2| = |z_3| = |z_4| \neq 0$.

Suppose $z_1 + z_2 = -z_3 - z_4 \neq 0$. The relation (1) gives $z_1 z_2 = z_3 z_4$, so $\{z_1, z_2\} = \{-z_3, -z_4\}$. On the other hand, if $z_1 + z_2 = 0$, then $z_3 + z_4 = 0$. In both cases, the numbers z_1, z_2, z_3, z_4 form two pairs of equal sum, whence the conclusion.

(b) Let $z_1 = \cos x + i\sin x$, $z_2 = \cos y + i\sin y$, $z_3 = \cos z + i\sin z$, and $z_4 = \cos t + i\sin t$ to get $z_1 + z_2 + z_3 + z_4 = 0$ and $|z_1| = |z_2| = |z_3| = |z_4| = 1$. As before, the numbers z_1, z_2, z_3, z_4 form two pairs of opposite numbers, so the same goes for numbers $z_1^{2n+1}, z_2^{2n+1}, z_3^{2n+1}, z_4^{2n+1}$. Therefore,

$$z_1^{2n+1} + z_2^{2n+1} + z_3^{2n+1} + z_4^{2n+1} = 0,$$

implying the claim.

6.2.6 More on the nth Roots of Unity

Problem 11. *For all positive integers k, define*

$$U_k = \{z \in \mathbb{C} \mid z^k = 1\}.$$

Prove that for every pair of integers m and n with $0 < m < n$, we have

$$U_1 \cup U_2 \cup \cdots \cup U_m \subset U_{n-m+1} \cup U_{n-m+2} \cup \cdots \cup U_n.$$

(Romanian Mathematical Regional Contest "Grigore Moisil," 1997)

Solution. Let $p = 1, 2, \ldots, m$ and let $z \in U_p$. Then $z^p = 1$.

Note that $n-m+1, n-m+2, \ldots, n$ are m consecutive integers, and since $p \le m$, there is an integer $k \in \{n-m+1, n-m+2, \ldots, n\}$ such that p divides k.

Let $k = k'p$. It follows that $z^k = (z^p)^{k'} = 1$, so $z \in U_k \subset U_{n-m+1} \cup U_{n-m+2} \cup \ldots \cup U_n$, as claimed.

Remark. An alternative solution can be obtained from the fact that

$$\frac{(a^n - 1)(a^{n-1} - 1) \cdots (a^{n-k+1} - 1)}{(a^k - 1)(a^{k-1} - 1) \cdots (a - 1)}$$

is an integer for all positive integers $a > 1$ and $n > k$.

Problem 12. *Let a, b, c, d, α be complex numbers such that $|a| = |b| \neq 0$ and $|c| = |d| \neq 0$. Prove that all roots of the equation*

$$c(bx + a\alpha)^n - d(ax + b\overline{\alpha})^n = 0, \ n \ge 1,$$

are real numbers.

Solution. Rewrite the equation as

$$\left(\frac{bx + a\alpha}{ax + b\overline{\alpha}} \right)^n = \frac{d}{c}.$$

Since $|c| = |d|$, we have $\left| \frac{d}{c} \right| = 1$. Consider

$$\frac{d}{c} = \cos t + i \sin t, \quad t \in [0, 2\pi).$$

It follows that

$$\frac{bx_k + a\alpha}{ax_k + b\overline{\alpha}} = u_k, \tag{1}$$

where
$$u_k = \cos \frac{t + 2k\pi}{n} + i \sin \frac{t + 2k\pi}{n}, \quad k = 0, 1, \ldots, n - 1.$$

The relation (1) implies that

$$x_k = \frac{\bar{b}\bar{\alpha}u_k - a\alpha}{b - au_k}, \quad k = 0, 1, \ldots, n - 1.$$

To prove that the roots $x_k, k = \overline{0, n-1}$ are real numbers, it suffices to show that $x_k = \overline{x_k}$ for all $k = 0, 1, \ldots, n - 1$.

Set $|a| = |b| = r$. Then

$$\overline{x_k} = \frac{\bar{b}\alpha\overline{u_k} - \bar{a}\bar{\alpha}}{\bar{b} - \bar{a}\overline{u_k}} = \frac{\frac{r^2}{b} \cdot \alpha \cdot \frac{1}{u_k} - \frac{r^2}{a} \cdot \bar{\alpha}}{\frac{r^2}{b} - \frac{r^2}{a} \cdot \frac{1}{u_k}}$$

$$= \frac{\alpha a - \bar{b}\bar{\alpha}u_k}{au_k - b} = x_k, \quad k = \overline{0, n-1},$$

as desired.

Problem 13. *Suppose that $z \neq 1$ is a complex number such that $z^n = 1$, $n \geq 1$. Prove that*
$$|nz - (n + 2)| \leq \frac{(n + 1)(2n + 1)}{6}|z - 1|^2.$$

(Crux Mathematicorum, 2003)

Solution. Differentiating the familiar identity

$$\sum_{k=0}^{n} z^k = \frac{x^{n+1} - 1}{x - 1}$$

with respect to x, we get

$$\sum_{k=1}^{n} kx^{k-1} = \frac{nx^{n+1} - (n + 1)x^n + 1}{(x - 1)^2}.$$

Multiplying both sides by x and differentiating again, we arrive at

$$\sum_{k=1}^{n} k^2 x^{k-1} = g(x),$$

where

$$g(x) = \frac{n^2 x^{n+2} - (2n^2 + 2n - 1)x^{n+1} + (n + 1)^2 x^n - x - 1}{(x - 1)^3}.$$

Taking $x = z$ and using $|z| = 1$ (which we were given), we obtain

$$|g(z)| \le \sum_{k=1}^{n} k^2 |z|^{k-1} = \frac{n(n+1)(2n+1)}{6}. \tag{1}$$

On the other hand, taking into account that $z^n = 1, z \ne 1$, we get

$$g(z) = \frac{n(nz^2 - 2(n+1)z + n + 2)}{(z-1)^3} = \frac{n(nz - (n+2))}{(z-1)^2}. \tag{2}$$

From (1) and (2), we therefore conclude that

$$|nz - (n+2)| \le \frac{(n+1)(2n+1)}{6} |z - 1|^2.$$

Problem 14. *Let M be a set of complex numbers such that if x, $y \in M$, then $\dfrac{x}{y} \in M$. Prove that if the set M has n elements, then M is the set of the nth roots of 1.*

Solution. Setting $x = y \in M$ yields $1 = \dfrac{x}{y} \in M$. For $x = 1$ and $y \in M$, we obtain $\dfrac{1}{y} = y^{-1} \in M$.

If x and y are arbitrary elements of M, then $x, y^{-1} \in M$, and consequently,

$$\frac{x}{y^{-1}} = xy \in M.$$

Let $x_1, x_2, \ldots, x_n$ be the elements of set M and take at random an element $x_k \in M, k = \overline{1, n}$. Since $x_k \ne 0$ for all $k = \overline{1, n}$, the numbers $x_k x_1, x_k x_2, \ldots, x_k x_n$ are distinct and belong to the set M, whence

$$\{x_k x_1, x_k x_2, \ldots, x_k x_n\} = \{x_1, x_2, \ldots, x_n\}.$$

Therefore, $x_k x_1 \cdot x_k x_2 \cdots x_k x_n = x_1 x_2 \cdots x_n$, and hence $x_k^n = 1$; that is, x_k is an nth root of 1.

The number x_k was arbitrary; hence M is the set of the nth roots of 1, as claimed.

Problem 15. *A finite set A of complex numbers has the following property: $z \in A$ implies $z^n \in A$ for every positive integer n.*

(a) *Prove that $\sum_{z \in A} z$ is an integer.*

(b) *Prove that for every integer k, one can choose a set A that satisfies the above condition and $\sum_{z \in A} z = k$.*

(Romanian Mathematical Olympiad—Final Round, 2003)

Solution.

(a) We will denote by $S(X)$ the sum of the elements of a finite set X. Suppose $0 \neq z \in A$. Since A is finite, there exist positive integers $m < n$ such that $z^m = z^n$, whence $z^{n-m} = 1$. Let d be the smallest positive integer k such that $z^k \in 1$. Then $1, z, z^2, \ldots, z^{d-1}$ are distinct, and the dth power of each is equal to 1; therefore, these numbers are the dth roots of unity. This shows that $A \backslash \{0\} = \bigcup\limits_{k=1}^{m} U_{n_k}$, where $U_p = \{z \in \mathbb{C} | z^p = 1\}$. Since $S(U_p) = 0$ for $p \geq 2$, $S(U_1) = 1$ and $U_p \cap U_q = U_{(p,q)}$, we get

$$S(A) = \sum_k S(U_{n_k}) - \sum_{k<l} S(U_{n_k} \cap U_{n_l})$$

$$+ \sum_{k<l<s} S(U_{n_k} \cap U_{n_l} \cap U_{n_s}) + \cdots = \text{ an integer.}$$

(b) Suppose that for some integer k there exists $A = \bigcup\limits_{k=1}^{m} U_{n_k}$ such that $S(A) = k$. Let $p_1, p_2, \ldots, p_6$ be the distinct primes that are not divisors of any n_k. Then

$$S(A \cup U_{p_1}) = S(A) + S(U_{p_1}) - S(A \cap U_{p_1}) = k - S(U_1) = k - 1.$$

Also,

$$S(A \cup U_{p_1 p_2 p_3} \cup U_{p_1 p_4 p_5} \cup U_{p_2 p_4 p_6} \cup U_{p_3 p_5 p_6})$$

$$= S(A) + S(U_{p_1 p_2 p_3}) + S(U_{p_1 p_4 p_5}) + S(U_{p_2 p_4 p_6}) + S(U_{p_3 p_5 p_6})$$

$$- S(A \cap U_{p_1 p_2 p_3}) - \cdots + S(A \cap U_{p_1 p_2 p_3} \cap U_{p_1 p_4 p_5})$$

$$+ \cdots - S(A \cap U_{p_1 p_2 p_3} \cap U_{p_1 p_4 p_5} \cap U_{p_2 p_4 p_6} \cap U_{p_3 p_5 p_6})$$

$$= k + 4 \cdot 0 - 4S(U_1) - \sum_{k=1}^{6} S(U_{p_k}) + 10S(U_1) - 5S(U_1) + S(U_1)$$

$$= k - 4 + 10 - 5 + 1 = k + 2.$$

Hence, if there exists A such that $S(A) = k$, then there exist B and C such that $S(B) = k - 1$ and $S(C) = k + 2$. The conclusion now follows easily.

Problem 16. *Let $n \geq 3$ be an odd integer. Evaluate* $\sum\limits_{k=1}^{\frac{n-1}{2}} \sec \dfrac{2k\pi}{n}$.

(Mathematical Reflections)

Solution. We will prove that

$$\sum_{k=1}^{\frac{n-1}{2}} \sec\frac{2k\pi}{n} = \begin{cases} \dfrac{n-1}{2}, & \text{if } n \equiv 1 \pmod 4 \\ -\dfrac{n+1}{2}, & \text{if } n \equiv 3 \pmod 4 \end{cases}$$

Let T_n denote Chebyshev's polynomial of the first kind of degree n, which is defined by the formula

$$T_n(\cos\theta) = \cos(n\theta).$$

Since $T_n'(\cos\theta) = n\sin(n\theta)/\sin\theta$, we conclude that

$$\{\cos(k\pi/n) : 1 \le k \le n-1\}$$

are the $n-1$ distinct zeros of T_n', which is then of degree $n-1$. This proves that there exists a constant λ such that

$$T_n'(X) = \lambda \prod_{1 \le k < n} (X - \cos(k\pi/n)),$$

and consequently,

$$\frac{T_n''(X)}{T_n'(X)} = \sum_{k=1}^{n-1} \frac{1}{X - \cos(k\pi/n)}.$$

Noting that $\cos\dfrac{k\pi}{n} = \cos\dfrac{(n-k)\pi}{n}$, we see that

$$\frac{T_n''(X)}{T_n'(X)} = \frac{1}{2}\sum_{k=1}^{n-1}\left(\frac{1}{X - \cos(k\pi/n)} + \frac{1}{X + \cos(k\pi/n)}\right)$$

$$= \sum_{k=1}^{n-1}\frac{X}{X^2 - \cos^2(k\pi/n)},$$

so

$$\frac{T_n''(X)}{T_n'(X)} = \sum_{k=1}^{n-1}\frac{2X}{2X^2 - 1 - \cos(2k\pi/n)},$$

and by substituting $X = \cos\theta$, we get that

$$\frac{T_n''(\cos\theta)}{T_n'(\cos\theta)} = \sum_{k=1}^{n-1}\frac{2\cos\theta}{\cos(2\theta) - \cos(2k\pi/n)}.$$

On the other hand, from $T_n'(\cos\theta) = n\sin(n\theta)/\sin\theta$, we see that

$$-(\sin\theta)\frac{T_n''(\cos\theta)}{T_n'(\cos\theta)} = n\cot(n\theta) - \cot\theta.$$

Therefore, we conclude that

$$\sum_{k=1}^{n-1} \frac{1}{\cos(2\theta) - \cos(2k\pi/n)} = \frac{1}{2\sin^2\theta} - \frac{n\cot(n\theta)}{\sin(2\theta)},$$

and for odd n, this is equivalent to

$$\sum_{k=1}^{\frac{n-1}{2}} \frac{1}{\cos(2k\pi/n) - \cos(2\theta)} = \frac{n\cot(n\theta)}{2\sin(2\theta)} - \frac{1}{4\sin^2\theta}.$$

In particular, taking $\theta = \pi/4$, we obtain

$$\sum_{k=1}^{\frac{n-1}{2}} \frac{1}{\cos(2k\pi/n)} = \frac{n\cot(n\pi/4) - 1}{2} = \frac{n(-1)^{(n-1)/2} - 1}{2},$$

which is the desired conclusion.

Problem 17. *Let n be an odd positive integer and let z be a complex number such that $z^{2n-1} - 1 = 0$. Evaluate*

$$\prod_{k=0}^{n-1} \left(z^{2^k} + \frac{1}{z^{2^k}} - 1 \right).$$

(Mathematical Reflections)

Solution. Let

$$Z_n = \prod_{k=0}^{n-1} \left(z^{2^k} + \frac{1}{z^{2^k}} - 1 \right).$$

We have that

$$\left(z + \frac{1}{z} + 1 \right) Z_n = \left(z^2 + \frac{1}{z^2} + 1 \right) \left(z^2 + \frac{1}{z^2} - 1 \right) \cdots \left(z^{2^{n-1}} + \frac{1}{z^{2^{n-1}}} + 1 \right)$$

$$= \left(z^{2^n} + \frac{1}{z^{2^n}} + 1 \right).$$

However, from the given condition, we have that $z^{2^n} = z$. Finally,

$$\left(z + \frac{1}{z} + 1 \right) Z_n = \left(z + \frac{1}{z} + 1 \right).$$

Hence $Z_n = 1$.

Problem 18. *The expression $\sin 2° \sin 4° \sin 6° \ldots \sin 90°$ is equal to $p\sqrt{5}/2^{50}$, where p is an integer. Find p.*

Solution 1. All trigonometric arguments are in degrees. Notice that

$$\sin(90x) = \text{Im}[(\cos x + i \sin x)^{90}]$$

$$= \sum_{n=0}^{45} (-1)^n \binom{90}{2n+1} \sin^{2n+1}(x) \cos^{90-2x+1}(x)$$

$$= \sin(x)\cos(x) \sum_{n=0}^{44} \binom{90}{2n+1} \sin^{2n}(x)[\sin^2(x) - 1]^{44-n}.$$

Then

$$\frac{\sin(90x)}{\sin(x)\cos(x)} = P(\sin(x))$$

is a polynomial in $\sin(x)$ of degree 88, and it has roots at

$$\sin(x) = \pm\sin(2°), \pm\sin(4°), \ldots, \pm\sin(88°),$$

so it follows that these are exactly the roots of the polynomial. Observe that the constant term of $P(x)$ is 90, while the leading term has coefficient

$$\sum_{n=0}^{44} \binom{90}{2n+1} = \frac{(1+1)^{90} - (1-1)^{90}}{2} = 2^{89}.$$

It follows that

$$\frac{90}{2^{89}} = \prod_{n=-44, n\neq 0}^{44} \sin(2n) = (-1)^{44} \left(\prod_{n=1}^{44} \sin(2n) \right)^2,$$

and thus

$$\sin(90) \prod_{n=1}^{44} \sin(2n) = \sqrt{\frac{45}{2^{88}}} = \frac{3\sqrt{5}}{2^{44}}.$$

Then $p = 3 \cdot 2^6 = 192$.

Solution 2. Let $\omega = \cos\dfrac{2\pi}{90} + i\sin\dfrac{2\pi}{90}$. We have

$$\prod_{n=1}^{45} \sin(2n) = \sum_{n=1}^{45} \frac{\omega^n - 1}{2i\omega^{n/2}}.$$

By the symmetry of the sine (and the fact that $\sin(90) = 1$),

$$\prod_{n=1}^{45} \sin(2n) = \prod_{n=46}^{89} \sin(2n),$$

so

$$\left|\prod_{n=1}^{45} \sin(2n)\right|^2 = \sum_{n=1}^{89} \frac{|\omega^n - 1|}{2} = \frac{90}{2^{89}},$$

where we have used the usual geometric-series sum for roots of unity. The product is clearly positive and real, so it is equal to

$$\frac{\sqrt{45}}{2^{44}} = \frac{3\sqrt{5}}{2^{44}};$$

hence $p = 3 \cdot 2^6 = 192$.

Problem 19. *The polynomial $P(x) = (1 + x + x^2 + \ldots + x^{17})^2 - x^{17}$ has 34 complex roots of the form*

$$z_k = r_k[\cos(2\pi a_k) + i\sin(2\pi a_k)], \quad k = 1, 2, 3, \ldots, 34,$$

with $0 < a_1 \le a_2 \le a_3 \le \ldots \le a_{34} < 1$ and $r_k > 0$. Given that $a_1 + a_2 + a_3 + a_4 + a_5 = m/n$, where m and n are relatively prime positive integers, find $m + n$.

(2004 AIME I, Problem 13)

Solution. We see that the expression for the polynomial P is very difficult to work with directly, but there is one obvious transformation to make, sum the geometric series:

$$P(x) = \left(\frac{x^{18} - 1}{x - 1}\right)^2 - x^{17} = \frac{x^{36} - 2x^{18+1} + 1}{x^2 - 2x + 1} - x^{17}$$

$$= \frac{x^{36} - x^{19} - x^{17} + 1}{(x - 1)^2} = \frac{(x^{19} - 1)(x^{17} - 1)}{(x - 1)^2}.$$

This expression has roots at every 17th root and 19th root of unity other than 1. Since 17 and 19 are relatively prime, this means that there are no duplicate roots. Thus, a_1, a_2, a_3, a_4, a_5 are the five smallest fractions of the form $\frac{m}{19}$ or $\frac{n}{17}$ for $m, n > 0$.

Now, $\frac{3}{17}$ and $\frac{4}{19}$ can both be seen to be larger than any of $\frac{1}{19}, \frac{2}{19}, \frac{3}{19}, \frac{1}{17}, \frac{2}{17}$, so these latter five are the numbers we want to add:

$$\frac{1}{19} + \frac{2}{19} + \frac{3}{19} + \frac{1}{17} + \frac{2}{17} = \frac{6}{19} + \frac{3}{17} = \frac{6 \cdot 17 + 3 \cdot 1}{17 \cdot 19} = \frac{159}{323},$$

and so the answer is $159 + 323 = 482$.

Problem 20. *The sets $A = \{z : z^{18} = 1\}$ and $B = \{w : w^{48} = 1\}$ are both sets of complex roots of unity. The set $C = \{zw : a \in A \text{ and } w \in B\}$ is also a set of complex roots of unity. How many distinct elements are in C?*

Solution 1. The least common multiple of 18 and 48 is 144, so define

$$n = \cos \frac{2\pi}{144} + i \sin \frac{2\pi}{144}.$$

We can write the numbers of the set A as $\{n^8, n^{18}, \ldots, n^{144}\}$ and of the set B as $\{n^3, n^6, \ldots, n^{144}\}$. Now, n^x can yield at most 144 different values. All solutions for zw will be in the form of $n^{8k_1 + 3k_2}$.

Note that 8 and 3 are relatively prime, and it is well known that for two relatively prime integers a, b, the largest number that cannot be expressed as the sum of multiples of a, b is $ab - a - b$. For 3, 8, this is 13; however, we can easily see that the numbers 145 to 157 can be written in terms of 3, 8. Since the exponents are of roots of unity, they reduce modulo 144, so all numbers in the range are covered. Thus the answer is 144.

Solution 2. The 18th and 48th roots of 1 can be found by de Moivre's theorem. They are $\operatorname{cis}\left(\dfrac{2k_1\pi}{18}\right)$ and $\operatorname{cis}\left(\dfrac{2k_2\pi}{48}\right)$ respectively, where $\operatorname{cis}\theta = \cos\theta + i\sin\theta$ and k_1 and k_2 are integers from 0 to 17 and 0 to 47, respectively:

$$zw = \operatorname{cis}\left(\frac{k_1\pi}{9} + \frac{k_2\pi}{24}\right) = \operatorname{cis}\left(\frac{8k_1\pi + 3k_2\pi}{72}\right).$$

Since the trigonometric functions are periodic with period 2π, there are at most $72 \cdot 2 = 144$ distinct elements in C. As above, all of these will work.

Problem 21. *Let $n \geq 3$ be an integer and $z = \cos\dfrac{2\pi}{n} + i\sin\dfrac{2\pi}{n}$. Consider the sets*

$$A = \{1, z, z^2, \ldots, z^{n-1}\}$$

and

$$B = \{1, 1+z, 1+z+z^2, \ldots, 1+z+\ldots+z^{n-1}\}.$$

Determine $A \cap B$.

(Romanian Mathematica Olympiad—District Round, 2008)

Solution. Clearly, $1 \in A \cap B$. Let $w \in A \cap B$, $w \neq 1$. As a member of B,

$$w = 1 + z + \ldots + z^k = \frac{1 - z^{k+1}}{1 - z}$$

for some $k = 1, 2, \ldots, n - 1$. Since $w \in A$, we get

$$|w| = 1 \quad \text{and} \quad |1 - z^{k+1}| = |1 - z|.$$

The latter equality yields

$$\sin\frac{(k+1)\pi}{n} = \sin\frac{\pi}{n}, \quad \text{or} \quad \frac{(k+1)\pi}{n} = \pi - \frac{\pi}{n},$$

which implies $k = n - 2$, so

$$w = \frac{1 - \dfrac{1}{z}}{1 - z} = -\frac{1}{z}.$$

because $w \in A$, we must have $w^n = 1$, which means that n has to be even. So the answer is $A \cap B = \{1\}$ for odd n, and $A \cap B = \left\{1, -\dfrac{1}{z}\right\}$ for even n.

6.2.7 Problems Involving Polygons

Problem 12. *Prove that there exists a convex 1990-gon with the following two properties:*

(a) all angles are equal;
(b) the lengths of the sides are the numbers $1^2, 2^2, 3^2, \ldots, 1989^2, 1990^2$ in some order.

(31st IMO)

Solution. Suppose that such a 1990-gon exists and let $A_0, A_1, \ldots, A_{1989}$ be its vertices. The sides $A_k A_{k+1}$, $k = 0, 1, \ldots, 1989$ define the vectors $\overrightarrow{A_k A_{k+1}}$, which can be represented in the complex plane by the numbers

$$z_k = n_k w^k, \quad k = 0, 1, \ldots, 1989,$$

where $w = \cos\dfrac{2\pi}{1990} + i\sin\dfrac{2\pi}{1990}$. Here $A_{1990} = A_0$, and $n_0, n_1, \ldots, n_{1989}$ represents a permutation of the numbers $1^2, 2^2, \ldots, 1990^2$.

Because $\sum\limits_{k=0}^{1989} \overrightarrow{A_k A_{k+1}} = 0$, the problem can be restated as follows: find a permutation $(n_0, n_1, \ldots, n_{1989})$ of the numbers $1^2, 2^2, \ldots, 1990^2$ such that

$$\sum_{k=0}^{1989} n_k w^k = 0.$$

Observe that $1990 = 2 \cdot 5 \cdot 199$. The strategy is to add vectors after a suitable grouping of 2, 5, 199 vectors such that these partial sums can be directed toward a suitable result.

To begin, let us consider the pairing of numbers

$$(1^2, 2^2), (3^2, 4^2), \ldots, (1988^2, 1989^2)$$

and assign these lengths to pairs of opposite vectors respectively:

$$(w_k, w_{k+995}), \quad k = 0, \ldots, 994.$$

By adding the obtained vectors, we obtain 995 vectors of lengths

$$2^2 - 1^2 = 3; 4^2 - 3^2 = 7; 6^2 - 5^2 = 11; \ldots; 1989^2 - 1988^2 = 3979,$$

which divide the unit circle of the complex plane into 995 equal arcs.

Let $B_0 = 1, B_1, \ldots, B_{994}$ be the vertices of the regular 995-gon inscribed in the unit circle. We intend to assign the lengths $3, 7, 11, \ldots, 3979$ to the unit vectors $\overrightarrow{OB_0}, \overrightarrow{OB_1}, \ldots, \overrightarrow{OB_{994}}$ such that the sum of the obtained vectors is zero.

We divide 995 lengths into 199 groups of size 5:

$$(3, 7, 11, 15, 19), (23, 27, 31, 35, 39), \ldots, (3963, 3967, 3971, 3975, 3979).$$

Let $\zeta = \cos \dfrac{2\pi}{5} + i \sin \dfrac{2\pi}{5}, \omega = \cos \dfrac{2\pi}{199} + i \sin \dfrac{2\pi}{199}$ be the primitive roots of unity of order 5 and 199, respectively. Let $\mathcal{P}_1$ be the pentagon with vertices 1, $\zeta, \zeta^2, \zeta^3, \zeta^4$. Then we rotate $\mathcal{P}_1$ about the origin O with coordinates through angles $\theta_k = \dfrac{2k\pi}{199}, k = 1, \ldots, 198$, to obtain new pentagons $\mathcal{P}_2, \ldots, \mathcal{P}_{198}$, respectively. The vertices of $\mathcal{P}_{k+1}$ are $\omega^k, \omega^k\zeta, \omega^k\zeta^2, \omega^k\zeta^3, \omega^k\zeta^4, k = 0, \ldots, 198$. We assign to unit vectors defined by the vertices $\mathcal{P}_k$ of the respective lengths (Fig. 6.7)

$$2k + 3, 2k + 7, 2k + 11, 2k + 15, 2k + 19(k = 0, \ldots, 198).$$

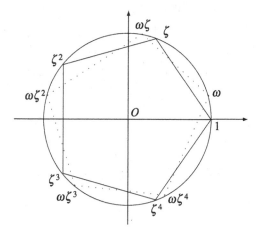

Figure 6.7.

Thus, we have to evaluate the sum

$$\sum_{k=0}^{198}[(2k+3)\omega^k+(2k+7)\omega^k\zeta+(2k+11)\omega^k\zeta^2+(2k+15)\omega^k\zeta^3+(2k+19)\omega^k\zeta^4]$$

$$=\sum_{k=0}^{198}2k\omega^k(1+\zeta+\zeta^2+\zeta^3+\zeta^4)+(3+7\zeta+11\zeta^2+15\zeta^3+19\zeta^4)\sum_{k=0}^{198}\omega^k.$$

Since $1+\zeta+\zeta^2+\zeta^3+\zeta^4=0$ and $1+\omega+\omega^2+\cdots+\omega^{198}=0$, it follows that the sum equals zero.

Problem 13. *Let A and E be opposite vertices of a regular octagon. Let a_n be the number of paths of length n of the form $(P_0, P_1, \ldots, P_n)$, where P_i are vertices of the octagon, and the paths are constructed using the rule $P_0 = A$, $P_n = E$, P_i and P_{i+1} are adjacent vertices for $i = 0, \ldots, n-1$, and $P_i \neq E$ for $i = 0, \ldots, n-1$.*
Prove that $a_{2n-1} = 0$ and $a_{2n} = \frac{1}{\sqrt{2}}(x^{n-1} - y^{n-1})$, for all $n = 1, 2, 3, \ldots,$ where $x = 2 + \sqrt{2}$ and $y = 2 - \sqrt{2}$.

(21st IMO)

Solution. It is convenient to take a regular octagon inscribed in a circle and label its vertices as follows:

$$A = A_0, A_1, A_2, A_3, A_4 = E, A_{-3}, A_{-2}, A_{-1}.$$

We imagine a step in the path to be rotation of angle $\dfrac{2\pi}{8} = \dfrac{\pi}{4}$ about the center O of the circumscribed circle of the octagon. In this way, a path is a sequence of such rotations, subject to certain conditions. If the rotation is counterclockwise, we add the angle $\dfrac{\pi}{4}$. If the rotation is clockwise, we add the angle $-\dfrac{\pi}{4}$. The starting point is A_0, which is represented by the complex number $z_0 = \cos 0 + i \sin 0$. Each vertex A_k of the octagon is represented by $z_k = \cos\dfrac{2k\pi}{8} + i\sin\dfrac{2k\pi}{8}$. It is convenient to work only with the angles $\dfrac{2k\pi}{8}, -4 \leq k \leq 4$. But these k's are integers considered modulo 8, so that $z_4 = z_{-4}$ and $A_4 = A_{-4}$ (Fig. 6.8).

We may associate to a path of length n, say $(P_0 P_1 \cdots P_n)$, an ordered sequence $(u, u_2, \ldots, u_n)$ of integers that satisfy the following conditions:

(a) $u_k = \pm 1$ for $k = 1, 2, \ldots, n$; more precisely, $u_i = \pm 1$ if the arc $(P_{k-1}P_k)$ is $\dfrac{\pi}{4}$, and $u_k = -1$ if the arc $(P_{k-1}P_k)$ is $-\dfrac{\pi}{4}$;
(b) $u_1 + u_2 + \cdots + u_k \in \{-3, -2, -1, 0, 1, 2, 3\}$ for all $k = 1, 2, \ldots, n-1$;
(c) $u_1 + u_2 + \cdots + u_n = \pm 4$.

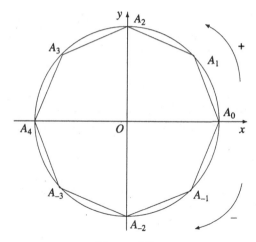

Figure 6.8.

For example, the sequence associated with the path $(A_0, A_{-1}, A_0, A_1, A_2, A_3, A_4)$ is $(-1, 1, 1, 1, 1, 1)$. From now on, we consider only sequences that satisfy (a)–(c). It is obvious that conditions (a)–(c) define a bijective function between the set of paths and the set of sequences.

For every sequence $u_1, u_2, \ldots, u_n$ and every k, $1 \leq k \leq n$, we call the sum $s_k = u_1 + u_2 + \cdots + u_k$ a partial sum of the sequence. It is easy to see that for each k, s_k is an even number if and only if k is even. Thus, $a_{2n-1} = 0$. Thus we have to prove the formula for even numbers. For small n, we have $a_2 = 0$, $a_4 = 2$; for example, only sequences $(1, 1, 1, 1)$ and $(-1, -1, -1, -1)$ of length 4 satisfy conditions (a)–(c).

In the following, we will prove a recurrence relation between the numbers a_n, n even. The first step is to observe that if $s_n = \pm 4$, then $s_{n-2} = \pm 2$. Moreover, if $(u_1, u_2, \ldots, u_{n-2})$ is a sequence that satisfies (a), (b), and $s_{n-2} = \pm 2$, there are only two ways to extend it to a sequence that satisfies (c) as well: either the sequence $(u_1, u_2, \ldots, u_{n-2}, +1, +1)$ or the sequence $(u_1, u_2, \ldots, u_{n-2}, -1, -1)$. So if we denote by x_n the number of sequences that satisfy (a), (b), and $s_n = \pm 2$, then n is even, and $a_n = x_{n-2}$.

Let y_n denote the number of sequences that satisfy (a), (b), and $s_n = 0$. Then n is even, and we have the equality

$$y_n = x_{n-2} + 2y_{n-2}. \tag{1}$$

This equality comes from the following constructions.

A sequence $(u_1, \ldots, u_{n-2})$ for which $s_{n-2} = \pm 2$ gives rise to a unique sequence of length n with $s_n = 0$ by extending it either to $(u_1, \ldots, u_{n-2}, 1, 1)$ or to $(u_1, u_2, \ldots, u_{n-1}, -1, -1)$. Also, a sequence $(u_1, \ldots, u_{n-2})$ with $s_{n-2} = 0$ gives rise either to the sequence $(u_1, \ldots, u_{n-2}, 1, -1)$ or to

$(u_1, \ldots, u_{n-2}, -1, 1)$. Finally, every sequence of length n with $s_n = 0$ ends in one of the following "terminations": $(-1, -1), (1, 1), (1, -1), (-1, 1)$.

The following equality is also satisfied:

$$x_n = 2x_{n-2} + 2y_{n-2}. \tag{2}$$

This corresponds to the property that every sequence of length n for which $s_n = \pm 2$ can be obtained either from a similar sequence of length $n - 2$ by adding the termination $(1, -1)$ or the termination $(-1, 1)$, or from a sequence of length $n - 2$ for which $s_{n-2} = 0$ by adding the termination $(1,1)$ or the termination $(-1, -1)$.

Now the problem is to derive $a_n = x_{n-2}$ from relations (1) and (2). By subtracting (1) from (2), we obtain $x_{n-2} = x_n - y_n$, for all $n \geq 4, n$ even. Thus, $y_{n-2} = x_{n-2} - x_{n-4}$. Substituting the last equality in (2), we obtain the recurrence relation $x_n = 4x_{n-2} - 2x_{n-4}$, for all $n \geq 4, n$ even. Taking into account that $x_n = a_{n+2}$, we obtain the linear recurrence relation

$$a_{n+2} = 4a_n - 2a_{n-2}, \quad n \geq 4, \tag{3}$$

with the initial values $a_2 = 0, a_4 = 2$.

The sequence $(a_n), n \geq 2, n$ even is uniquely defined by $a_2 = 0, a_4 = 2$ and the relation (3). Therefore, to answer the question, it is sufficient to prove that the sequence $(c_{2n})_{n \geq 1}, c_{2n} = \dfrac{1}{\sqrt{2}}((2 + \sqrt{2})^{n-1} - (2 - \sqrt{2})^{n-1})$ obeys the same conditions. This is a straightforward computation.

Problem 14. *Let A, B, C be three consecutive vertices of a regular polygon and let us consider a point M on the major arc AC of the circumcircle.*

Prove that

$$MA \cdot MC = MB^2 - AB^2.$$

Solution. Consider the complex plane with origin at the center of the polygon. Without loss of generality we may assume that the coordinates of A, B, C are $1, \varepsilon, \varepsilon^2$, respectively, where $\varepsilon = \cos \dfrac{2\pi}{n} + i \sin \dfrac{2\pi}{n}$.

Let $z_M = \cos t + i \sin t, t \in [0, 2\pi)$ be the coordinate of point M. From the hypothesis, we derive that $t > \dfrac{4\pi}{n}$. Then

$$MA = |z_M - 1| = \sqrt{(\cos t - 1)^2 + \sin^2 t} = \sqrt{2 - 2\cos t} = 2\sin \frac{t}{2};$$

$$MB = |z_M - \varepsilon| = \sqrt{2 - 2\cos \left(t - \frac{2\pi}{n} \right)} = 2\sin \left(\frac{t}{2} - \frac{\pi}{n} \right);$$

$$MC = |z_M - \varepsilon^2| = \sqrt{2 - 2\cos \left(t - \frac{4\pi}{n} \right)} = 2\sin \left(\frac{t}{2} - \frac{2\pi}{n} \right);$$

$$AB = |\varepsilon - 1| = \sqrt{2 - 2\cos\frac{2\pi}{n}} = 2\sin\frac{\pi}{n}.$$

We have

$$MB^2 - AB^2 = 4\sin^2\left(\frac{t}{2} - \frac{\pi}{n}\right) - 4\sin^2\frac{\pi}{n}$$

$$= 2\left(\cos\frac{2\pi}{n} - \cos\left(t - \frac{2\pi}{n}\right)\right)$$

$$= -2 \cdot 2\sin\frac{\frac{2\pi}{n} - \left(t - \frac{2\pi}{n}\right)}{2}\sin\frac{\frac{2\pi}{n} + \left(t - \frac{2\pi}{n}\right)}{2}$$

$$= 2\sin\frac{t}{2} \cdot 2\sin\left(\frac{t}{2} - \frac{2\pi}{n}\right) = MA \cdot MC,$$

as desired.

Problem 15. Let $A_1 A_2 \cdots A_n$ be a regular polygon inscribed in a circle C of radius 1. Find the maximum value of $\prod_{j=1}^{n} PA_j$, where P is an arbitrary point on circle C.

(Romanian Mathematical Regional Contest "Grigore Moisil," 1992)

Solution. Rotate the polygon $A_1 A_2 \cdots A_n$ so that the coordinates of its vertices are the complex roots of unity of order n: $\varepsilon_1, \varepsilon_2, \ldots, \varepsilon_n$. Let z be the coordinate of point P located on the circumcircle of the polygon and note that $|z| = 1$.

The equality

$$z^n - 1 = \prod_{j=1}^{n}(z - \varepsilon_j)$$

yields

$$|z^n - 1| = \prod_{j=1}^{n}|z - \varepsilon_j| = \prod_{j=1}^{n} PA_j.$$

Since $|z^n - 1| \le |z|^n + 1 = 2$, it follows that the maximal value of $\prod_{j=1}^{n} PA_j^2$ is 2 and is attained for $z^n = -1$, i.e., for the midpoints of arcs $A_j A_{j+1}, j = 1, \ldots, n$, where $A_{n+1} = A_1$.

Problem 16. Let $A_1 A_2 \cdots A_{2n}$ be a regular polygon with circumradius equal to 1 and consider a point P on the circumcircle. Prove that

$$\sum_{k=0}^{n-1} PA_{k+1}^2 \cdot PA_{n+k+1}^2 = 2n.$$

Solution. Without loss of generality, assume that points A_k have coordinates ε^{k-1} for $k = 1, \ldots, 2n$, where

$$\varepsilon = \cos\frac{\pi}{n} + i\sin\frac{\pi}{n}.$$

Let α be the coordinate of the point P, $|\alpha| = 1$. We have

$$PA_{k+1} = |\alpha - \varepsilon^k|$$

and

$$PA_{n+k+1} = |\alpha - \varepsilon^{n+k}| = |\alpha + \varepsilon^k|,$$

for $k = 0, \ldots, n - 1$. Then

$$\sum_{k=0}^{n-1} PA_{k+1}^2 \cdot PA_{n+k+1}^2 = \sum_{k=0}^{n-1} |\alpha - \varepsilon^k|^2 \cdot |\alpha + \varepsilon^k|^2$$

$$= \sum_{k=0}^{n-1} [(\alpha - \varepsilon^k)(\overline{\alpha} - \overline{\varepsilon}^k)][(\alpha + \varepsilon^k)(\overline{\alpha} + \overline{\varepsilon}^k)]$$

$$= \sum_{k=0}^{n-1} (2 - \alpha\overline{\varepsilon}^k - \overline{\alpha}\varepsilon^k)(2 + \alpha\overline{\varepsilon}^k + \overline{\alpha}\varepsilon^k)$$

$$= \sum_{k=0}^{n-1} (2 - \alpha^2\overline{\varepsilon}^{2k} - \overline{\alpha}^2\varepsilon^{2k}) = 2n - \alpha^2 \sum_{k=0}^{n-1} \overline{\varepsilon}^{2k} - \overline{\alpha}^2 \cdot \sum_{k=0}^{n-1} \varepsilon^{2k}$$

$$= 2n - \alpha \cdot \frac{\overline{\varepsilon}^{2n} - 1}{\overline{\varepsilon}^2 - 1} - \overline{\alpha}^2 \cdot \frac{\varepsilon^{2n} - 1}{\varepsilon^2 - 1} = 2n,$$

as desired.

Problem 17. Let $A_1 A_2 \ldots A_n$ be a regular n-gon inscribed in a circle with center O and radius R. Prove that for each point M in the plane of the n-gon, the following inequality holds:

$$\prod_{k=1}^{n} MA_k \leq (OM^2 + R^2)^{\frac{n}{2}}.$$

(Mathematical Reflections, 2009)

Solution. Let us work in the complex plane with O as the origin and without loss of generality, $R = 1$. Let

$$\omega = \cos\frac{2\pi}{n} + i\sin\frac{2\pi}{n},$$

and let the complex numbers $\omega, \omega^2, \ldots, \omega^n, x$ correspond to the points $A_1, A_2, \ldots, A_n, M$, respectively. Then our inequality is equivalent to

$$\prod_{k=1}^{n} |x - \omega^k| \leq \sqrt{(|x|^2 + 1)^n}.$$

Since $\omega, \omega^2, \ldots, \omega^n$ are the roots of $z^n - 1 = 0$, we have

$$\prod_{k=1}^{n} |x - \omega^k| = |x^n - 1| \leq |x|^n + 1,$$

by the triangle inequality. Hence it remains to show that

$$(|x|^n + 1)^2 \leq (|x|^2 + 1)^n \Leftrightarrow 2|x|^n \leq \sum_{k=1}^{n-1} \binom{n}{k} |x|^{2k},$$

which follows from AM-GM inequality, since $n \geq 3$ and

$$\sum_{k=1}^{n-1} \binom{n}{k} |x|^{2k} \geq n|x|^2 + n|x|^{2n-2} \geq 2n|x|^n + 2|x|^n.$$

Equality holds iff $|x| = 0$ i.e., when $M \equiv O$.

6.2.8 Complex Numbers and Combinatorics

Problem 11. *Calculate the sum* $s_n = \sum_{k=0}^{n} \binom{n}{k}^2 \cos kt$, *where* $t \in [0, \pi]$.

Solution. Let us consider the complex number $z = \cos t + i \sin t$ and the sum $t_n = \sum_{k=0}^{n} \binom{n}{k}^2 \sin kt$. Observe that

$$s_n + it_n = \sum_{k=0}^{n} \binom{n}{k}^2 (\cos kt + i \sin kt) = \sum_{k=0}^{n} \binom{n}{k}^2 (\cos t + i \sin t)^k.$$

In the product $(1 + X)^n (1 + zX)^n = (1 + (z + 1)X + zX^2)^n$, we set the coefficient of X^n equal to obtain

$$\sum_{\substack{0 \leq k, s \leq n \\ k+s=n}} \binom{n}{k}\binom{n}{s} z^s = \sum_{\substack{0 \leq k, s, r \leq n \\ k+s+r=n \\ s+2r=n}} \frac{n!}{k! s! r!} (z + 1)^s z^r. \qquad (1)$$

The above relation is equivalent to

$$\sum_{k=0}^{n} \binom{n}{k}^2 z^k = \sum_{k=0}^{[\frac{n}{2}]} \binom{n}{2k} \binom{2k}{k} (z+1)^{n-2k} z^k. \tag{2}$$

The trigonometric form of the complex number $1 + z$ is given by

$$1 + \cos t + i \sin t = 2 \cos^2 \frac{t}{2} + 2i \sin \frac{t}{2} \cos \frac{t}{2} = 2 \cos \frac{t}{2} \left(\cos \frac{t}{2} + i \sin \frac{t}{2} \right),$$

since $t \in [0, \pi]$. From (2), it follows that

$$s_n + it_n = \sum_{k=0}^{[\frac{n}{2}]} \binom{n}{2k} \binom{2k}{k} \left(2 \cos \frac{t}{2} \right)^{n-2k} \left(\cos \frac{nt}{2} + i \sin \frac{nt}{2} \right);$$

hence

$$s_n = \sum_{k=0}^{[\frac{n}{2}]} \binom{n}{2k} \binom{2k}{k} \left(2 \cos \frac{t}{2} \right)^{n-2k} \cos \frac{nt}{2},$$

$$t_n = \sum_{k=0}^{[\frac{n}{2}]} \binom{n}{2k} \binom{2k}{k} \left(2 \cos \frac{t}{2} \right)^{n-2k} \sin \frac{nt}{2}.$$

Remark. Here we have a few particular cases of (2).

(1) If $z = 1$, then

$$\sum_{k=0}^{n} \binom{n}{k}^2 = \sum_{k=0}^{[\frac{n}{2}]} \binom{n}{2k} \binom{2k}{k} 2^{n-2k} = \binom{2n}{n}.$$

(2) If $z = -1$, then

$$\sum_{k=0}^{n} (-1)^k \binom{n}{k}^2 = \begin{cases} 0 & \text{if } n \text{ is odd,} \\ (-1)^{\frac{n}{2}} \binom{n}{n/2}, & \text{if } n \text{ is even.} \end{cases}$$

(3) If $z = -\frac{1}{2}$, then

$$\sum_{k=0}^{n} (-1)^k \binom{n}{k}^2 2^{n-k} = \sum_{k=0}^{[\frac{n}{2}]} (-1)^k \binom{n}{2k} \binom{2k}{k} 2^k.$$

Problem 12. *Prove the following identities:*

(1) $\dbinom{n}{0} + \dbinom{n}{4} + \dbinom{n}{8} + \cdots = \frac{1}{4}\left(2^n + 2^{\frac{n}{2}+1}\cos\frac{n\pi}{4}\right).$

(Romanian Mathematical Olympiad—Second Round, 1981)

(2) $\dbinom{n}{0} + \dbinom{n}{5} + \dbinom{n}{10} + \cdots$

$$= \frac{1}{5}\left[2^n + \frac{(\sqrt{5}+1)^n}{2^{n-1}}\cos\frac{n\pi}{5} + \frac{(\sqrt{5}-1)^n}{2^{n-1}}\cos\frac{2n\pi}{5}\right].$$

Solution.

(1) In Problem 4, consider $p = 4$ to obtain

$$\dbinom{n}{0} + \dbinom{n}{4} + \dbinom{n}{8} + \cdots = \frac{2^n}{4}\left(1 + 2\left(\cos\frac{\pi}{4}\right)^n\cos\frac{n\pi}{4}\right)$$

$$= \frac{1}{4}\left(2^n + 2^{\frac{n}{2}+1}\cos\frac{n\pi}{4}\right).$$

(2) Let us consider $p = 5$ in Problem 4. We find that

$$\dbinom{n}{0} + \dbinom{n}{4} + \dbinom{n}{8} + \cdots$$

$$= \frac{2^n}{5}\left(1 + 2\left(\cos\frac{\pi}{5}\right)^n\cos\frac{n\pi}{5} + 2\left(\cos\frac{2\pi}{5}\right)^n\cos\frac{2n\pi}{5}\right).$$

Using the well-known relations

$$\cos\frac{\pi}{5} = \frac{\sqrt{5}+1}{4} \quad \text{and} \quad \cos\frac{2\pi}{5} = \frac{\sqrt{5}-1}{4},$$

the desired identity follows.

Problem 13. *Consider the integers A_n, B_n, C_n defined by*

$$A_n = \dbinom{n}{0} - \dbinom{n}{3} + \dbinom{n}{6} - \cdots,$$

$$B_n = -\dbinom{n}{1} + \dbinom{n}{4} - \dbinom{n}{7} + \cdots,$$

$$C_n = \dbinom{n}{2} - \dbinom{n}{5} + \dbinom{n}{8} - \cdots.$$

The following identities hold:

(1) $A_n^2 + B_n^2 + C_n^2 - A_nB_n - B_nC_n - C_nA_n = 3^n;$
(2) $A_n^2 + A_nB_n + B_n^2 = 3^{n-1}.$

Solution.

(1) Let ε be a cube root of unity different from 1. We have

$$(1 - \varepsilon)^n = A_n + B_n\varepsilon + C_n\varepsilon^2, (1 - \varepsilon^2)^n = A_n + B_n\varepsilon^2 + C_n\varepsilon.$$

Hence

$$A_n^2 + B_n^2 + C_n^2 - A_n B_n - B_n C_n - C_n A_n = (A_n + B_n\varepsilon + C_n\varepsilon^2)(A_n + B_n\varepsilon^2 + C_n\varepsilon)$$

$$= (1 - \varepsilon)^n(1 - \varepsilon^2)^n = (1 - \varepsilon - \varepsilon^2 + 1)^n = 3^n.$$

(2) It is obvious that $A_n + B_n + C_n = 0$. Replacing $C_n = -(A_n + B_n)$ in the previous identity, we get $A_n^2 + A_n B_n + C_n^2 = 3^{n-1}$.

Problem 14. *Let $p \geq 3$ be a prime and let m, n be positive integers divisible by p such that n is odd. For each m-tuple $(c_1, \ldots, c_m)$, $c_i \in \{1, 2, \ldots, n\}$, with the property that $p | \sum_{i=1}^{m} c_i$, let us consider the product $c_1 \cdots c_m$. Prove that the sum of all these products is divisible by $\left(\dfrac{n}{p}\right)^m$.*

Solution. For $k \in \{0, 1, \ldots, p - 1\}$, consider $x_k = \sum c_1 \cdots c_m$, the sum of all products $c_1 \cdots c_m$ such that $c_i \in \{1, 2, \ldots, n\}$ and $\sum_{i=1}^{m} c_i \equiv k \pmod{p}$.

If $\varepsilon = \cos\dfrac{2\pi}{p} + i\sin\dfrac{2\pi}{p}$, then

$$(\varepsilon + 2\varepsilon^2 + \cdots + n\varepsilon^n)^m = \sum_{c_1, \ldots, c_m \in \{12, \ldots, n\}} c_1 \ldots c_m \varepsilon^{c1 + \cdots + cm} = \sum_{k=0}^{p-1} x_k \varepsilon^k.$$

Taking into account the relation

$$\varepsilon + 2\varepsilon^2 + \cdots + n\varepsilon^n = \frac{n\varepsilon^{n+2} - (n+1)\varepsilon^{n+1} + \varepsilon}{(\varepsilon - 1)^2} = \frac{n\varepsilon}{\varepsilon - 1}$$

(see Problem 9 in Sect. 5.4 or Problem 13 in Sect. 5.5), it follows that

$$\frac{n^m}{(\varepsilon - 1)^m} = \sum_{k=0}^{p-1} x_k \varepsilon^k. \tag{1}$$

On the other hand, from $\varepsilon^{p-1} + \cdots + \varepsilon + 1 = 0$, we obtain that

$$\frac{1}{\varepsilon - 1} = -\frac{1}{p}(\varepsilon^{p-2} + 2\varepsilon^{p-3} + \cdots + (p-2)\varepsilon + p - 1);$$

hence

$$\frac{n^m}{(\varepsilon - 1)^m} = \left(-\frac{n}{p}\right)(\varepsilon^{p-2} + 2\varepsilon^{p-3} + \cdots + (p-2)\varepsilon + p - 1)^m.$$

Put

$$(X^{p-2} + 2X^{p-3} + \cdots + (p-2)X + p - 1)^m = b_0 + b_1 X + \cdots + b_{m(p-2)} X^{m(p-2)},$$

and obtain

$$\frac{n^m}{(\varepsilon - 1)^m} = \left(-\frac{n}{p}\right)^m (y_0 + y_1\varepsilon + \cdots + y_{p-1}\varepsilon^{p-1}), \tag{2}$$

where $y_j = \displaystyle\sum_{k \equiv j (\mathrm{mod}\ p)} b_k$.

From (1) and (2), we get

$$x_0 - ry_0 + (x_1 - ry_1)\varepsilon + \cdots + (x_{p-1} - ry_{p-1})\varepsilon^{p-1} = 0,$$

where $r = \left(-\dfrac{n}{p}\right)^m$. From Proposition 4 in Sect. 2.2.2, it follows that $x_0 - ry_0 = x_1 - ry_1 = \cdots = x_{p-1} - ry_{p-1} = k$. Now it is sufficient to show that $r|k$. But

$$pk = x_0 + \cdots + x_{p-1} - r(y_0 + \cdots + y_{p-1})$$
$$= (1 + 2 + \cdots + n)^m - r(b_0 + \cdots + b_{m(p-2)})$$
$$= (1 + 2 + \cdots + n)^m - r(1 + 2 + \cdots + (p-1))^m,$$

and we obtain

$$pk = \left(\frac{n(n+1)}{2}\right)^m - r\left(\frac{p(p-1)}{2}\right)^m.$$

Since the right-hand side is divisible by pr, it follows that $r|k$.

Problem 15. *Let k be a positive integer and $a = 4k - 1$. Prove that for every positive integer n, the integer*

$$s_n = \binom{n}{0} - \binom{n}{2}a + \binom{n}{4}a^2 - \binom{n}{6}a^3 + \cdots \text{ is divisible by } 2^{n-1}.$$

(Romanian Mathematical Olympiad—Second Round, 1984)

Solution. Expanding $(1 + i\sqrt{a})^n$ by the binomial theorem and then separating the even and odd terms, we obtain

$$(1 + i\sqrt{a})^n = s_n + i\sqrt{a}t_n. \tag{1}$$

Passing to conjugates in (1), we get

$$(1 - i\sqrt{a})^n = s_n - i\sqrt{a}t_n. \tag{2}$$

From (1) and (2), it follows that

$$s_n = \frac{1}{2}[(1 + i\sqrt{a})^n + (1 - i\sqrt{a})^n]. \tag{3}$$

The quadratic equation with roots $z_1 = 1 + i\sqrt{a}$ and $z_2 = 1 - i\sqrt{a}$ is $z^2 - 2z + (a + 1) = 0$. It is easy to see that for every positive integer n, the following relation holds:

$$s_{n+2} = 2s_{n+1} - (1 + a)s_n. \tag{4}$$

Now we proceed by induction by step 2. We have $s_1 = 1$ and $s_2 = 1 - a = 2 - 4k = 2(1 - 2k)$, and hence the desired property holds. Assume that $2^{n-1}|s_n$ and $2^n|s_{n+1}$. From (4), it follows that $2^{n+1}|s_{n+2}$, since $1 + a = 4k$ and $2^{n+1}|(1 + a)s_n$.

Problem 16. *Let m and n be integers greater than 1. Prove that*

$$\sum_{\substack{k_1+k_2+\dots+k_n=m \\ k_1,k_2,\dots,k_n \geq 0}} \frac{1}{k_1!k_2!\dots k_n!} \cos(k_1 + 2k_2 + \dots + nk_n)\frac{2\pi}{n} = 0.$$

(Mathematical Reflections, 2009)

Solution. Let L denote the left-hand side of the proposed identity. We observe that L is the real part of the complex number

$$Z = \sum_{\substack{k_1+k_2+\dots+k_n=m \\ k_1,k_2,\dots,k_n \geq 0}} \frac{\omega^{k_1+2k_2+\dots+nk_n}}{k_1!k_2!\dots k_n!} = \sum_{\substack{k_1+k_2+\dots+k_n=m \\ k_1,k_2,\dots,k_n \geq 0}} \frac{\omega^{k_1}(\omega^2)^{k_2}\dots(\omega^n)^{k_n}}{k_1!k_2!\dots k_n!},$$

where $\omega = \cos\dfrac{2\pi}{n} + i\sum\dfrac{2\pi}{n}$.

Now, using the multinomial theorem, we have

$$Z = (\omega + \omega^2 + \dots + \omega^{n-1} + 1)^m = \left(\frac{\omega^n - 1}{\omega - 1}\right)^m = 0$$

(since $\omega^n = 1$). Thus, $L = \mathrm{Re}(Z) = 0$.

Problem 17. *Given an integer $n \geq 2$, let a_n, b_n, c_n be integers such that*

$$(\sqrt[3]{2} - 1)^n = a_n + b_n\sqrt[3]{2} + c_n\sqrt[3]{4}.$$

Show that $c_n \equiv 1 \pmod{3}$ if and only if $n \equiv 2 \pmod{3}$.

(Romanian IMO Team Selection Test, 2013)

Solution. The binomial expansion of $(\sqrt[3]{2} - 1)^n$ yields

$$c_n = \sum_{\substack{k \equiv 2 \\ (\text{mod } 3)}} (-1)^{n-k} \cdot 2^{(k-2)/3} \binom{n}{k} \equiv (-1)^n \sum_{\substack{k \equiv 2 \\ (\text{mod } 3)}} \binom{n}{k} \quad (\text{mod } 3).$$

Since

$$\sum_{\substack{k \equiv 2 \\ (\text{mod } 3)}} \binom{n}{k} = \frac{1}{3}((1+1)^n + \varepsilon(1+\varepsilon)^n + \varepsilon^2(1+\varepsilon^2)^n)$$

$$= \frac{1}{3}\left(2^n + 2\cos(n+3)\frac{\pi}{3}\right),$$

where $1 + \varepsilon + \varepsilon^2 = 0$, the condition $n \equiv 2 \pmod 3$ may be restated as

$$3c_n = (-1)^n \left(2^n + 2\cos(n+2)\frac{\pi}{3}\right) \equiv 3 \pmod 9.$$

Consideration of n modulo 6 yields $3c_n \equiv 3 \pmod 9$ if $n \equiv 2$ or $5 \pmod 6$, and $3c_n \equiv 0 \pmod 9$ otherwise. The conclusion follows.

6.2.9 Miscellaneous Problems

Problem 12. *Solve in complex numbers the system of equations*

$$\begin{cases} x|y| + y|x| = 2z^2, \\ y|z| + z|y| = 2x^2, \\ z|x| + x|z| = 2y^2. \end{cases}$$

Solution. Using the triangle inequality, we have

$$2|z|^2 = |x|y| + y|x|| \le |x||y| + |y||x|,$$

so $|z|^2 \le |x| \cdot |y|$. Likewise,

$$|y|^2 \le |x| \cdot |z| \text{ and } |z|^2 \le |y||x|.$$

Summing these inequalities yields

$$|x|^2 + |y|^2 + |z|^2 \le |x||y| + |y||z| + |z||x|.$$

This implies that

$$|x| = |y| = |z| = a.$$

If $a = 0$, then $x = y = z = 0$ is a solution of the system. Consider $a > 0$. The system may be written as

$$\begin{cases} x + y = \dfrac{2}{a}z^2, \\[2mm] y + z = \dfrac{2}{a}x^2, \\[2mm] z + x = \dfrac{2}{a}y^2, \end{cases}$$

Subtracting the last two equations gives

$$x - y = \frac{2}{a}(y^2 - x^2), \text{ i.e., } (y - x)\left(y + x + \frac{2}{a}\right) = 0.$$

Case 1. If $x = y$, then $x = y = \dfrac{z^2}{a}$. The last equation implies

$$z + \frac{z^2}{a} = 2\frac{z^4}{a^3}.$$

This is equivalent to

$$2\left(\frac{z}{a}\right)^3 = \frac{z}{a} + 1;$$

hence

$$\frac{z}{a} = 1 \quad \text{or} \quad \frac{z}{a} = \frac{-1 \pm i}{2}.$$

If $z = a$, then $x = y = z = a$ is a solution of the system. If $\dfrac{z}{a} = \dfrac{-1 \pm i}{2}$, then

$$1 = \left|\frac{z}{a}\right| = \left|\frac{-1 \pm i}{2}\right| = \frac{\sqrt{2}}{2},$$

which is a contradiction.

Case 2. If $x + y = -\dfrac{2}{a}$, then $-\dfrac{2}{a} = \dfrac{2}{a}z^2$. We obtain $z = \pm i$ and $a = |z| = 1$. Consider $z = i$; then

$$x = (x + y) - (y + z) + z = 2z^2 - 2x^2 + z = -2 + i - 2x^2,$$

or equivalently,

$$2x^2 + x + 2 - i = 0.$$

Then $x = i$ or $x = -\dfrac{1}{2} - i$. Since $|x| = a = 1$, we have $x = i$. Then $y = 2x^2 - z = -2 - i$ and $|y| = \sqrt{5} \neq a = 1$, so the system has no solution. The case $z = -i$ has the same conclusion.

Therefore, the solutions are $x = y = z = a$, where $a \geq 0$ is a real number.

Problem 13. *Solve in complex numbers the following:*

$$\begin{cases} x(x-y)(x-z) = 3, \\ y(y-x)(y-z) = 3, \\ z(z-x)(z-y) = 3. \end{cases}$$

(Romanian Mathematical Olympiad—Second Round, 2002)

Solution. In every solution (x, y, z), we have $x \neq 0$, $y \neq 0$, $z \neq 0$ and $x \neq y$, $y \neq z$, $z \neq x$. We can divide each equation by another and obtain new equations:

$$\begin{aligned} x^2 + y^2 &= yz + zx, \\ y^2 + z^2 &= xy + zx, \\ z^2 + x^2 &= xy + yz. \end{aligned} \tag{1}$$

By adding them, one obtains the equality

$$x^2 + y^2 + z^2 = xy + yz + zx. \tag{2}$$

After subtracting the second equation in (1) from the first, one obtains $x + y + z = 0$. By squaring this identity, one obtains an improvement of (2):

$$x^2 + y^2 + z^2 = xy + yz + zx = 0. \tag{3}$$

Using (3) in (1), one obtains

$$x^2 = zy, \ y^2 = zx, \ z^2 = xy, \tag{4}$$

and also

$$x^3 = y^3 = z^3 = xyz.$$

It follows that x, y, z are distinct roots of the same complex number $a = xyz$. From $x^3 = y^3 = z^3 = xyz = a$ we obtain

$$x = \sqrt[3]{a}, \ t = \varepsilon \sqrt[3]{a}, \ z = \varepsilon^2 \sqrt[3]{a}, \tag{5}$$

where $\varepsilon^2 + \varepsilon + 1 = 0$, $\varepsilon^3 = 1$. When we introduce the relations (5) in the first equation of the original system, we obtain $a^3(1 - \varepsilon)(1 - \varepsilon^2) = 3$. Taking into account the computation

$$(1 - \varepsilon)(1 - \varepsilon^2) = 1 - \varepsilon - \varepsilon^2 + 1 = 3,$$

we have $a^3 = 1$. Hence, we obtain, using (5), that (x, y, z) is a permutation of the set $\{1, \varepsilon, \varepsilon^2\}$.

Problem 14. *Let X, Y, Z, T be four points in the plane. The segments $[XY]$ and $[ZT]$ are said to be connected if there is some point O in the plane such that the triangles OXY and OZT are right isosceles triangles in O.*

Let $ABCDEF$ be a convex hexagon such that the pairs of segments $[AB]$, $[CE]$, and $[BD]$, $[EF]$ are connected. Show that the points A, C, D, and F are the vertices of a parallelogram and that the segments $[BC]$ and $[EA]$ are connected.

(Romanian Mathematical Olympiad—Final Round, 2002)

Solution. Suppose that the triangles OXY and OZT are oriented counterclockwise. Let x, y, z, t be the coordinates of the points X, Y, Z, T, and let m be the coordinate of O. Since these are right isosceles triangles, we have $x - m = i(y - m), z - m = i(t - m)$. It follows that $m(1 - i) = x - iy = z - it$. We deduce that $x - z = i(y - t)$.

Conversely, if $x - iy = z - it$, the coordinate of O is $m = \dfrac{x - iy}{1 - i}$, and the triangles OXY and OZT are right and isosceles.

Let a, b, c, d, e, f be the coordinates of the given hexagon in that order. We can write $a - ib = c - ie, b - id = e - if$. It follows that $a + d = c + f$, i.e., $ACDF$ is a parallelogram.

Multiplying the first equality by i, we obtain $b - ic = e - ia$, i.e., BC and AE are connected.

Problem 15. *Let $ABCDE$ be a cyclic pentagon inscribed in a circle with center O that has angles $\hat{B} = 120°, \hat{C} = 120°, \hat{D} = 130°, \hat{E} = 100°$. Show that the diagonals BD and CE meet at a point belonging to the diameter AO.*

(Romanian IMO, Team Selection Test, 2002)

Solution. By standard computations, we find that on the circumscribed circle, the sides of the pentagon subtend the following arcs: $\overset{\frown}{AB} = 80°, \overset{\frown}{BC} = 40°, \overset{\frown}{CD} = 80°, \overset{\frown}{DE} = 20°$ and $\overset{\frown}{EA} = 140°$. It is then natural to consider all these measures as multiples of $20°$ that correspond to the primitive 18th roots of unity, say $\omega = \cos\dfrac{2\pi}{18} + i\sin\dfrac{2\pi}{18}$. We thus assign to each vertex, starting from $A(1)$, the corresponding root of unity: $B(\omega^4), C(\omega^6), D(\omega^{10}), E(\omega^{11})$. We shall use the following properties of ω:

$$\omega^{18} = 1, \quad \omega^9 = -1, \quad \overline{\omega}^k = \omega^{18-k}, \quad \omega^6 - \omega^3 + 1 = 0. \tag{A}$$

We need to prove that the coordinate of the common point of the lines BD and CE is a real number (Fig. 6.9).

The equation of the line BD is

$$\begin{vmatrix} z & \bar{z} & 1 \\ \omega^4 & \bar{\omega}^4 & 1 \\ \omega^{10} & \bar{\omega}^{10} & 1 \end{vmatrix} = 0, \tag{1}$$

and the equation of the line CE is

$$\begin{vmatrix} z & \bar{z} & 1 \\ \omega^6 & \bar{\omega}^6 & 1 \\ \omega^{11} & \bar{\omega}^{11} & 1 \end{vmatrix} = 0. \tag{2}$$

Equation (1) can be written as follows:

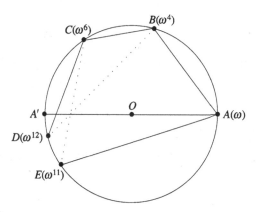

Figure 6.9.

$$z(\omega^{14} - \omega^8) - \bar{z}(\omega^4 - \omega^{10}) + (\omega^{12} - \omega^6) = 0,$$

or

$$z\omega^8(\omega^6 - 1) + \bar{z}\omega^4(\omega^6 - 1) + \omega^6(\omega^6 - 1) = 0.$$

Using the properties of ω, we derive a simplified version of (1):

$$z\omega^4 + \bar{z} + \omega^2 = 0. \tag{1'}$$

In the same way, (2) becomes

$$z\omega + \bar{z} - \omega^3(\omega^4 - 1) = 0. \tag{2'}$$

From (1') and (2') we obtain the following expression for z:

$$z = \frac{-\omega^7 + \omega^3 - \omega^2}{\omega^4 - \omega} = \frac{-\omega^6 + \omega^2 - \omega}{\omega^6} = -1 + \frac{\omega - 1}{\omega^5}.$$

To prove that z is real, it will suffice to prove that it coincides with its conjugate. It is easy to see that

$$\frac{\omega - 1}{\omega^5} = \frac{\bar{\omega} - 1}{\bar{\omega}^5}$$

is equivalent to

$$\bar{\omega}^4 - \bar{\omega}^5 = \omega^4 - \omega^5,$$

i.e., $\omega^{14} - \omega^{13} = \omega^4 - \omega^5$, which is true by the properties of ω given in (A).

Problem 16. *A function f is defined on the complex numbers by*

$$f(z) = (a + bi)z,$$

where a and b are positive numbers. This function has the property that the image of each point in the complex plane is equidistant from that point and the origin. Given that $|a + bi| = 8$ and that $b^2 = m/n$, where m and n are relatively prime positive integers, find $m + n$.

(1999 AIME, Problem 9)

Solution 1. Suppose we pick an arbitrary point in the complex plane, say $(1, 1)$. According to the definition of

$$f(x) = f(1 + i) = (a + bi)(1 + i) = (a - b) + (a + b)i,$$

this image must be equidistant to $(1, 1)$ and $(0, 0)$. Thus the image must lie on the line with slope -1 that passes through $\left(\dfrac{1}{2}, \dfrac{1}{2}\right)$, so its graph is $x + y = 1$. Substituting $x = (a - b)$ and $y = (a + b)$, we get $2a = 1 \Rightarrow a = \dfrac{1}{2}$. By the Pythagorean theorem, we have

$$\left(\frac{1}{2}\right)^2 + b^2 = 8^2 \Rightarrow b^2 = \frac{255}{4},$$

and the answer is 259.

Solution 2. We are given that $(a + bi)z$ is equidistant from the origin and z. This translates to

$$|(a + bi)z - z| = |(a + bi)z|,$$
$$|z(a - 1) + bzi| = |az + bzi|,$$
$$|z||(a - 1) + bi| = |z||a + bi|,$$
$$(a - 1)^2 + b^2 = a^2 + b^2, \Rightarrow a = \frac{1}{2}.$$

Since $|a + bi| = 8$, $a^2 + b^2 = 64$. But $a = \dfrac{1}{2}$, and thus $b^2 = \dfrac{255}{4}$. So the answer is 259.

Solution 3. Let P and Q be the points in the complex plane represented by z and $(a + bi)z$, respectively. Then $|a + bi| = 8$ implies $OQ = 8OP$. Also, we are given $OQ = PQ$, so OPQ is isosceles with base OP. Notice that the base angle of this isosceles triangle is equal to the argument θ of the complex number $a + bi$, because $(a + bi)z$ forms an angle of θ with z. Drop the altitude/median from Q to the base OP, and you end up with a right triangle showing that

$$\cos\theta = \frac{\frac{1}{2}OP}{8OQ} = \frac{\frac{1}{2}|z|}{8|z|} = \frac{1}{16}.$$

Since a and b are positive, z lies in the first quadrant, and $\theta < \pi/2$. Hence by right triangle trigonometry,

$$\sin\theta = \frac{\sqrt{255}}{16}.$$

Finally,

$$b = |a + bi|\sin\theta = 8\frac{\sqrt{255}}{16} = \frac{\sqrt{255}}{2} \text{ and } b^2 = \frac{255}{4},$$

so the answer is 259.

Problem 17. Let $F(z) = \dfrac{z+i}{z-i}$ for all complex numbers $z \neq i$, and let

$$z_n = F(z_{n-1})$$

for all positive integers n. Given that $z_0 = \dfrac{1}{137} + i$ and $z_{2002} = a + i$, where a and b are real numbers, find $a + b$.

(2002 AIME I, Problem 12)

Solution. Integrating F we get

$$F(z) = \frac{z+i}{z-i},$$

$$F(F(z)) = \frac{\dfrac{z+i}{z-i}+i}{\dfrac{z+i}{z-i}-i} = \frac{(z+i)+i(z-i)}{(z+i)-i(z-i)} = \frac{z+i+zi+1}{z+i-zi-1} = \frac{(z+1)(i+1)}{(z-1)(1-i)}$$

$$= \frac{(z+1)(i+1)^2}{(z-1)(1^2+1^2)} = \frac{(z+1)(2i)}{(z-1)(2)} = \frac{z+1}{z-1}i,$$

$$F(F(F(z))) = \frac{\dfrac{z+1}{z-1}i+i}{\dfrac{z+1}{z-1}i-i} = \frac{\dfrac{z+1}{z-1}+1}{\dfrac{z+1}{z-1}-1} = \frac{(z+1)+(z-1)}{(z+1)-(z-1)} = \frac{2z}{2} = z.$$

From this, it follows that $z_{k+3} = z_k$ for all k. Thus

$$z_{2002} = z_{3\cdot667+1} = z_1 = \frac{z_0+i}{z_0-i} = \frac{\left(\dfrac{1}{137}+i\right)+i}{\left(\dfrac{1}{137}+i\right)-i} = \frac{\dfrac{1}{137}+2i}{\dfrac{1}{137}} = 1 + 274i.$$

Thus $a + b = 1 + 274 = 275$.

Problem 18. *Given a positive integer n, it can be shown that every complex number of the form $r + si$, where r and s are integers, can be uniquely expressed in the base $-n + i$ using the integers $1, 2, \ldots, n^2$ as digits. That is, the equation*

$$r + si = a_m(-n+i)^m + a_{m-1}(-n+i)^{m-1} + \ldots + a_1(-n+i) + a_0$$

is valid for a unique choice of nonnegative integer m and digits $a_0, a_1, \ldots, a_m$ chosen from the set $\{0, 1, 2, \ldots, n^2\}$, with $a_m \neq 0$. We write

$$r + si = (a_m a_{m-1} \ldots a_1 a_0)_{-n+i}$$

to denote the base-$(-n+i)$ expansion of $r + si$. There are only finitely many integers $k + 0i$ that have four-digit expansions

$$k = (a_3 a_2 a_1 a_0)_{-3+i}, \quad a_3 \neq 0.$$

Find the sum of all such k.

(1989 AIME, Problem 14)

Solution. First, we find the first three powers of $-3 + i$:

$$(-3+i)^1 = -3+i, \quad (-3+i)^2 = 8 - 6i, \quad (-3+i)^3 = -18 + 26i.$$

So we need to solve the Diophantine equation

$$a_1 - 6a_2 + 26a_3 = 0 \Rightarrow a_1 - 6a_2 = -26a_3.$$

The least possible value of the left-hand side is -54, so $a_3 \leq 2$. We try cases:

Case 1. $a_3 = 2$. The only solution is $(a_1, a_2, a_3) = (2, 9, 2)$.
Case 2. $a_3 = 1$. The only solution is $(a_1, a_2, a_3) = (4, 5, 1)$.
Case 3. $a_3 = 0$. It is impossible for a_3 to equal 0, for otherwise, we would not have a four-digit number.

So we have the four-digit integers $(292a_0)_{-3+i}$ and $(154a_0)_{-3+i}$, and we need to find the sum of all integers k that can be expressed by one of these.

For $(292a_0)_{-3+i}$, we plug the first three digits in base 10 to get $30 + a_0$. The sum of the integers k in that form is 345.

For $(154a_0)_{-3+i}$, we plug the first three digits into base 10 to get $10 + a_0$. Thus the sum of the integers k in that form is 145. The answer is $345 + 145 = 490$.

Problem 19. *There is a complex number z with imaginary part 164 and a positive integer n such that*

$$\frac{z}{z+n} = 4i.$$

Find n.

(2009 AIME, Problem 2)

Solution. Let $z = a + 164i$. Then

$$\frac{a + 164i}{a + 164i + n} = 4i \text{ and } a + 164i = (4i)(a + n + 164i) = 4i(a + n) - 656.$$

By comparing coefficients, equating the real terms on the leftmost and rightmost sides of the equation, we conclude that $a = -656$.

By equating the imaginary terms on each side of the equation, we conclude that

$$164i = 4i(a + n) = 4i(-656 + n).$$

We now have an equation for n:

$$4i(-656 + n) = 164i,$$

and this equation shows that $n = 697$.

Problem 20. *Let u, v, w be complex numbers of modulus 1. Prove that one can choose signs $+$ and $-$ such that*

$$|\pm u \pm v \pm w| \leq 1.$$

(Romanian Mathematical Olympiad—District Round, 2007)

Solution. Denote by uppercase letters the points having as complex coordinates the corresponding lowercase letters. We have that $u + v + w$ is the complex coordinate of the orthocenter H of the triangle UVW.

If UVW is acute or right, we take all signs to be $+$, and this gives the solution, because H is interior to UVW, and so interior to the circumcircle.

Otherwise, one angle is obtuse, say W. Then for $w' = -w$, we get the acute triangle UVW', reducing the problem to the first case.

Problem 21. *Consider a complex number z, $z \neq 0$ and the real sequence*

$$a_n = \left| z^n + \frac{1}{z^n} \right|, \quad n \geq 1.$$

(a) Show that if $a_1 > 2$, then

$$a_{n+1} < \frac{a_n + a_{n+2}}{2}, \quad \text{for all } n \in \mathbb{N}^*.$$

(b) Prove that if there exists $k \in \mathbb{N}^*$ such that $a_k \leq 2$, then $a_1 \leq 2$.

(Romanian Mathematical Olympiad—District Round, 2010)

Solution 1.

(a) We easily observe that

$$2\left|z^{n+1} + \frac{1}{z^{n+1}}\right| < \left|z + \frac{1}{z}\right| \cdot \left|z^{n+1} + \frac{1}{z^{n+1}}\right|$$

$$= \left|z^n + \frac{1}{z^n} + z^{n+2} + \frac{1}{z^{n+2}}\right| \leq \left|z^n + \frac{1}{z^n}\right| + \left|z^{n+2} + \frac{1}{z^{n+2}}\right|.$$

(b) Suppose for the sake of obtaining a contradiction that $a_1 > 2$. Then (a) implies that the sequence $a_{n+1} - a_n$ is strictly increasing, so $a_{n+1} - a_n > a_2 - a_1$. But

$$a_2 = \left|z^2 + \frac{1}{z^2}\right| = \left|\left(z + \frac{1}{z}\right)^2 - 2\right| \geq \left(z + \frac{1}{z}\right)^2 - 2 = a_1^2 - 2 > a_1,$$

and therefore, the sequence $(a_n)_n$ is strictly increasing. Hence $a_k \geq a_1 > 2$ for all k, a contradiction.

Solution 2. Consider the sequence $(\alpha_n)_{n \geq 1}$ given by

$$\alpha_n = z^n + \frac{1}{z^n}.$$

Extend it to the left with the term $\alpha_0 = z^0 + \frac{1}{z^0} = 2$, and set $\alpha = \alpha_1$. Clearly, $a_n = |\alpha_n|$. We have

$$\alpha\alpha_n = \left(z + \frac{1}{z}\right)\left(z^n + \frac{1}{z^n}\right)$$

$$= \left(z^{n+1} + \frac{1}{z^{n+1}}\right) + \left(z^{n-1} + \frac{1}{z^{n-1}}\right) = \alpha_{n+1} + \alpha_{n-1}$$

for all $n \geq 1$, so the sequence $(\alpha_n)_{n \geq 0}$ satisfies the linear recurrence relation

$$\alpha_{n+1} = \alpha\alpha_n - \alpha_{n-1}.$$

Then for $|\alpha| > 2$, we have

$$a_n = |\alpha_n| = \left|\frac{\alpha_{n+1} + \alpha_{n-1}}{\alpha}\right| \leq \frac{|\alpha_{n+1}| + |\alpha_{n-1}|}{|\alpha|} < \frac{a_{n+1} + a_{n-1}}{2},$$

i.e., the sequence $(a_n)_{n \geq 0}$ is convex.

But then if $a_1 = |\alpha| > 2 = a_0$, every convex sequence is (strictly) increasing, since from $a_n > a_{n-1}$ follows

$$a_{n+1} > 2a_n - a_{n-1} = a_n + (a_n - a_{n-1}) > a_n,$$

and the assertion is proved by simple induction. Conversely, if there exists $k \in \mathbb{N}^*$ such that $a_k \leq 2$, then $a_1 \leq 2$. Therefore, the proof of (b) comes directly from (a), and the nature of the sequence is no longer relevant.

Problem 22. *Consider the set $M = \{z \in \mathbb{C} \mid |z| = 1, \text{ Re} z \in \mathbb{Q}\}$. Prove that the complex plane contains an infinity of equilateral triangles with vertices in M.*

(Romanian Mathematical Olympiad—Final Round, 2012)

Solution. Let $z = a + bi$ be a complex number of modulus 1 such that $a \in \mathbb{Q}$:

$$a^2 + b^2 = 1.$$

An equilateral triangle having z as one vertex that will satisfy the given condition has the other two vertices at the points $z(-1/2 \pm (i\sqrt{3})/2)$.

The real parts of these numbers are $-a/2 \pm (b\sqrt{3})/2$. Because $a \in \mathbb{Q}$, we have $-a/2 \pm (b\sqrt{3})/2) \in \mathbb{Q}$ if and only if $b\sqrt{3} \in \mathbb{Q}$. Let $q = b/\sqrt{3} \in \mathbb{Q}$.

To conclude the solution, we have to prove that the equation

$$a^2 + 3q^2 = 1$$

has an infinity of solutions $(a, q) \in \mathbb{Q} \times \mathbb{Q}$, i.e., that the equation $m^2 + 3n^2 = p^2$ admits an infinity of solutions $(m, n, p) \in \mathbb{N} \times \mathbb{N} \times \mathbb{N}$.

Since $3n^2 = (p - m)(p + m)$, we look for solutions such that $p - m = 3$ and $p + m = n^2$. we have $n^2 = 2m + 3$, so n is odd. Putting $n = 2k + 1$, $k \in \mathbb{N}^*$, we obtain $m = 2k^2 + 2k - 1$ and $p = 2k^2 + 2k + 2$. Then

$$a = (2k^2 + 2k - 1)/(2k^2 + 2k + 2), \quad b = ((2k + 1)\sqrt{3})/(2k^2 + 2k + 2),$$

and $z = a + bi$ is of modulus 1 with $a, b > 0$. So the triangle with one vertex in z is uniquely determined. Since $k \in \mathbb{N}$ is arbitrary, the conclusion follows.

Problem 23. *Let $(a_n)_{n \geq 1}$ be a sequence of nonnegative integers such that $a_n \leq n$ for all $n \geq 1$ and $\sum_{k=1}^{n-1} \cos \frac{\pi a_k}{n} = 0$ for all $n \geq 2$. Find a closed formula for the general term of the sequence.*

(Romanian Mathematical Olympiad—District Round, 2012)

Solution. Observe that $a_1 = 1$ and $\cos \frac{\pi a_1}{3} + \cos \frac{\pi a_2}{3} = 0$ implies $a_2 = 2$. Induct on n to prove that $a_n = n$, $n \geq 1$. Suppose $a_k = k$ for all $k = 1, 2, \ldots, n - 1$. The given relation can be rewritten as

$$\cos \frac{\pi a_n}{n + 1} = -\sum_{k=1}^{n-1} \cos \frac{\pi k}{n + 1}.$$

Set

$$z = \cos \frac{\pi}{n+1} + i \sin \frac{\pi}{n+1}$$

and observe that

$$z + z^2 + z^3 + \ldots + z^n = \frac{z - z^{n+1}}{1 - z} = \frac{1 + z}{1 - z}.$$

Use $\bar{z} = \dfrac{1}{z}$ to get

$$\overline{\left(\frac{1+z}{1-z} \right)} = -\frac{1+z}{1-z}$$

and hence $\mathrm{Re} \dfrac{1+z}{1-z} = 0$, and consequently,

$$\sum_{k=1}^{n} \cos \frac{\pi k}{n+1} = 0.$$

From $\cos \dfrac{\pi a_n}{n+1} = \cos \dfrac{\pi n}{n+1}$ and $a_n \leq n$ we get $a_n = n$, as claimed.

Problem 24. *Let a and b be two rational numbers such that the absolute value of the complex number $z = a + bi$ is equal to 1. Prove that the absolute value of the complex number $z_n = 1 + z + z^2 + \ldots + z^{n-1}$ is a rational number for all odd integers n.*

(Romanian Mathematical Olympiad—District Round, 2012)

Solution. Set $z = \cos t + i \sin t$, $t \in [0, 2\pi)$, and observe that $\sin t$, $\cos t$ are both rational numbers. For $z = 1$, the claim holds. For $z \neq 1$, write

$$|z_n| = |1 + z + z^2 + \ldots + z^{n-1}| = |(z^n - 1)/(z - 1)|.$$

Let $n = 2k + 1$, $k \in \mathbb{N}$. Then

$$|(z^n - 1)/(z - 1)| = \left| \sin \frac{(2k+1)t}{2} \Big/ \sin \frac{t}{2} \right|.$$

It is sufficient to prove that $x_k = \sin \dfrac{(2k+1)t}{2} \Big/ \sin \dfrac{t}{2}$ is a rational number. Observe that $x_{k+1} - x_k = 2\cos(k+1)t$, $k \in \mathbb{N}$ and $x_0 = 1 \in \mathbb{Q}$. Since

$$\cos(k+1)r = \mathrm{Re} z^{k+1} = \mathrm{Re}(a + bi)^{k+1} \in \mathbb{Q},$$

by induction we get that x_k is rational for all $k \in \mathbb{N}$.

Glossary

Antipedal triangle of point M: The triangle determined by perpendicular lines from vertices A, B, C of triangle ABC to MA, MB, MC, respectively.

Area of a triangle: The area of triangle with vertices with coordinates z_1, z_2, z_3 is the absolute value of the determinant

$$\Delta = \frac{i}{4} \begin{vmatrix} z_1 & \overline{z_1} & 1 \\ z_2 & \overline{z_2} & 1 \\ z_3 & \overline{z_3} & 1 \end{vmatrix}.$$

Area of pedal triangle of point X **with respect to the triangle** ABC:

$$\text{area}[PQR] = \frac{\text{area}[ABC]}{4R^2}|x\overline{x} - R^2|.$$

where x is the coordinate of X and R is the circumradius of the triangle.

Argument of a complex number: If the polar representation of complex number z is $z = r(\cos t^* + i \sin t^*)$, then $arg(z) = t^*$.

Barycenter of set $\{A_1, \ldots, A_n\}$ **with respect to weights** $m_1, \ldots, m_n$: The point G with coordinate $z_G = \frac{1}{m}(m_1 z_1 + \cdots + m_n z_n)$, where $m = m_1 + \cdots + m_n$.

Barycentric coordinates: In triangle ABC, the unique real number μ_a, μ_b, μ_c such that

$$z_P = \mu_a a + \mu_b b + \mu_c c, \text{ where } \mu_a + \mu_b + \mu_c = 1.$$

Basic invariants of triangle: semiperimeter s, inradius r, circumradius R.

Binomial equation: An algebraic equation of the form $Z^n + a = 0$, where $a \in \mathbb{C}^*$.

Blundon's inequalities: The necessary and sufficient conditions for three positive real numbers to be the semiperimeter s, the circumradius R, and inradius r, of a triangle.

Ceva's theorem: Let AD, BE, CF be three cevians of triangle ABC. Then lines AD, BE, CF are concurrent if and only if

$$\frac{AF}{FB} \cdot \frac{BD}{DC} \cdot \frac{CE}{EA} = 1.$$

Cevian of a triangle: Any segment joining a vertex to a point on the opposite side.

Concyclicity condition: If points $M_k(z_k)$, $k = 1, 2, 3, 4$, are not collinear, then they are concyclic if and only if

$$\frac{z_3 - z_2}{z_1 - z_2} : \frac{z_3 - z_4}{z_1 - z_4} \in \mathbb{R}^*.$$

Collinearity condition: $M_1(z_1)$, $M_2(z_2)$, $M_3(z_3)$ are collinear if and only if $\dfrac{z_3 - z_1}{z_2 - z_1} \in \mathbb{R}^*$.

Complex coordinate of point A with Cartesian coordinates (x, y): The complex number $z = x + yi$. We use the notation $A(z)$.

Complex coordinate of the midpoint of segment $[AB]$: $z_M = \dfrac{a + b}{2}$, where $A(a)$ and $B(b)$.

Complex coordinates of important centers of a triangle: Consider the triangle ABC with vertices with coordinates a, b, c. If the origin of complex plane is in the circumcenter of triangle ABC, then:

- The centroid G has coordinate $z_G = \dfrac{1}{3}(a + b + c)$.

- The incenter I has coordinate $z_I = \dfrac{\alpha a + \beta b + \gamma c}{\alpha + \beta + \gamma}$, where α, β, γ are the side lengths of triangle ABC.

- The orthocenter H has coordinate $z_H = a + b + c$.

- The Gergonne point J has coordinate $z_J = \dfrac{r_\alpha a + r_\beta b + r_\gamma c}{r_\alpha + r_\beta + r_\gamma}$, where r_α, r_β, r_γ are the radii of the three excircles of the triangle.

- The Lemoine point K has coordinate $z_K = \dfrac{\alpha^2 a + \beta^2 b + \gamma^2 c}{\alpha^2 + \beta^2 + \gamma^2}$.

- The Nagel point N has coordinate

$$z_N = \left(1 - \frac{\alpha}{s}\right) a + \left(1 - \frac{\beta}{s}\right) b + \left(1 - \frac{\gamma}{s}\right) c.$$

- The center O_9 of nine-point circle has coordinate $z_{O_9} = \dfrac{1}{2}(a + b + c)$.

Complex number: A number z of the form $z = a + bi$, where a, b are real numbers and $i = \sqrt{-1}$.

Complex product of complex numbers a and b: $a \times b = \dfrac{1}{2}(\bar{a}b - a\bar{b})$.

Conjugate of a complex number: The complex number $\bar{z} = a - bi$, where $z = a + bi$.

Cyclic sum: Let n be a positive integer. Given a function f of n variables, define the cyclic sum of variables $(x_1, x_2, \ldots, x_n)$ as

$$\sum_{\text{cyc}} f(x_1,\, x_2,\, \ldots,\, x_n) = f(x_1,\, x_2,\, \ldots,\, x_n) + f(x_2,\, x_3,\, \ldots,\, x_n,\, x_1)$$

$$+ \cdots + f(x_n,\, x_1,\, x_2,\, \ldots,\, x_{n-1}).$$

De Moivre's formula: For an angle α and integer n,

$$(\cos\alpha + i\sin\alpha)^n = \cos n\alpha + i\sin n\alpha.$$

Distance between points $M_1(z_1)$ and $M_2(z_2)$: $M_1 M_2 = |z_2 - z_1|$.

Equation of a circle: $z \cdot \overline{z} + \alpha \cdot z + \overline{\alpha} \cdot \overline{z} + \beta = 0$, where $\alpha \in \mathbb{C}$ and $\beta \in \mathbb{R}$.

Equation of a line: $\overline{\alpha} \cdot \overline{z} + \alpha z + \beta = 0$, where $\alpha \in \mathbb{C}^*$, $\beta \in \mathbb{R}$ and $z = x + iy \in \mathbb{C}$.

Equation of a line determined by two points: If $P_1(z_1)$ and $P_2(z_2)$ are distinct points, then the equation of the line $P_1 P_2$ is

$$\begin{vmatrix} z_1 & \overline{z_1} & 1 \\ z_2 & \overline{z_2} & 1 \\ z & \overline{z} & 1 \end{vmatrix} = 0.$$

Euler's formula: Let O and I be the circumcenter and incenter, respectively, of a triangle with circumradius R and inradius r. Then

$$OI^2 = R^2 - 2Rr.$$

Euler line of triangle: The line determined by the circumcenter O, the centroid G, and the orthocenter H.

Extend law of sines: In a triangle ABC with circumradius R and sides α, β, γ the following relations hold:

$$\frac{\alpha}{\sin A} = \frac{\beta}{\sin B} = \frac{\gamma}{\sin C} = 2R.$$

Heron's formula: The area of triangle ABC with sides α, β, γ is equal to

$$\text{area}[ABC] = \sqrt{s(s-\alpha)(s-\beta)(s-\gamma)},$$

where $s = \dfrac{1}{2}(\alpha + \beta + \gamma)$ is the semiperimeter of the triangle.

Isometric transformation: A mapping $f : \mathbb{C} \to \mathbb{C}$ preserving the distance.

Lagrange's theorem: Consider the points $A_1,\, \ldots,\, A_n$ and the nonzero real numbers $m_1,\, \ldots,\, m_n$ such that $m = m_1 + \cdots + m_n \neq 0$. The following relation holds for every point M in the plane:

$$\sum_{j=1}^{n} m_j M A_j^2 = mMG^2 + \sum_{j=1}^{n} m_j GA_j^2,$$

where G is the barycenter of the set $\{A_1, \ldots, A_n\}$ with respect to weights $m_1, \ldots, m_n$.

Modulus of a complex number: The real number $|z| = \sqrt{a^2 + b^2}$, where $z = a + bi$.

Morley's theorem: The three points of adjacent trisectors of angles form an equilateral triangle.

Nagel line of a triangle: The line I, G, N.

nth roots of a complex number z_0: Any solution Z of the equation

$$Z^n - z_0 = 0.$$

nth roots of unity: The complex numbers

$$\varepsilon_k = \cos \frac{2k\pi}{n} + i \sin \frac{2k\pi}{n}, \ k \in \{0, 1, \ldots, n-1\}.$$

The set of all these complex numbers for a given n is denoted by U_n.

Orthogonality condition: If $M_k(z_k)$, $k = 1, 2, 3, 4$, then lines M_1M_2 and M_3M_4 are orthogonal if and only if $\dfrac{z_1 - z_2}{z_3 - Z_4} \in i\mathbb{R}^*$.

Orthopolar triangles: Consider triangle ABC and points X, Y, Z situated on its circumcircle. Triangles ABC and XYZ are orthopolar (or S-triangles) if the Simson–Wallace line of point X with respect to triangle ABC is orthogonal to line YZ.

Pedal triangle of point X: The triangle determined by projections of X on the sides of triangle $A\,B\,C$.

Polar representation of a complex number $z = x + yi$: The representation $z = r(\cos t^* + i \sin t^*)$, where $r \in [0, \infty)$ and $t^* \in [0, 2\pi)$.

Primitive nth root of unity: An nth root $\varepsilon \in U_n$ such that $\varepsilon^m \neq 1$ for all positive integers $m < n$.

Quadratic equation: The algebraic equation $ax^2 + bx + c = 0$, $a, b, c \in \mathbb{C}$, $a \neq 0$.

Real product of complex numbers a and b: $a \cdot b = \dfrac{1}{2}(\overline{a}b + a\overline{b})$.

Reflection across a point: The mapping $s_{z_0} : \mathbb{C} \to \mathbb{C}$, $s_{z_0}(z) = 2z_0 - z$.

Reflection across the real axis: The mapping $s : \mathbb{C} \to \mathbb{C}$, $s(z) = \overline{z}$.

Rotation: The mapping $r_a : \mathbb{C} \to \mathbb{C}$, $r_a(z) = az$, where a is a given complex number.

Rotation formula: Suppose that $A(a)$, $B(b)$, $C(c)$ and C is the rotation of B with respect to A by the angle α. Then $c = a + (b - a)\varepsilon$, where $\varepsilon = \cos \alpha + i \sin \alpha$.

Similar triangles: Triangles $A_1A_2A_3$ and $B_1B_2B_3$ with the same orientation are similar if and only if

$$\frac{a_2 - a_1}{a_3 - a_1} = \frac{b_2 - b_1}{b_3 - b_1}.$$

Simson-Wallace line: For a point M on the circumcircle of triangle ABC, the projections of M on lines BC, CA, AB are collinear.

Translation: The mapping $t_{z_0} : \mathbb{C} \to \mathbb{C}$, $t_{z0}(z) = z + z_0$.

Trigonometric identities:

$$\sin^2 x + \cos^2 x = 1,$$
$$1 + \cot^2 x = \csc^2 x,$$
$$\tan^2 x + 1 = \sec^2 x;$$

addition and subtraction formulas:

$$\sin(a \pm b) = \sin\ a\ \cos b \pm \cos\ a\ \sin b,$$
$$\cos(a \pm b) = \cos\ a\ \cos b \mp \sin\ a\ \sin b,$$
$$\tan(a \pm b) = \frac{\tan a \pm \tan b}{1 \mp \tan a \tan b},$$
$$\cot(a \pm b) = \frac{\cot a \cot b \mp 1}{\cot a \pm \cot b};$$

double-angle formulas:

$$\sin 2a = 2\sin\ a\ \cos a = \frac{2 \tan a}{1 + \tan^2 a},$$

$$\cos 2a = 2\cos^2 a - 1 = 1 - 2\sin^2 a = \frac{1 - \tan^2 a}{1 + \tan^2 a},$$

$$\tan 2a = \frac{2 \tan a}{1 - \tan^2 a};$$

triple-angle formulas:

$$\sin 3a = 3\sin a - 4\sin^3 a,$$
$$\cos 3a = 4\cos^3 a - 3\cos a,$$
$$\tan 3a = \frac{3\tan a - \tan^3 a}{1 - 3\tan^2} a;$$

half-angle formulas:

$$\sin^2 \frac{a}{2} = \frac{1 - \cos a}{2},$$
$$\cos^2 \frac{a}{2} = \frac{1 + \cos a}{2},$$
$$\tan \frac{a}{2} = \frac{1 - \cos a}{\sin a} = \frac{\sin a}{1 + \cos a};$$

sum-to-product formulas:

$$\sin a + \sin b = 2 \sin \frac{a+b}{2} \cos \frac{a-b}{2},$$

$$\cos a + \cos b = 2 \cos \frac{a+b}{2} \cos \frac{a-b}{2},$$

$$\tan a + \tan b = \frac{\sin(a+b)}{\cos a \cos b};$$

difference-to-product formulas:

$$\sin a - \sin b = 2 \sin \frac{a-b}{2} \cos \frac{a+b}{2},$$

$$\cos a - \cos b = -2 \sin \frac{a-b}{2} \sin \frac{a+b}{2},$$

$$\tan a - \tan b = \frac{\sin(a-b)}{\cos a \cos b};$$

product-to-sum formulas:

$$2 \sin a \cos b = \sin(a+b) + \sin(a-b),$$

$$2 \cos a \cos b = \cos(a+b) + \cos(a-b),$$

$$2 \sin a \sin b = -\cos(a+b) + \cos(a-b).$$

Viète's theorem: Let x_1, x_2, ..., x_n be the roots of the polynomial

$$P(x) = a_n x^n + a_{n-1} x^{n-1} + \cdots + a_1 x + a_0,$$

where $a_n \neq 0$ and $a_0, a_1, \ldots, a_n \in \mathbb{C}$. Let s_k be the sum of the products of the x_i taken k at a time. Then

$$s_k = (-1)^k \frac{a_{n-k}}{a_n},$$

that is,

$$x_1 + x_2 + \cdots + x_n = \frac{a_{n-1}}{a_n},$$

$$x_1 x_2 + \cdots + x_i x_j + x_{n-1} x_n = \frac{a_{n-2}}{a_n},$$

$$\cdots$$

$$x_1 x_2 \cdots x_n = (-1)^n \frac{a_0}{a_n}.$$

References

[1] Adler, I., *A New Look at Geometry*, John Day, New York, 1966.

[2] Andreescu, T., editor, *Mathematical Reflections—The First Two Years*, XYZ Press, Dallas, 2011.

[3] Andreescu, T., editor, *Mathematical Reflections—The Next Two Years*, XYZ Press, Dallas, 2012.

[4] Andreescu, T., Andrica, D., *360 Problems for Mathematical Contests*, GIL Publishing House, Zalău, 2003.

[5] Andreescu, T., Andrica, D., *Proving some geometric inequalities by using complex numbers*, Mathematical Education, Vol. **1**, No. 2(2005), 19–26.

[6] Andreescu, T., Dospinescu, G., *Problems from the Book*, XYZ Press, Dallas, 2010.

[7] Andreescu, T., Dospinescu, G., *Straight from the Book*, XYZ Press, Dallas, 2012.

[8] Andreescu, T., Enescu, B., *Mathematical Treasures*, Birkhäuser, Boston, 2003.

[9] Andreescu, T., Feng, Z., *Mathematical Olympiads 1998–1999, Problems and Solutions from Around the World*, The Mathematical Association of America, 2000.

[10] Andreescu, T., Feng, Z., *Mathematical Olympiads 1999–2000, Problems and Solutions from Around the World*, The Mathematical Association of America, 2002.

[11] Andreescu, T., Feng, Z., Lee, G. Jr., *Mathematical Olympiads 2000–2001, Problems and Solutions from Around the World*, The Mathematical Association of America, 2003.

[12] Andreescu, T., Gelca, R., *Mathematical Olympiad Challenges*, Birkhäuser, Boston, 2000.

[13] Andreescu, T., Kedlaya, K., *Mathematical Contests 1996–1997, Olympiads Problems and Solutions from Around the World*, American Mathematics Competitions, 1998.

[14] Andreescu, T., Kedlaya, K., *Mathematical Contests 1997–1998, Olympiads Problems and Solutions from Around the World*, American Mathematics Competitions, 1999.

[15] Andrica, D., Barbu, C., *A geometric proof of Blundon's inequalities*, Mathematical Inequalities & Applications, Vol. **15**, No. 2(2012), 361–370.

[16] Andrica, D., Barbu, C., Minculete, N., *A geometric way to generate Blundon type inequalities*, Acta Universitatis Apulensis, No. 31/2012, 93–106.

[17] Andrica, D., Bişboacă, N., *Complex Numbers from A to Z* (Romanian), Millennium, Alba Iulia, 2001.

[18] Andrica, D., Bogdan, I., *A formula for areas in terms of complex numbers* (Romanian), Revista de Matematică Transylvania, **3**(1999), 3–14.

[19] Andrica, D., Nguyen, K.L., *A note on the Nagel and Gergonne points*, Creative Math. & Inf., **17**(2008).

[20] Andrica, D., Varga, C., Văcăreţu, D., *Selected Topics and Problems in Geometry* (Romanian), PLUS, Bucharest, 2002.

[21] Baptist, Peter, *Die Entwicklung der Neueren Dreiecksgeometrie*, Wissenschaftsverlag, Mannheim, 1992.

[22] Baker, H. F., *Principles of Geometry*, Vol. 1–3, University Press, Cambridge, 1943.

[23] Bălună, M., Becheanu, M., *Romanian Mathematical Competitions*, Romanian Mathematical Society, Bucharest, 1997.

[24] Becheanu, M., *International Mathematical Olympiads 1959–2000. Problems. Solutions. Results*, Academic Distribution Center, Freeland, USA, 2001.

[25] Berger, M., *Géométrie*, CEDUC Nathan Paris, 1977–1978.

[26] Berger, M. et al., *Problèmes de géométrie commentés et redigés*, Paris, 1982.

[27] Brânzei, D., *Notes on Geometry*, Paralela 45, Pitești, 1999.

[28] Brumfiel, C. E. et al., *Geometry*, Addison-Wesley, Reading, MA, 1975.

[29] Coxeter, H. S. M., *Introduction to Geometry*, John Wiley & Sons, New York, 1969.

[30] Coxeter, H. S. M., Greitzer, S. L., *Geometry Revisited*, Random House, New York, 1967.

[31] Deaux, R., *Introduction to the Geometry of Complex Numbers*, Ungar, New York, 1956. (Deaux, R., *Introduction à la géométrie des nombres complexes*, Brussels, 1947.)

[32] Dincă, M., Chiriță, M., *Complex Numbers in High School Mathematics* (Romanian), All Educational, Bucharest, 1996.

[33] Dunham, William, *Euler: The Master of Us All*, Mathematical Association of America, 1999.

[34] Engel, A., *Problem-Solving Strategies*, Springer-Verlag, New York, 1998.

[35] Fano, G., *Complementi di geometria*, Felice Gilli, Turin, 1935.

[36] Fenn, R., *Geometry*, Springer-Verlag, New York, 2001.

[37] Gleason, A. M., Greenwood, R. E., Kelly, L. M., *The William Lowell Putnam Mathematical Competition. Problems and Solutions: 1938–1964*. The Mathematical Association of America, 1980.

[38] Gelca, R., Andreescu, T., *Putnam and Beyond*, Springer, New York, 2007.

[39] Hahn, L., *Complex Numbers & Geometry*, The Mathematical Association of America, 1994.

[40] Johnson, R. A., *Advanced Euclidean Geometry*, New York, 1960.

[41] Kedlaya, K. S., Poonen, B., Vakil, R., *The William Lowell Putnam Mathematical Competition 1985–2000*. The Mathematical Association of America, 2002.

[42] Kutepov, A., Rubanov, A., *Problems in Geometry*, MIR, Moscow, 1975.

[43] Lalescu, T., *La géométrie du triangle*, Librairie Vuibert, Paris, 1937.

[44] Lozansky, E., Rousseau, C., *Winning Solutions*, Springer-Verlag, New York, 1996.

[45] Mihalca, D. et al., *Quadrilateral Geometry* (Romanian), Teora, Bucharest, 1998.

[46] Mihalescu, C., *The Geometry of Remarkable Elements* (Romanian), Editura Tehnică, Bucharest, 1957.

[47] Mihăileanu, N. N., *Using Complex Numbers in Geometry* (Romanian), Editura Tehnică, Bucharest, 1968.

[48] Modenov, P. S., *Problems in Geometry*, MIR, Moscow, 1981.

[49] Modenov, P. S., Parkhomenko, A. S., *Geometric Transformations*, Academic Press, New York, 1965.

[50] Moisotte, L., *1850 exercices de mathématique*, Bordas, Paris, 1978.

[51] Nahin, P. J., *An Imaginary Tale. The Story of $\sqrt{-1}$* (Romanian), Theta, Bucharest, 2000.

[52] Nicula, V., *Complex Numbers* (Romanian), Scorpion 7, Bucharest, 1999.

[53] Pedoe, D., *A Course of Geometry for Colleges and Universities*, Cambridge University Press, Cambridge, 1970.

[54] Pompeiu, D., *The Mathematical Works* (Romanian), Academiei, Bucharest, 1959.

[55] Prasolov, V. V., *Problems of Plane Geometry*, 2 volumes, Nauka, Moscow, 1986.

[56] Retali, V., Biggiogero, G., *La geometria del triangolo* (cap. XXIV din *Enciclopedia delle matematiche elementari*, vol. II, parte I, Milan, 1937).

[57] Sălăgean, Gr. S., *The Geometry of the Complex Plane* (Romanian), Promedia-Plus, Cluj-Napoca, 1997.

[58] Schwerdtfeger, H., *Geometry of Complex Numbers*, University of Toronto Press, Toronto, 1962.

[59] Sergyeyev, I. N., *Foreign Mathematical Olympiads*, Nauka, Moscow, 1987.

[60] Stanilov, G., Kuchnov, Y., Gjorgjev, V., *Vectors and Plane Geometrical Transformations*, Narodna Prosveta, Sofia, 1979.

[61] Tomescu, I. et al., *Problems from High School Mathematical Olympiads (1950–1990)* (Romanian), Editura Ştiinţifică, Bucharest, 1992.

[62] Tomescu, I. et al., *Balkan Mathematical Olympiads 1984–1994* (Romanian), Gil, Zalău, 1996.

[63] Tonov, I. K., *Complex Numbers* (Bulgarian), Narodna Prosveta, Sofia, 1979.

[64] Yaglom, I. M., *Complex Numbers in Geometry*, Academic Press, New York, 1968.

Author Index

T. Andreescu and D. Andrica, *Complex Numbers from A to ... Z*,   387
DOI 10.1007/978-0-8176-8415-0, © Springer Science+Business Media New York 2014

Subject Index